LAND AND ITS USES – ACTUAL AND POTENTIAL

An Environmental Appraisal

NATO CONFERENCE SERIES

I	Ecology
II	Systems Science
III	Human Factors
IV	Marine Sciences
V	Air–Sea Interactions
VI	Materials Science

I ECOLOGY

LAND AND ITS USES – ACTUAL AND POTENTIAL

An Environmental Appraisal

Edited by

F. T. Last
Institute of Terrestrial Ecology
Penicuik, Midlothian, Scotland

M. C. B. Hotz
Commission on Lead in the Environment
Toronto, Ontario, Canada

and

B. G. Bell
Institute of Terrestrial Ecology
Penicuik, Midlothian, Scotland

Published in cooperation with NATO Scientific Affairs Division

PLENUM PRESS · NEW YORK AND LONDON

Library of Congress Cataloging in Publication Data

NATO Seminar on Land and Its Uses—Actual and Potential: an Environmental Appraisal
  1982: Edinburgh, Scotland)
  Land and its uses—actual and potential: an environmental appraisal.

  (NATO conference series. I, Ecology; v. 10)
  "Proceedings of a NATO Seminar on Land and Its Uses—Actual and Potential: An En-
vironmental Appraisal, held September 19–October 1, 1982, in Edinburgh, Scotland"
—T.p. verso.
  "Published in cooperation with NATO Scientific Affairs Division."
  Includes bibliographical references and index.
  1. Environmental protection—Congresses. 2. Land use—Environmental aspects—
Congresses. 3. Landscape protection—Congresses. I. Last, F. T. II. Hotz, M. C. B. III.
Bell, B. G. IV. North Atlantic Treaty Organization. Scientific Affairs Division. V. Title. VI.
Series.
TD169.N385 1982   333.73   85-28111
ISBN-13:978-1-4612-9278-4   e-ISBN-13:978-1-4613-2169-9
DOI: 10.1007/978-1-4613-2169-9

Proceedings of a NATO Seminar on Land and Its Uses—Actual and Potential:
An Environmental Appraisal, held September 19–October 1, 1982, in
Edinburgh, Scotland

© 1986 Plenum Press, New York
Softcover reprint of the hardcover 1st edition 1986

A Division of Plenum Publishing Corporation
233 Spring Street, New York, N.Y. 10013

PREFACE

During its existence the Ecosciences Panel of the North Atlantic
Treaty Organisation was constantly concerned with (i) the communication gap between the generators of ecological/environmental information and those who use it and (ii) the narrow interpretation of
'environmental' which too frequently was taken as being synonymous
with pollution.

Because of this concern, and because the panel recognised that
land-use is perhaps the overriding facet of environmental policy it
was decided to arrange the Seminar recorded in this volume :-

Land and its Uses : Actual and Potential
An Environmental Appraisal

The development of this Seminar was chaired by Professor F. T.
Last who was enthusiastically supported by B. G. Bell (U.K.),
Drs S. Bie (Norway), O. W. Heal (U.K.), R. Herrmann (Federal Republic
of Germany), M.C.B.Hotz (formerly of NATO, Belgium, but now in
Canada), L. Munn (Canada) and N. Yassoglou (Greece). Together, they
decided that the participants should include (i) planners/decision-
makers and (ii) scientists generating ecological/environmental information, in the hope that they would gain a better understanding of
each others problems and attitudes and as a result identify how
information can be prepared in a more usable form.

Regretably a considerable volume of ecological/environmental
information is lying idle simply because it is not in a form that
can be used by planners/decision-makers. This situation must be
corrected in the hope that the likely impacts of different policy
scenarios can be predicted in a rational manner, eliminating potential
damaging options before implementation.

The work of the Seminar was greatly advanced by the participation of: J. C. Baral (Nepal), L. Canadas-Cruz (Equador), M. Elahi (Pakistan), M. M. Gure (Somalia), S. Mingramm (Mexico), J. L. Mwangala (Zambia), T. A. Okusami (Nigeria), and H. Rodrigues (Colombia) who were generously funded by the United Nations Environment Programme (UNEP).

F. T. LAST
M. C. B. HOTZ
B. G. BELL

CONTENTS

CONTENTS

C REMOTE SENSING

D CASE STUDIES

INTRODUCTION

F. T. Last, M. C. B. Hotz, and B. G. Bell

The degree of 'environmental' awareness among populations differs in different countries but throughout the world it seems to be increasing, an increase that surprisingly seems to be encouraged rather than deterred by the effects of economic recession. Even in difficult times the concept of 'quality of life' seems to be holding its place in the forefront, a position undoubtedly attributable in some parts of the world to the effects, actual and perceived, of atmospheric pollutants on freshwater and terrestrial ecosystems, on buildings and also on the clarity or haziness of the air.

Pollutants such as sulphur dioxide, oxides of nitrogen, photo-chemical oxidants and acid rain, although of immense relevance are, however, only one of the results of man's activities, though admit-tedly one that attracts widespread publicity. By series of excellent films we are being increasingly informed of the dangers of deforestation, including the damage done to resources of trees, to associated assemblages of animals and other plants, to water and to the maintenance of soil fertility, with soil erosion adversely affecting water quality and imparing aquatic biota. But these influences, bad as they may be, are overshadowed at present by the catastrophic effects of a succession of droughts, notably on the African Continent.

What do these major and minor catastrophies have in common ? They all reflect to a greater or lesser extent the inept and usually avaricious manner in which most of the human race attempts to use its natural resources. The World is often divided into two parts (i) the 'developed' and (ii) the 'developing', with the implied suggestion that the former has learned how to sustain or steward its resources. Nothing could be further from the truth. In reality,

1

'development' tends to be synonymous with 'exploitation'. Can we
really claim that oil is being extracted and used prudently ? Is
our land being used effectively and wisely ?

While hoping that standards of living in developing countries
will be greatly improved while those in developed countries do not
deteriorate, it is clear that we are all in danger of overlooking
the lessons and/or principles that should be learnt from primitive
societies which knew, if nothing else, how to husband and sustain
their resources. Curiously, in developed countries, with a relatively strong interventionist approach to the management of resources,
we have a situation in which individuals who are technically informed
about natural resources are rarely those in a position to take, or
make, 'political' decisions about those resources. On the one hand
there are 'decision-makers' (politicians, financiers, planners etc.)
while on the other there are the technically informed scientists.
Although the situation is undoubtedly improving, there is still an
unacceptable 'divide'. How many scientists, environmental or otherwise, really understand the nature of the decisions confronting
planners, financiers ....? Very few indeed, judging by the manner in
which scientists prepare 'briefs' for decision-makers. Similarly,
how many planners/financiers are sympathetic to the science underlying
the technically-related judgements that they have to make ? Admit-
tedly the situation is improving, but it will not be changed fundamentally until educators recognize that socially-related subjects
(including culture, history, geography and economics) are not optional
alternatives to the physical and biological sciences: instead they
complement each other in a system aiming to "educate for living".

To tackle the more limited objective of improving the rapport
between decision-makers and their scientific counterparts, the North
Atlantic Treaty Organisation provided funds for a seminar restricted
to a consideration of :-
Land and its Uses : Actual and Potential : an Environmental Appraisal
This book records the main contributions to that Workshop, which
was attended by planners concerned with the integrated consideration
of urban and rural environments, both tropical and temperate, and
scientists trained as geologists, hydrologists, foresters, freshwater
ecologists, agriculturists, soil experts ......who, without exception,
were trying to relate their individual skills to an integrated
'holistic' approach to the wise development of land resources, in
essence 'landscape ecology'. Fundamentally all ecosystems are
dynamic, with deaths being offset by recruits. Our aim should be to
conserve, not preserve, but how should scientists, generating an
increasing flood of information, condense and crystallize it so that
it is in a form that can be comprehended and used by decision-makers?
The scientists generating environmental data must be discriminatory;
on the other hand the decision-makers must be sympathetic to science.
We invite our readers to judge whether the participants in the Workshop, who were committed to the 'cause', have been able to give a
lead.

A
IDENTIFICATION OF ECOLOGICAL FACTORS
CHARACTERISING THE RANGE
OF *TERRESTRIAL* HABITATS
i - Urban

PLANNING* AND THE PHYSICAL ENVIRONMENT

P. Johnson-Marshall

Department of Urban Design and Regional Planning
University of Edinburgh
Edinburgh EH1 1JZ, Scotland

THE WORLD OBSERVED

During the last decade there has been a growing public awareness in many countries that the world is one place with finite limits, and that any discoveries of new lands must be made in space. The possibilities of colonizing other planets do not seem to be very great at the present time, so it is, therefore, necessary to take stock of the physical resources of the one we all live in for the benefit of all its inhabitants in the future.

Ever since the United Nations (UN) was first created some of its specialist agencies, notably the Food and Agriculture Organization and the World Health Organization, have been reporting on certain aspects of the terrestrial situation with growing concern. A number of dramatic developments concerned with such phenomena as pollution and the depletion of wild life, and in particular endangered species such as the whales, led to the organization by the United Nations of the first World Conference on the Environment at Stockholm in 1972. Although considered by many cynics as something of a five day wonder, the published conclusions and recommendations were in many aspects as important as the original UN Declaration of Human Rights. Shortly after the Stockholm conference it was proposed that a similar meeting should be held on urban problems. This resulted in the UN Conference on Human Settlements, which was held at Vancouver in 1976, and produced recommendations of comparable significance.

---

* The word planning is used in its original sense of preparing plans of the physical environment.

The Stockholm Conference had initiated the United Nations Environmental Programme (UNEP) to deal with broad problems concerned with
the environment as a whole, and after the Vancouver Conference, the
United Nations Centre for Human Settlements (UNCHS) was formed to
deal specifically with urban problems. Both institutions were established in Nairobi, and provided the foundation for necessary
international action.

In spite of the recognition of the gravity of the problems
revealed by the two Conferences and the official UN actions taken
to deal with them, too many politicians in too many countries
allowed the vital challenge to fade into comparative obscurity, and
Ward (1982) drew attention to this dangerous relapse shortly before
her death. It is of extreme importance that the fundamental
problems of the environment be continually brought to the notice of
both the public and its political representatives, and today pop-
ularly available means exist to do so.

First, it is necessary to emphasise the dangers that lie ahead,
and indeed not very far ahead, for the human species as a whole.
These dangers should be regarded as equivalent to a traditional war
situation, and action should be taken both at national and international levels to deal with them with the same degree of urgency.
In fact, such action should replace the present concentration of
resources on military organization, and continuous pressure should
be brought to bear on the decision makers, from the grass roots
level, to change their policies towards ecological survival and
cooperation.

Some of the problems can be set out for popular information in
a number of different ways. Of first importance are those revealed
by UN and other world statistical sources. More accurate figures
are becoming available which indicate food and dietary deficiencies,
unsatisfactory health conditions, population movements with
resulting demographic imbalance, and widespread housing shortages,
particularly in the larger human settlements in developing countries
many of these are responsible for almost insoluble problems.

From an ecological viewpoint, the incorrect use of the earth's
surface is leading to a rapid destruction of its usable area. In
accordance with Malthus' classic dictum, population increases are
still tending to outstrip food supply, and temporary solutions sometimes exacerbate the problem in ecological terms. For instance, the
widespread use of chemicals in agriculture sometimes does more harm
than good, and primitive husbandry, on the one hand, or commercially
dominated farming methods on the other, often lead to soil erosion,
soil degradation, and eventually desertification.

The uneven distribution of mineral resources has led to imbalances
and irrational political/economic tensions which distort rational

assessments of need and distribution, even within individual
countries. Of these, oil is only the most glaring example. Fortunately, many of these problems can be highlighted by new techniques
such as satellite photography, and the stark problems of our world
can be illustrated to the people in a vivid and comprehendable way.
Three of the most urgent problems thus revealed are (i) the
increase of desertification, (ii) the overconcentration of population in large cities, and (iii) the decay of rural areas.

At the UN Conference on Desertification (see Walls, 1980), held
in Nairobi in 1977, several valuable case studies were presented
on desert recovery projects. Some were of limited aspect, and in
this context it is very important that all such projects should be
considered as part of broader regional planning and development
programmes, and not as isolated first-aid operations.

The concentration of population in large conurbations was highlighted by the International Congress on Metropolitan Cities held
in Mexico City in 1981, but so far effective solutions for this
problem, particularly in developing countries, have not been found.
The idea of reducing large urban populations, for instance, militates
against powerful economic, commercial and even psychological forces,
in spite of the grossly indadequate living conditions of so many of
the inhabitants.

The decay of rural areas still awaits major international action,
although there have been several UN sponsored regional conferences
and a number of practical initiatives, such as the Ujaama village
programme in Tanzania, which indicated some progress but require monitoring at the international level.

A general conclusion can be reached that a fundamentally new
"client", or political, approach is required, based on terrestrial
ecological balance and the conservation of energy. However, a
number of manifest dangers to world civilization, and even to human
existence, must first be recognized as such, and kept constantly
before the international public until they have been eradicated.
Outstanding among them are, first, that the traditional concept of
the human right to unlimited and thoughtless exploitation of world
resources is no longer valid, and is now endangering what Ward
(1982) called "the seed corn"; second, that the unscientific
removal and degradation of the vegetal cover of the soil, coupled
with pollution, must be arrested; and third, that uncontrolled
population expansion and movement, with the mass migration from
rural areas to large cities, can no longer be tolerated.

SIGNS OF HOPE

In terms of the limited field of physical planning, any hope of

success implies a major reassessment in the basic approach to managing the use of the environment, and giving even more emphasis to the
need for coordination at all levels of physical planning and development. As always, it is important to cite precedents where they exist;
and the theory of the 'mini-paradise'*, whereby a good environment is
created over a limited area is, therefore, an important concept in
terms of demonstration for everyone.

Fortunately, signs of hope are appearing in many countries, and
there are enough demonstrations to provide some guidance for future
action. At the basic level of desertification, there have been
considerable achievements, notably the case studies which were
reported at the UN Conference on Desertification.

Some countries have accepted conservation policies, and in increasing areas of the earth's surface ecologically sound solutions are
being adopted, although short term political and military considerations tend to frustrate the best intentions. Nevertheless, a strong
"grass roots" movement is growing up, based on an ecological approach,
energy conservation, and appropriate technology. All these activities
need to be recorded, monitored, and publicised, as widely as possible.
It may even be, that when enough people become aware of the critical
state of the earth, conventional politics will wither, and be replaced
by a concensus based on ecology.

In terms of physical planning, it should be remembered that planning policies have hitherto been based largely on the assumption of
unlimited energy and resources, and their maximum technological
exploitation. The planners need new instructions from their clients
based on conservation instead of exploitation. In the meantime
physical planning will continue, since it is now an accepted part
of the social/technical services in most countries, and every year
useful experiences are recorded.

Several countries have already accepted the idea of physical planning at the national level. These include the Netherlands, Poland,
and Ghana, while even in Scotland, National Planning Guidelines,
prepared by the Scottish Development Department, already exist.

It is now almost forgotten that the USA had a carefully worked
out national physical plan prepared for the New Deal's National
Resources Committee (1935) and it played a significant role in the
selection of the Tennessee Valley as a demonstration region for
remedial action and recovery.

---

*The Persian poet Firdawsi described Shah Jahan's palace in Delhi
 in lyrical terms, "If there is a heaven on earth it is here...."

Plans may be of two basic types, definitive or remedial. Few examples of definitive regional planning exist, although one could say that Wakefield's concepts for Australia and New Zealand were germinal of their kind; and, in the case of Adelaide (Dutton, 1960), led to an outstanding new town plan for which the city has never ceased to be grateful. In contemporary terms, the Netherlands polder schemes (Rijswaterstraat Communications, 1964) are the nearest to definitive regional planning, since the land was reclaimed from the sea and was laid out with a comprehensive plan which included proposals for human settlements, transport, agriculture, and recreation. As with the new towns in Britain it is now possible to visit these regions to see how the plans have been implemented and to learn some lessons accordingly.

For developing countries the Geziera scheme (Gaitskell, 1959) in the Sudan is the nearest to a definitive solution, although the recent projects in Sinkiang (Science Press, 1977) in north-west China may prove to be equally significant. For examples of remedial regional plans, the Tennessee Valley Authority* operations are still, after nearly fifty years, the most outstanding, although they lack the comprehensiveness which the inclusion of human settlements would provide.

Most of the large metropolitan cities have had not one but many remedial regional plans, but the problem has nearly always been a failure of political will and continuity of policy. The Greater London Plan of 1944 (Abercrombie, 1944), one of the best ever prepared for a metropolis, is a good example. The initial momentum, which began with the implementation of decentralization, conservation of the green belt and the creation of eight New Towns, was gradually lost through a failure of political will. In Scotland the new Regional Authorities have all prepared Regional Plans which are well worth study, although, with the exception of the Highlands Region with the highlands and Islands Development Board, the British tendency to separate planning from development may well limit their effectiveness.

At the urban level a considerable number of definitive solutions exist in various parts of the world, and may be visited and studied. The immediate and obvious achievements in developed countries are the thirty New Towns in the United Kingdom, but important examples exist in France, Sweden, Finland, the Netherlands, and Poland; while in developing countries there are, among others, Brasilia (Acropole, 1960) in Brazil and Chandigarh (Evenson, 1966) in India, as useful case studies. Sadly, Australia, which could have been the scene for a fine achievement at Monarto, near Adelaide in South Australia, changed its government policy at a critical moment. The USA, which

---

*The Annual Reports still make good reading

made a false start during the New Deal (Stein, 1966) has in recent
years only created two, Reston and Columbia, near Washington D.C.,
both by private enterprise, and both having to confront the difficulties that are normally encountered when an inappropriate method
is used for a task as complex as establishing and developing a new
town. Nevertheless, new urban settlements offer valuable case
studies for planners and clients alike, as they present an opportunity to apply new concepts, new standards, and new criteria, over
the whole range of urban components. As the plans are realized,
the lessons can be learnt accordingly, but over a time span of
several generations.

The situation in regard to existing settlements is far more
complex. Most have grown without comprehensive physical plans, and
planning has therefore been of a remedial character. Added to this,
it is often difficult to assess the results of planning proposals
unless they are monitored over a considerable period. For instance,
the consistent planning policy whereby Paris was remodelled from a
medieval city to become the world prototype of a renaissance city,
took over three hundred years (Couperie, 1968); while Edinburgh,
which decided in the eighteenth century to proceed by planned urban
expansion rather than renewal, took over one hundred years to complete its New Town (Youngson, 1966). Most cities which have adopted
planning effectively, such as Stockholm, have used both methods
(Bergvall, 1952), although recently the recognition of past environmental quality has led to a powerful movement for the conservation
of historic urban areas, and sometimes of whole urban settlements.
In fact, it is now being recognized that some historic areas are
more effective in terms of energy saving and liveability than the
large scale high energy consuming urban developments which evolved
from the concepts of early 20th century innovators such as Le
Corbusier (1929).

The components of human settlements consist largely of buildings
designed by architects, and both they and their clients need to
reappraise their approach to design in the light of a policy of
energy conservation. One obvious implication is a reduction in the
use of resource and energy intensive machinery, and the introduction
of as many energy saving materials and techniques as possible. The
effect on infrastructure could also be profound. If the private
motor vehicle is seen as a luxury in the future on account of the
shortage of fossil fuel, then towns should be laid out with a view
to walking or cycling wherever possible, or at least of making
maximum use of public transport, or a combination of both. The
enormous energy use involved in underground rail services would also
be questioned, and in this context it could be said that when a city
needs an underground service it needs another city. The whole
problem of waste disposal requires review, negative action being
required in many countries when it is a polluting agent, and positive
action in terms of its potential organic use, following up the exp-

eriments with bio-mass in countries such as Switzerland and Sweden (Mourer, 1979). Only if the people of developed countries are willing to scale down their high resource and energy using life styles, and if the people of developing countries accept the fact that it is impossible for them to achieve a "Las Vegas" type of environment, can we begin to dream in realistic terms of a world for tomorrow.

A DREAM FOR THE FUTURE

It is important to begin with dreams, and to follow in an evolutionary way the well-trodden path of such environmental visionaries as Leonardo da Vinci (Rosci, 1978), Thomas More (1965), Robert Owen (Morton, 1969), William Morris (1907), Ebenezer Howard (1946), and Patrick Geddes (Boardman, 1978).

In democratic societies it is essential to find out what the public wants, but it is a truism to say that the people like what they know, rather than know what they like. The experts should, therefore, set out a panorama of environmental ideas based on ideals. Such a dream could begin with the concept of a good environment for everyone, a world paradise. In the land that encompassed the ancient Palestine and Mesopotamia, where so many philosophical ideas originated, and much of which seems to have been generally brown and dry, the ideal physical environment was seen as essentially a green and pleasant land, flowing with milk and honey. If emotional ideas could be converted into practical projects, this would mean the recovery of the wilderness by irrigation systems, regular water supply, shelterbelts, and good husbandry, based on the best scientific knowledge available. What is interesting is that the Biblical ideal, transmitted through the works of Italian painters in the sixteenth, and French painters in the seventeenth, centuries, almost became a reality in Britain during the 18th century, where good husbandry and the humanized landscape combined to form a land both green and pleasant, which was at the same time very productive.

Each part of the earth's surface needs its environmental dream, one that responds to climate, terrain, and culture. When brought together in a terrestrial ideal, these dreams could form the basis for action at all levels. The nations, at last united for a common purpose, would commission a long term plan for the biosphere (and remember that the British 18th century landscape improvers had no hope of seeing the results of their visions).

The world long term plan would indicate priorities where the most acute environmental dangers exist (Doxiades and Papaioannou, 1972). It would be essential that they took precedence over any other considerations, and that remedial action should be effective. Perhaps water is the most critical commodity, and should receive paramount attention, but not in a narrow sectoral way.

When Xenophon led his ten thousand to the aid of a Persian contender for the throne, he was astonished at the number of obstacles in the form of water channels he had to cross. Today too many of these channels are mounds in the desert, and still await excavation. When the Egyptian high dam at Aswan became a political shuttlecock, it frustrated a rational approach to a Nile Valley plan, where a large number of small dams might have had an infinitely better ecological result. But water control is only one aspect of the comprehensively planned development programmes which are needed.

THE CITY OF TOMORROW

Planners today no longer see an urban settlement as an isolated island, nor do they now view the task purely in urban terms. As the Geddesian concept of regional planning has evolved, so the idea of a city as an isolated phenomenon has changed to that of a network of settlements of various sizes, with their populations determined basically at regional level (Christaller, 1966). There is a vital need to keep close links between urban and non-urban land, so that town dwellers are always aware of their non-urban surroundings, not only in terms of urban parks for recreation, but also of agriculture, forestry, and other uses. A clear concept for this policy was set out by Geddes' contemporary Ebenezer Howard (1946), and it may well be that Howard's Garden City dream is nearer to our present needs in terms of energy conservation, good health, and well being, than so many of the recent technologically-based utopias. The fashionable catchwords of growth and change, and conversely of no growth and no change, need to be replaced by the concept of an ideal environment for both town and country that is based on ecological balance.

The approach to planning a city should begin with a client's brief. Since few clients as yet understand how to prepare this seminal document, the planners often become involved in assisting them. First, it is necessary to set out some basic criteria, and the following are essential for any human settlement - health, safety, efficiency, flexibility, continuity, balance, identity, and beauty. Second, minimum space standards must be set for all buildings and land uses. Third, Geddes' traditional SURVEY/ANALYSIS/ PLAN method should be adopted, but in a balanced way, and with a fourth task of IMPLEMENTATION added. Fourth, although it may be a truism to say so, planning must be undertaken for, and with, the people, as planning is a social service for all the citizens. As it is impossible to consult with all the people all the time, it is essential for the public to have representatives who really do represent them, who understand complex problems and issues, and who bring a philosophic approach to their task as collective clients.

Perhaps the key requirement for city fathers is nobility of intent over generations.

This is the spirit that created Venice, Florence, and a few other cities of high environmental quality, and it is one which should animate us all in our quest for civilisation.

REFERENCES

Abercrombie, P., 1944, "The Greater London Plan", HMSO, London.

Acropole, 1960, "Brasilia", Acropole special issue no.256-257.

Bergvall, J., ed., 1952, "General Plan for Stockholm", City of Stockholm, Stockholm.

Boardman, P., 1978, "The World of Patrick Geddes", Routledge, London.

Christaller, W., 1966, "Central Places in Southern Germany", Prentice Hall, Englewood Cliffs, New Jersey.

Couperie, P., 1968, "Paris through the Ages," Barrie and Jenkins, London.

Doxiades, C.A., and Papaioannou, J.G., 1972, The concept of Ecumenopolis, Ekistics, 33:199.

Dutton, G., 1960, "Founder of a City", Rigby, Sydney.

Evenson, N., 1966, "Chandigarh", University of California, Berkeley.

Gaitskell, A., 1959, "Geziera", Faber, London.

Howard, E., 1946, "Garden Cities of Tomorrow", Faber, London.

Johnson-Marshall, P., 1972, "The Planning and Management of Human Settlements for Environmental Quality", United Nations, New York.

Le Corbusier, 1929, "The City of Tomorrow", Architectural Press, London.

Maurer, J. ed., 1979, Proceedings of International Society of City and Regional Planners (ISOCARP) conference, Strasburg, ORL Institut Ethz, Zurich.

More, T., 1965. "Utopia", Dent, London.

Morris, W., 1907, "News from Nowhere", Longmans, London.

Morton, A., 1969, "The Life and Ideas of Robert Owen", Wishart, London.

National Resources Committee, 1935, "Regional Factors in National Planning and Development", U.S.Government, Washington.

Rijswaterstraat Communications, 1964, "A Structure Plan for the South Ijsselmeerpolders", Netherlands Government, The Hague.

Rosci, M. 1978, "The Hidden Leonardo", Phaidon, London.

Science Press, 1977, "China Tames her Deserts", Government of China, Peking.

Stein, C., 1966, "Towards New Towns for America", M.I.T.Press, Cambridge, Mass.

Walls, J., 1980, "Land Man and Sand", Macmillan, London.

Ward, B., 1982, Introduction to Eckholm, E., "Down to Earth; Environment and Human Needs", Pluto, London.

Youngson, A.E., 1966, "The Making of Georgian Edinburgh", Edinburgh University Press, Edinburgh.

2

PLANNED URBAN UNIT DEVELOPMENT: A CASE STUDY: LOUVAIN-LA-NEUVE

P. Laconte

Université Catholique de Louvain
Avenue G. Lemaitre 13
b-1348 Louvain-la-Neuve, Belgium

ABSTRACT

In contrast to the overall de-urbanization of Belgium, the
planning and development of the new university town of Louvain-la-
Neuve, in the metropolitan area of Brussels, is an attempt to create
a high density-lowrise mixture of urban functions in the European
urban tradition. Nine hundred hectares were acquired by the University of Louvain, and an eventual population of up to 50 000 can be
accommodated in the town site. Only the central part of the site
has been developed; the remaining portion is being kept as open
space. The concept emphasizes small groups of town houses and terraced streets, and the division of the site into small plots, each
developed separately by different architects, has led to both variety
and economy.

The new town is highly 'pedestrianized'; that is, its road
system discourages the use of cars for short trips and it has a
railway station to encourage outward journeys by train. The aim is
to develop the "town and gown" interaction found in the old university
towns, with a social mix both of university and town, as well as
university and industry. Moreover, a science park for high technology research and development firms is situated close to an industrial research park. As to resource conservation, the site, located
in a water catchment area, is provided with a dual sewerage system.
All storm waters converge into a 5.5 ha reservoir, which is only
partly sealed in order to allow replenishment of the water table.
This lake is to be filled by the end of 1985. Another resource conservation experiment is the intended use of a nearby sanitary landfill, for energy recovery, agricultural and landscape purposes.

INTRODUCTION

Louvain-la-Neuve is a new university town designed and developed
by the Université Catholique de Louvain (U.C.L.). The University of
Louvain was founded in 1425 in the city of Louvain (Leuven in Flemish),
located 25 km east of Brussels. It consisted of two universities:
one Flemish speaking, the Katholieke Universiteit Leuven (K.U.L.) and
one French speaking, the Université Catholique de Louvain. In 1968
as part of a language-based regionalization process, it was decided
to move the French-speaking U.C.L. to a rural site in the southern
part of Belgium, some 30 km south of Brussels, mainly within the
municipality of Ottignies (Figure 1).

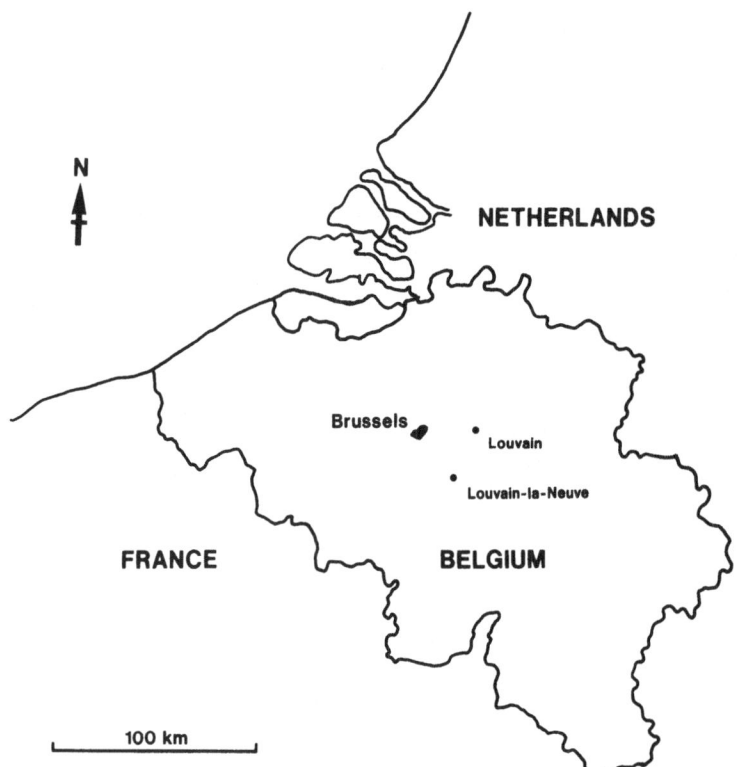

Figure 1: Map of Belgium showing position of Louvain-la-Neuve.

The University authorities decided to develop a university integrated into a new town, rather than an isolated campus. It was
believed that this goal could only be achieved by the development of
a new town by the U.C.L. itself, following the pattern of the earlier
English new towns.

THE PLANNING CONCEPT: AIMS AND OBJECTIVES

The main objective pursued in planning the new university was to recreate the kind of interaction which existed between the academic and urban activities in old Louvain, creating more than a mere university campus, but also a new town with commercial, residential, and industrial activities. Another basic objective was to create an attractive environment for people to live in and several principles were advanced with this in mind.

The presence of a university often tends to create a closed community, so every effort was to be made to attract a diversified population to the new town. This was very obviously dependent upon the variety of employment possibilities; the industrial area would have to provide work opportunities in many different fields.

The town was planned to reflect the human dimension at every stage of development and in each neighbourhood. Low-rise buildings and pedestrian walkways were emphasized. Identification of the individual with his home or place of residence and perception of the community as a whole were considered to be essential elements in the successful development of the town. There had to be interaction between the town and the university, but both were to be allowed to develop separately.

OVERALL PLAN

The long-term planning objective was to have a balanced community of up to 50 000 people, with a maximum of 15 000 students. To achieve that result the university bought 900 ha of land and appointed the Groupe Urbanisme Architecture (Planning and Architecture Group), directed by R. Lemaire, J. P. Blondel and P. Laconte, to produce a master plan.

This master plan was approved in 1970 and allocated an area of 350 ha in the centre of the site for university and urban development, 150 ha for an industrial research and development area, and the remainder was to be allocated for farming and forestry. The first buildings were completed in 1972.

The plan is based on a pedestrian walkway forming a long linear backbone around which urban development can take place (Figure 2). Most of the community facilities and shops are located there and it is linked by pedestrian walkways to the four mainly residential districts which surround it. These districts contain a mixture of academic buildings, student housing, flats, houses, shops and entertainment facilities - an arrangement which should allow complete integration of university and town activities.

Figure 2:   Part of the pedestrian walkway which runs through Louvain-
            la-Neuve forming the backbone around which urban develop-
            ment has taken place.

The master plan also allowed for the growth of urban services parallel to population and housing growth. By mid-1980 there were 130 shops and restaurants.

LAND ACQUISITION

The university was able to acquire this land by means of a loan from the Belgian government at the low interest rate of 3.18 per cent per year over 40 years, including capital repayments. The conditions of the government loan stipulate that the university can make no profit from the sale of land. Its policy is, therefore, to retain ownership of it and to sell long-term leases (from 55 to 99 years) on the land not required for university development to private developers.

PLANNING

The land to be developed was divided into a large number of small plots (200 to 400 sq m). Some were designated for university buildings and the remainder put on the market and the leasehold sold to individuals or small developers.

The buildings were designed by a large number of architects (over fifty for the university buildings alone), and constructed by several small contractors. This had the result of increasing the competition between the contractors and reduced architects cost estimates by up to 30 per cent. The increased burden of co-ordination borne by the university and the occasional bankruptcy of small contractors have not outweighed this advantage.

The division into small units encouraged visual variety within the framework fixed by the master plan, which laid down rules for building size - there is a maximum of three stories - and imposed a degree of uniformity on the building materials used.

HOUSING

The high density-lowrise concept is based on terraced forms of attached double-family houses, town houses, semi-detached houses and lowrise apartment houses. The format chosen depends in each case on the topography, the proximity to the centre of the town or to other residential areas, the growth of the town and so forth. The main objective is always the integration of the town into the landscape by taking advantage of natural contours.

The emphasis throughout has been on small groups of town houses. A layout of groups of about seven houses covering 120 to 200 sq m

was found to be a particularly economical way to build, as it taps
the substantial market of building materials for large single-family
houses. If the number of dwellings in a building unit is greater
than seven, one enters the market of large projects, and it becomes
more economical to build ever larger units, which become inhuman in
their scale and give rise to vertical transportation problems.

The use of terraced forms of housing with roofs and attics which
might accommodate solar heating devices confers a possible advantage
in the future. In addition, attics provide storage space and thus
act as the "memory of the family".

The need for elevators has been reduced because buildings are
low. Moreover, the contour of the site has been used in such a way
as to provide entrances at different levels. The few elevators which
have been installed are for goods deliveries and for the use of handi-
capped people.

COMMUNITY FACILITIES

Community facilities, including cultural activities, administrative services, shops, restaurants, and offices, are mainly located
in the town centre. The four residential areas connected to the
centre by pedestrian ways, nevertheless have their own local shops
to meet everyday needs. Other community facilities, such as schools,
churches, sport and recreational facilities are distributed throughout the town, so that they can all be reached on foot.

The centre includes 33 000 sq m of commercial space located above
the car parks, roads, and the railway station.

The buildings assigned to sports and physical education (67 000
sq m) are located near the centre of Hocaille, one of the residential
areas.

The university's open spaces and sporting facilities serve as
meeting places for the population of Louvain-la-Neuve and the surrounding areas.

TRANSPORTATION NETWORKS

There is no through traffic in the centre of Louvain-la-Neuve,
save for the railway which penetrates the very heart of the town.
The road design is based on a ring road and culs-de-sac. All motor
traffic coming towards the town first follows this outer road, from
which three others give access to the centre and the underground

parking lots. Residential areas are connected to the outer road and
from it to the town centre.

The railway station is in the centre of town and there is ample
parking space nearby. This encourages trip to Brussels, for instance,
by train (30 minutes) rather than by car, and links Louvain-la-Neuve
to the main Brussels-Luxembourg-Basel line. The new station is a
terminal, and beyond it is a non edificandi area where no building
can take place. This allows for the future construction of a loop
railway which could serve other settlements.

After several years of operation, the number of railway passengers
is already 100 per cent higher than expected and the frequency and
length of trains has increased.

THE INDUSTRIAL RESEARCH PARK

Besides commercial, administrative and cultural activities developed in the centre of the new university town, an industrial research park, was built to attract non-polluting, technology-based
firms and to diversify the social mix of the population of Louvain-
la-Neuve. The close proximity (less than 1 km) of most firms to the
social facilities, shops and restaurants of Louvain-la-Neuve, has
proved to be the most effective way of encouraging contracts. As of
mid-1980, 30 firms had settled there - for example, IBM, Monsanto,
BP Chemicals, Abbott and Cyanamid - and have created over 1 000 jobs.

RESOURCE MANAGEMENT

Sewage and water management

The water collected on the site replenishes the groundwater
reserves. All rainwater is collected in a 5.5 ha artificial lake
at the lowest point on the site which serves as a storm basin. As
stormwater can be polluted, part of this lake area forms a temporary
holding area.

The bottom of the lake is only partially sealed so that natural
infiltration allows recharge of the aquifer. In dry weather, the
lake level will be kept constant by water pumped from the water table.
This basin should be an attractive as well as useful feature of the
new town. The fact that waste water is separated from the rainwater
means that a smaller quantity has to be treated. As treatment costs
are proportional to the quantity of water treated and not to its
degree of pollution, this represents an important economy. It has
thus been possible to reduce the diameter of the sewage pipes to
50 cm. The lake is now ready to be filled.

Heating

The high density-low-rise concept has the immediate consequence
of protecting the plateau from the unpleasant north winds and improving the microclimate. This is in itself a source of energy saving;
indeed, it is a feature of old towns that a curtain of continuous,
low-rise buildings acts as a barrier to the winds and preserves heat,
which is a major consideration, given the climate of Belgium. The
buildings retain heat because the houses have common walls and form
an aggregate which diminishes heat loss, unlike isolated apartment
blocks or detached houses. After five years a noticeable improvement in energy saving has occurred at Louvain-la-Neuve.

Methane recovery

Close to the site an existing sanitary landfill is used to
collect the garbage from the Brussels area. This has been an obvious
environmental nuisance, but a technique has been developed at the
university to recover the methane generated from the garbage to heat
the plants of the industrial park. A pilot operation is expected to
start in 1985.

Management and maintenance costs

Scaling down to small individual units is not, per se, a source
of economy, but it has been noticed that people take better care of
their own individual homes and undertake voluntary maintenance rather
than depend on the services provided by a large organization. The
variety of design also allows for greater flexibility in the future
and tends to reduce the rate of obsolescence of the buildings and the
need for renewal.

CONCLUSIONS

This case study of Louvain-la-Neuve suggests that a high density-
lowrise layout has some economic advantages not only saving energy
consumption per capita but also by lowering the cost per square metre.
It seems worth considering for new developments in any country, and
it appears to be suitable both for urban extensions and new settlements. Its flexibility facilitates the use of local materials, local
craftmanship, and local techniques.

The savings on the road investment needed in the linear pattern
of development allows for a "generative" growth which can, depending
on the economic circumstances, accelerate or stop altogether without
incurring undue debts for capital repayments.

In terms of social policy the high density-low-rise layout cannot
in itself determine life-styles and change consumer attitudes or

behaviour, but it does create the physical framework for informal contact among people and with their surroundings. It can produce a better harmony of scale between existing buildings, trees, and new types of construction and link urban amenities to our urban history which probably has more to teach us about man's requirements than a few decades of mono-functional allocation of urban space.

Such layouts do, however, depend on their small scale and on competition between a wide range of contractors. The implementation of such a project requires specific skills in developing working relations with each level of government (municipal, provincial, regional and national). The novel legal and administrative aspects of the implementation of the Louvain-la-Neuve project have generated comparative research projects on the factors influencing urban planning and design, and the institutional prerequisites of building an environment better adapted to its inhabitants.

3

FRAGMENTATION OF LAND - RURAL AND URBAN:

A MAJOR PROBLEM OF SOUTH EUROPEAN MEDITERRANEAN COUNTRIES

M. L. Da Costa Lobo

Avenida Miguel Bombarda 83
1000 Lisbon
Portugal

INTRODUCTION

Landscape has been affected by natural influences with resultant tectonic movements, erosion and the formation of river basins, with watersheds and deep waters. Many of these natural influences tend to fragment the landscape, a process accelerated by the activities of Man.

Man has developed an acquisitive instinct often leading to the possession of private property and consequently to the fragmentation of land, the boundaries being indicated by fences and hedges which, in turn, also contribute to changes in the landscape. Land is also fragmented, for the convenience of administrators, into administrative divisions. There are five reasons for the subdivision, or parcelling, of land:

    (i)   to facilitate different uses of adjoining land
    (ii)   the acquisition of private property
    (iii) the definition of administrative divisions
    (iv)   the identification of National frontiers
    (v)   the restriction of large engineering works, networks.

Within each parcel of land there is a clearly preferred use, the implementation of which is the prerogative of private or public owners. Strict legally binding rules regulate ownership by inheritance and/or acquisition; in some instances these rules stimulate further subdivision.

The administration subdivision of Portugal can be traced to the Romans who subdivided the "Peninsula Iberica" into Provinces, the

Portuguese Kingdom being founded in the XIII th century, when its
boundary was established as a result of conflicts, notably, with
the Moors and "Castilla", the latter being one of the Kingdoms of
Spain. Very soon "concelhos", series of muncipalities, were established for the exercise of power at the community level. Subsequent
development and demographic growth, led to the recognition, in
association with the Christian churches, of parishes. In due course
parishes were superceded by "freguesias", a subdivision of "concelhos".
Also, the need of a regional framework led to the adoption of
"provincias", and later, "distritos". As a result of the 1976
Constitution a new 'local' authority, the Administrative Region, was
foreseen, it will replace the "distritos". While this has been
happening, 296 "concelhos" have persisted and their powers greatly
widened by Acts "DL 79/77" and "DL 1/79" - Municipal Power (1979)
and Municipal Finances (1979).

Historically, the crown, the church, and ranges of Portuguese
foreign nobles have exercised authority over almost all of the
inhabited land. Following a succession of liberal movements and
battles in the XIX th century the monarch finally fell in 1910 with
the State acquiring the monasteries and other crown lands. At the
same time the large landowners saw their "latifundia" in the south
of Portugal (Alentejo) occupied, for some years, by 'workers'. In
return for working on the "latifundia" rural workers were able to
rent rural plots, "foros", on which they were able to build houses
while eking out their subsistence from the rest. Since the 1974
revolution, some of the "latifundia", have again been occupied by
rural workers. However, a new law has been enacted to limit the
amount of land owned by one landlord, the 'excess' land being
expropriated by the State and given to cooperatives for development. Some of these state owned farms have been subdivided into
small family farms the 'title' of which has been given to rural
workers. These administrative divisions are devices for optimizing
management but they also provoke political 'in fighting' e.g. the
strongly felt parochial views re boundaries are the causes of
friction between neighbouring municipalities and parishes.

FRAGMENTATION OF RURAL LAND

Four patterns of land fragmentation can be identified in
Portugal:

(i)   The "latifundia" of the Alentejo (to the east and south
       of Lisbon)
(ii)   The mountains
(iii) The "minifundia" of the Porto Region and Minho
(iv)   The "minifundia" of the Aveiro Region

(i)   The Alentejo latifundia effectively led to a separation of
rural and urban settlements. The rural settlements are centred on
"montes" - groups of farm buildings and workers' accommodation,
usually at the centre of a large estate, sometimes near to the landlord's residence. Whereas this arrangement is typical of the large
non-fragmented landscape where the heritage is associated with the
landlord's residence, in towns, land ownership has been determined
by the configuration of roads and the need for privacy, with heritage
being related to castles and churches. In recent times, urban bye-
laws have played an increasingly important role. Houses vary from
8 to 12 metres in depth with 10 m as an average, while a housing plot
is extended to 20 m by the inclusion of 6 metres for a backyard and
4 metres for a front garden. But this standard plot size, is likely
to be modified by the layout of roads and converging streets. With
a depth of 20 m, plots may be 6 m to 10 m wide for terraced housing
and 10 to 15, or 12 to 25 m for semi-detached, and detached, houses
respectively.

(ii)   The mountains: a mixture of large forest estates, subject
to private, state or communal (baldios) ownership with small holdings,
"minifundia", in the valleys, based on subsistence agriculture. In
these instances the goat is the prime domesticated animal providing
meat, milk, cheese, leather and bone (for handicrafts): on the other
hand, goats are responsible for irreparable damage to forests with
consequent land erosion. Typically, living conditions in the mountains are sub-standard and not surprisingly emigration is commonplace.

Towns in the mountainous regions tend to be small and very
densely populated. Their charm is associated with irregular streets
the location of which has been determined by history and topography.
Roads tend to follow watersheds or take what used to be the easiest
and possibly safest (vide banditry) route between neighbouring
settlements.

(iii)   The minifundia of the Porto Region and Minho: the fertile
land and generally favourable environment of this Region has led to
very large population densities, which in turn have made a demand on
other resources - namely industrial developments. Typically each
family augments its 'subsistence' farming by a secondary occupation.
Many crafts are strongly developed - tapestry, embroidery, jewellery,
furniture making, etc. - but inevitably the concentrated work force
has attracted the incursion of 'big' industries. The adoption of a
secondary occupation has given the population a degree of stability
but because of the system of inheritance, at the price of severe land
fragmentation. By tradition, bequeathed land of differing quality
(i.e. suitable for arable farming, forestry etc.) has been equally
shared among offspring which frequently means plots are not adjacent.
If two such inheritances come together - as a result of a marriage -

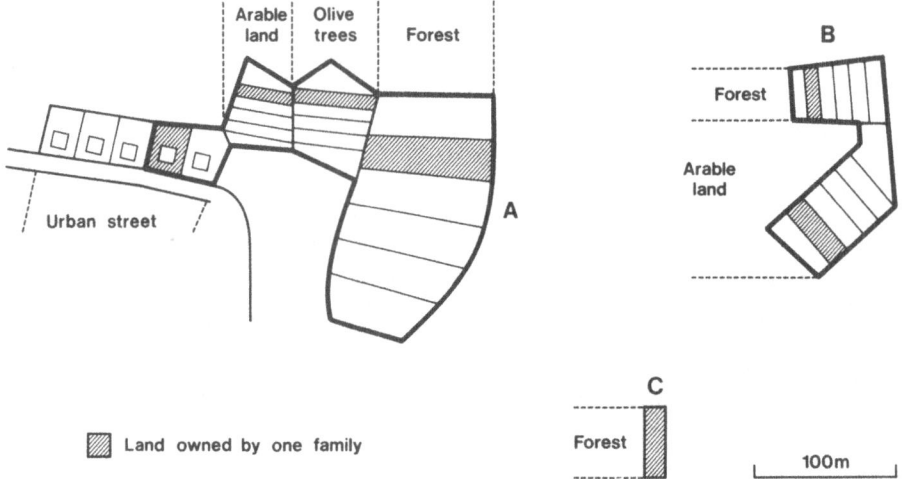

Figure 1: The seven separate plots of land comprising the frag-
           mented farm of one family in the northern Guimaraes
           region of Portugal. Four of the plots were inherited
           by the farmer as equal share (of six) of his father's
           land (Farm A), two were inherited similarly by his
           wife (Farm B). They subsequently bought plot C. The
           distance between A, B and C is not to the scale pro-
           vided which refers only to plot size. A, B and C are
           separated by several hundred metres.

the resultant 'farm' can consist of many plots of land often some
distance apart (Figure 1). Each family endeavours to have fertile
land suitable for arable crops and vegetables (sometimes irrigated),
sloping land for vineyards and less fertile land for olives and
pines.

Although the excessive degree of fragmentation may be regarded
as inefficient, the possession of land gives owners a sense of
security. Furthermore, it enables those members of a family who
are not employed elsewhere to contribute to the well-being of the
family with cultivation being regarded as an invaluable hobby. This
system is compatible with the individualistic nature of the population of the Porto region and, by enabling the production of fresh
food, fosters a strong sense of 'family'.

Even if it were desirable, the traditional method of inheritance
virtually precludes land rationalisation with the aim of obtaining
few large fields or areas of woodland. The problem of fragmentation
is likely to intensify as population pressures lead to land speculation for house construction. It will be difficult to prevent the
wider unplanned scatter of houses throughout the Region although the
Acts of 1965 and 1973, which insist that the subdivision of land must
await planning approval, may act as deterrents.

(iv) The minifundia of the Aveiro Region: the trends in this
Region are similar to those in the Porto Region and Minho but are
less intense. The population density is less and the degree of
fragmentation less extreme. As a result, living standards are
higher, the situation being aided by the traditional emigration of
surplus population to South America. While farming is still a
profitable activity, the production of wine is an important factor;
as is the availability of fish, salt, seaweeds (for fertiliser)
from the sea. Together, these enable most families to have a reasonable income but already the pressures of land speculation and
accelerated land fragmentation are evident with increasing urban
affluence.

Typically the towns developed from rural settlements and with the
addition of churches, shops and schools, development has tended to
be linear with discernible ribbon development linking adjoining
communities.

FRAGMENTATION OF URBAN LAND THROUGH HISTORY

The fragmentation of urban land through the ages can be conveniently followed by studying maps. From a formal orthogonal roman
lay-out, urban development became unsystematic, a random and unplanned arrangement being characteristic of the Middle Ages with the
designation of "mourarias" (the Arab quarters) and "judiarias" (the
quarters of the Jews and Christians, very often designed by the

Templars). This happened during the XII th century, as part of the
Christian strategy in the Portuguese Kingdom for reconquering the
south of the Iberian Peninsula. The Arab quarters typically had
labyrinthe streets; the houses of the rich having internal "pateos".
Subsequently urban development was determined by the Jesuits (XVI
and XVII centuries) and with an increasing military influence, the
walled towns were built for ease of defence.

The XVIII th century, mainly through "Marques de Pombal" a powerful prime-minister, saw the reassertion of civilian design notably
in the reconstruction of Lisbon after the 1755 earthquake. This
reconstruction was based on a "gridiron" with open squares for markets,
processions, public meetings etc, the blocks being about 25 m x 75 m.
In the XIX th century many of the present-day concepts of urban
development started to emerge with incipient ribbon development, conurbations of detached and semi-detached houses and blocks of flats,
the location of which was not rigidly determined by street alignments.
At the same time towns started to intrude into rural areas with the
consequent fragmentation of the rural fringe. These developments
tended to occur so rapidly that the organisations concerned with
urban planning were severely tested in the 1910s/1920s. In many of
the Mediterranean Countries of Southern Europe illegal subdivisions
and developments were commonplace with the construction of 'unsymp-
athetic' homes which conformed with neither traditional housing nor
by-law regulations, for example :

(i) the "gececondu" or the illegal development of usually flimsy
houses on private land without the permission of the landowner as
happened in Turkey. (vide the "bairros da lata" in Portugal).

and (ii) the "loteamentos clandestinos" developed by private
speculators in Portugal with total disregard for coordination.

These developments can have a profound effect. Thus, two
initially similar small towns (same size and shape) can, on development, acquire totally different forms. In one instance, the urban
perimeter may extend in a stepwise manner every year with the retention of an urban master, or structure, plan whereas in the other,
the random development of the rural surroundings will lead to a pro-
liferation of nearby satellites. Thereafter the steady spread of
population into intervening areas will lead to the emergence of a
consolidated but unstructured conurbation with all its attendant
faults.

ILLEGAL FRAGMENTATION OF LAND

Introduction

The Mediterranean countries of Spain, Greece, Portugal, Turkey,
Yugoslavia and southern Italy have features in common which differ

from those of other parts of Europe which were 'industrialised' sooner. Whereas urbanisation is expected to decline in most North American and Western European countries it is, on the contrary, expected to intensify in the Mediterranean countries where the agricultural population still accounts for a large part of the total labour force. Already rapid urban growth has seriously challenged the provision of land with services, the availability of housing and the establishment of institutional frameworks - these problems are likely to be exacerbated in the future. Past attempts to solve these problems have usually been inadequate and misconceived. They usually did not take due regard of planning and regulatory statutes and, as a consequence, were often wasteful of resources, particularly land. This triggered off conflict between rural and urban land allocation and heightened tensions between the public, and private, sectors.

The planned control of urban development is one of the most serious challenges facing Mediterranean countries. It is often thought of as an urban problem but in reality it is fundamentally a matter of land-use. Land taken for the urban sector automatically decreases the rural resource. How have the authorities attempted to resolve this problem ?

Illegal development

There are many different types of illegal development. Type A, is common to all the countries (except, perhaps, in Yugoslavia) - having obtained official approval for building construction on legally recognised plots of land ('lots'), construction is done with total disregard of official building regulations often at an inflated density (numbers of housing units per unit area). Type B involves building without approval of any sort and ipso facto occurs without due regard to water supplies, sewerage, electricity, etc. Shanty-towns with dwellings constructed overnight of easily perishable material are frequently located on publically-owned land where dwellers are subject to eviction. Type C relates to the illegal fragmentation and sale of land outside the geographical limits of already urban developments, a form of infringement characteristic of emigrants who wish to send money back to, and invest in, their country of birth.

These illegal developments and others pose many problems:

(i)   lack of basic services and infrastructure creating an unsatisfactory environment; poor standard, unsafe, insanitary living conditions where disease may flourish; minimal transport facilities for travel to work.

(ii)   social problems, resulting either in violence and unacceptably high crime rates or in family disruption and the loss of opportunities.

(iii)   political instability, with a lack of local involvement.
(iv)   conditions not conducive to economic development.

Illegal subdivision and housing in Portugal

Historical background. Although individuals may be living in
towns, rural traditions in the predominantly rural Mediterranean
countries persist. It is possibly for this reason that urban planning
has not been wholeheartedly accepted in Mediterranean countries -
Portugal is no exception. Before the mid-20 th century, there was
virtually no central government involvement in urban development
except to a limited extent in major towns. Since 1940, however·,
development plans have been systematically developed for the larger
towns. The 1944 "Urban Planning Act" provided a general town
planning directive but it was insufficient. Unplanned developments
continued to mushroom in spite of the Act. A crisis was reached
during the 1950s and 1960s: it was attributed to internal migration
from less, to more, favoured parts of the country and to the flow of
money to the country from emigrants to Northern Europe who remitted
part of their earnings for the construction of houses near their
places of birth. Never, in Portuguese history, were so many houses
built. In the peripheral areas of big cities, the illegal subdivision of sites was commonplace suggesting that the government was
either unable or did not wish to release land to fuel the massive
urban migration which accompanied industrialisation or to answer the
requirements of emigrants. There was a lack of clarity in municipal
and central government policies.

Control of illegal Subdivision of land. The law does not allow
the fragmentation of agricultural land below a predetermined size
without a municipal permit. Nonetheless, illegal speculative subdivisions persist. To counter these illegal acts it is essential
to have a realistic land policy which can be implemented simply and
efficiently. The policy needs to identify positively the land that
can be used instead of saying what can not be used.

In 1974, there were substantial political changes in Portugal
with the installation of a new regime and the granting of independence to former colonies. In attempting to widen the democratic basis
of government there was a noticeable increase in illegal urban sprawl,
especially in Porto and Lisbon and along the coastal strip. The
country was ill-prepared, with inadequate provisions for water supply,
power, and sewerage. These deficiencies have led inevitably to
residential developments lacking these amenities and, in turn, these
defects spawn a variety of economic and social problems. As a result,
the task of maintaining a stable urban environment in Portugal has
been greatly magnified. Irrational patterns of urban development are
common and beautiful landscapes and structures of architectural merit
have been destroyed or are in jeopardy. The situation necessitates
a new approach to the planning of urban developments.

Control of illegal Subdivision and Housing in Portugal. In
Portugal, as in every other Mediterranean country, illegal subdivision
and housing has attracted the serious attention of policy-makers.
After the political upheaval of 1974, the Portuguese government announced a move to decentralise some of its powers to authorities at
municipal and regional levels (the regionalization of Portugal is
still to be confirmed).

Some municipalities tried to cope with the new problems of rapid
urbanisation and political change. They have also tried to improve
the 'administration' by the appointment of technical specialists.
However, despite some successes, the image of urban planning in
Portugal is less than satisfactory with disputes between organisations
and expert groups, and between local authorities and central government (Ministry of Internal Affairs, Ministry of Finance and Planning,
and Ministry of Housing and Public Works) exacerbating a situation
already hampered by the more or less universal lack of financial
resources.

Case Study of Illegal Development: Municipality of Seixal

Seixal, one of the suburbs of Lisbon, with its own industrial
estates, has grown continuously from 1970 to 1981 (+138 %) with
industrial expansion and population growth being at the centre of the
problem. The shortage of housing and the inflated prices in Lisbon
has compelled the population to explore the nearby suburbs and inevitably illegal developments in Seixal have been located alongside the
main road from Seixal to Lisbon, mainly near the major road inter-
sections. To counter this trend a municipally organized local action
group with strong citizen participation has been instituted in an
attempt to implement a General Strategic Plan for Seixal.

The General Strategic Plan for Seixal

The General Strategic Plan includes a requirement for private
developers to cede a proportion of their land to the Municipality.
This requirement was introduced because it was foreseen that strong
municipal action would be required to achieve a good spatial location
of urban development based on a municipal stock of land - in Seixal's
case, 300 hectares are needed for urban development during the next
5 years. The Seixal Plan is also attempting to safeguard land of
'environmental' importance and that needed for future railways and
roads, electrical transmission, pipelines etc. The proportion of a
developer's land to be ceded to the Municipality for landscaping
ranges from 0 - 80 per cent depending on zonation. Of the land
retained by the developer, 30 - 50 per cent must be set aside for
the construction of public and social facilities and as a result of
these considerations the percentage of land actually ceded by a
developer to the Municipality ranges from 50 per cent to 83 per cent.
These requirements, which are applied without exception, are intended

to avoid ad hoc and hasty decisions based on arbitrary reports and
subjective opinions. They strengthen the element of urban control.

The Plan for Seixal will be implemented by the development of
Complementary Plans, for each of its seven sub-areas. The Complementary Plans will detail the land to be designated for services, roads,
environmental and cultural purposes etc. It is always important to
study new strategies and policies taking local needs and attitudes
into account - a plan must not be allowed to become discredited. The
rule should be - maximum individual freedom until that freedom incurs
unacceptable social costs (compromises and rules are therefore
needed).

Community Action. Starting in 1974, a system of planning and
management aid groups (GAPG) was established as a technical adjunct
to local authorities; the groups were formed to deal specifically
with the problem of illegal housing.

Summary and Conclusions

Predictably rapid urbanisation will continue in Mediterranean
countries during the 1980s - thereby exacerbating an existing problem.
There will be increased demands for land, services and housing and
social facilities while, at the same time, resources will be limited
with a still deficient ability to 'moderate', and implement, agreed
and agreeable plans.

To minimize friction and to facilitate the wise use of valuable
resources it is proposed to :

(i)   Design plans, and organise an administrative structure,
        capable of implementing the plans without involving the
        heavy hand of government.
(ii)   Make annual progress reports on the development of urban
        policies, and submit them for national and international
        review at least among OECD countries; and monitor the
        implementation of each plan seeking the collaboration of,
        and exchange of ideas between, experts with different
        backgrounds and points of view.
(iii)   Generalise the system of monitored case studies to include
        urban renewal and conservation.
(iv)   Develop the training and education of planners and
        related officials, not discounting the value of profess-
        ional education abroad.
(v)   Strengthen the professional bodies related to town plan-
        ning and to encourage them to take due regard to rural,
        in addition to urban, issues, and the rural/urban inter-
        face, noting that the urban population ipso facto makes
        demands on its rural environment. They are inter-related.

ECONOMICS OF LAND FRAGMENTATION

Essentially, prices per unit of land are larger for small, than for large, parcels. For plots of 5000 m^2 a price of 500$00 m^{-2} (in Portugal, there are 100 cents per Escudo ($)) would be considered reasonable but if this area were to be sold in units of 400 m^2, a price of 600$00 m^{-2} would be expected, an increase of 20 %. In this way, the seeds of inflation are sown. In Portugal, if it were proposed to develop a 20 ha site to house 5 000 individuals, an allowance of 25 m^2 per person would be expected for social services, green zones, play-grounds and other civic installations. That is a total area of 25 m^2 x 5 000 = 12.5 ha leaving a residue of 7.5 (= 20.0 - 12.5) ha for streets and housing plots. If 20 % or 1.5 ha were designated for streets, 6 ha would be available for housing. If each structure occupied 20 % of its ground allocation there would be a total ground floor area of 12 000 m^2. If each individual needs 25 m^2 of building area then to house 5000 individuals the buildings must have $\frac{25 \times 5000}{12000}$ = 10.42 storeys on average. If each m^2 were valued at 100$00 then 20 ha would cost 20 000 000$00. However as 12.5 ha will be set aside for services and social facilities, the cost price has to be concentrated on 7.5 ha to give a unit cost of 267$00 per sq.m and this disregards the costs of street-works and of infrastructure, administration, interest payments and the effects of scale.

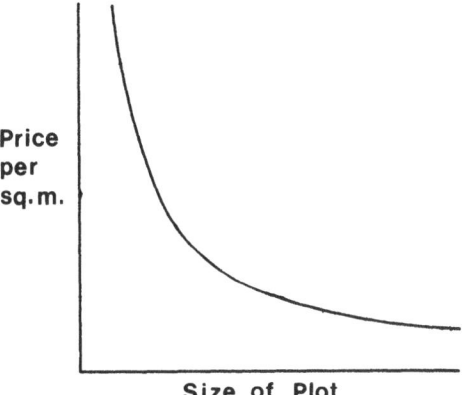

Size of Plot

Figure 2: The relationship of plot size to price as land is subdivided and sold as smaller plots for development, in
Portugal.

It is possible to predict the final value of land when it is subdivided into small plots - it follows a hyperbolic function (Figure 2). Assuming a plot of land (so) with the total area (sg) devoted to buildings, streets and social amenities being divisible into Sp and Sn, the areas required for single housing plots, and

other buildings and roads respectively (Figure 3), then a final land
evaluation can be calculated with knowledge of the following attributes.

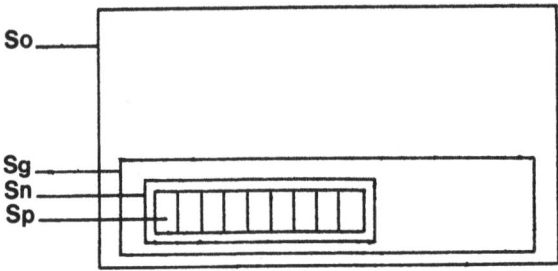

Figure 3: The proportion of land used for single housing plots (Sp),
           other buildings and roads (Sn), related to the total
           built-up area (Sg) and the initial plot of land (So).

l = cost of land per sq. m

l_o = h (So) where l_o = initial cost of land per sq. m

Sg = $K_{o/g}$ x So (i.e. area of land available for overall devel-
       opment from the initial area purchased (So) is determined
       by $K_{o/g}$, the coefficient transforming So to Sg)

Sni = Area of land of housing unit i

Sni = $K_{g/n}$ x Sg (i.e. area of land available for housing units
       from the area Sg is determined by $K_{g/n}$, the coefficient
       transforming Sg to Sn)

Sp = $K_{n/p}$ x Sn (i.e.area of land available for individual plots
       from the area available for housing Sn is determined by
       $K_{n/p}$, the coefficient transforming Sn to Sp)

$t_{g/n}$ = time elapsed in establishing secondary infrastructure (n)
       following that of general infrastructure (g)

$t_{n/p}$ = time elapsed in building after establishing general infra-
       structure ($t_{g/p}$ = $t_{g/n}$ + $t_{n/p}$)

$t_{o/p}$ = time elapsed between acquisition of land and completion of
       urban development

$t_{p/s}$ = time delay in selling plots following completion

K_d = a factor to account for the distribution of secondary infra-

structure costs of each level of land sale to reduce the
final cost $(K_d \simeq 0.5)$

C_1 = % of plot sales

C_2 = % of total administrative investment/year

f   = interest on loans

r   = risk factor (a percentage of each sale level)

The final price of the plot per sq. m is :

$$A = \frac{1}{K_{o/g} \cdot K_{g/n} \cdot K_{n/p}} \cdot 1_o \cdot (1 + f + C_2)^{t}o/g$$

$$B = (A + 1_g \cdot \frac{1}{K_{g/n} \cdot K_{n/p}} + 1_n \cdot \frac{K_d}{K_{n/p}}) \cdot (1 + f + C_x)^{t}g/n$$

$$C = B \cdot (1 + f + C_2)^{t}p/s \cdot (1 + C_1 + r)$$

$$= C \cdot h(Sp)$$

The value of h (Sp) is always >1 and increases by $\frac{1}{Sp}$ if Sp >S_{min}
where S_{min} = minimum size of plot.

To consider further the process of buying large areas of land
and selling sub-divided smaller plots we can take a hypothetical
example e.g. an area 50 x 50 km on the outskirts of a metropolis
where illegal sub-division has already taken place before urban
expansion reached the site. The farmland site was originally bought
by black market developers for 100$00 per sq.m. They divided the
land into 400 000 farms of 100 m x 50 m which each require a 25 %
land allocation for streets and amenities. The total land area
required for each farm is 6 250 sq. m. The developer then sold
these farms at 250$00 m^{-2}. His % profit, allowing for the loss on
the 25 % required for services etc., amounts to 100 %.

The 'secondary' developer who bought these farms further subdivided the land into units for house building of 200 sq. m and,
after spending 250$00 m^{-2} on infrastructure, sells the land at
1 500$00 m^{-2}. In fact, he can only sell 70 % of the land he bought,
because 30 % is used for the development of infrastructure.
Therefore his per cent profit is limited to $\frac{1500 - (250 + 250)}{500}$ x

100 x $\frac{70}{100}$ = 140 %

Speculation and land shortage could bring the price/sq. m to
3 750$00 which is an additional 150 % profit on the 1 500$00 m^{-2}
price.

The cost to the authorities of providing social amenities for
these smaller units with the concomitant increase in population
becomes a very serious problem. The acceptable number of people per
hectare on land designated as farmland with farms of 5 000 sq. m is
4.8 persons/hectare. In an area 50 x 50 km this would allow settlement of 1 200 000 persons. Each person requires 25 sq. m, a total of
30 x 10^6 sq. m. This land is ceded by the developer according to
Portuguese Law. However, if each 5 000 sq. m farm is subdivided into
10 plots of 400 sq. m (allowing for loss of land for streets etc.)
the 50 x 50 km area will break down into 4 000 000 plots. Assuming an
average number of people on each plot to be 3, there are 12 000 000
people on the land. This works out at 48 persons/hectare thus
reducing the amount of land 'available' for social amenities to 2.5
sq. m per person. On 200 sq. m plots this land is reduced to 1.25
sq. m/person (housing 24 000 000). In this situation the authorities,
working to a requirement of 25 sq. m/person for social amenities
need to find 24 x 10^6 x 25 = 600 x 10^6 sq. m leaving an additional
requirement of 570 x 10^6 sq. m. This land would cost 570 x 10^6 x
1500 = 855 000 million Esc. (at 1500 Esc./sq.m). This is a prime
cause of financial concern !

Although the example of a 50 x 50 km area of land is hypothetical,
in similar situations the developers have no difficulty in finding
buyers and the sequence of events just outlined is likely to be
followed. The buyers are likely to be emigrants, investors, developers who may each purchase four farms each 5 000 sq.m, leading to
up to 1 200 000 inhabitants occupying 2 500 sq. km. This is an
inefficient use of land. If the same number of persons were located
on 400 sq. m plots the density of persons per hectare would be 60
i.e. $3 \times \dfrac{10\ 000 - (0.2 \times 10\ 000)}{400}$ where 3 = average number of persons/
family; 10 000 (sq.m) = 1 hectare; 0.2 x 10 000 (sq.m) = area required
for streets/hectare; 400 (sq.m) = 1 plot.

Taking into account the need for space, for social amenities and
industry, the density approaches 50 persons/hectare. The land used
for urban purposes would be only $\dfrac{1\ 200\ 000}{50}$ = 24 000 ha which would
release 226 000 ha for agriculture, forestry and amenity in an area
50 x 50 km. Clearly this is a wiser use of available land resources.
If multi-storey blocks of flats were used, up to 130 persons/hectare
could be housed while maintaining the same area for agriculture etc.
In fact, when the process of land fragmentation proceeds over a
large area the local authority cannot possibly buy the land needed
for social amenities. The tendency is for them to buy land at the
last possible moment which is usually when fragmentation has taken
place and pressures already exist for amenities. Thus, when confronted by high land prices and when suitably located land is not
always available, the authorities reduce the social amenities to a
minimum and place them at considerable distances from housing. This
process has led to inadequate facilities bought at exhorbitant prices.

TOWN PLANNING ISSUES

The overriding issue is the ownership of land and the legal
consequences and limitations of that ownership. If this can be
resolved then it is possible to achieve unified planning and implementation, as happened when the Rotterdam Centre was developed after
World War Two and when the centre of Lisbon was developed in the
18th century after the disastrous earthquake. But, in Mediterranean
countries, this ideal solution can rarely, if ever, be obtained.
However, to mitigate against the inflationary and divisive influences
of land fragmentation, the use of a local-tax index might be considered, for instance:

Maximum floor space index for plots of <5000 sq. m   = 0.3
    "   "   "   "   "   "   " <5 hectares = 0.4
    "   "   "   "   "   "   " <50 hectares = 0.5

Attempts to wrestle with this problem were made in Great Britain
whose Act of 1947 attempted to abolish development value of land.
In Portugal there was an attempt in 1970 with the creation of the
concept of 'Programmed Expansions'. This system was based on the
progressive and compulsory acquisition of all land needed for urban
expansion (incidentally, at a price unrelated to the proposed land-
use). However, it was doomed to failure because of the undue restriction on enterprise. Another attempt is being made, the new Act
being based on a greater degree of collaboration between the private
and public sectors. After designating an area for development the
municipality can invite the land-owner to develop. If he/she is not
willing to do so, the municipality can offer the investment-cost of
infrastructural development paying the 'developer' 7 % of the value
of the final product. If the owner is still unwilling to develop,
the municipality can acquire the designated site by compulsory means.

A second problem is the "rehabilitation" of housing developments
where due regard is not paid to the space allowances for 'infrastructure' and 'social facilities'. Realistically, more land is
required, but usually this is not available without regrouping the
different parcels of land.

Rules and regulatory devices, including payments by tenants and
owners, can be developed with the best of intentions but, bearing
in mind temperament, they can be counter-productive. In the
Mediterranean countries there is still a need for a long period of
education before the social limitations of integrated planning are
acceptable and accepted.

A third issue concerns the design of building developments
including the siting of open spaces of different shapes and sizes.
The success or otherwise of a town or city depends on the fragmentation of space. It can change the micro-climate and alter appear-

ances. If we accept that green belts should be interspersed within
a town, the distance between neighbouring green belts will be of
importance. If the belts are not more than 3 km apart everyone is
likely to be within 1.5 km of a green zone with its associated influences on the quality of life, a topic that also revolves on the
prudent use and location of trees and hedges.

REGIONAL PLANNING ISSUES

    Land fragmentation is a central issue. Ideally it would be pre-
ferable to work with hexagonal packages of land with the distance
from the centre to the periphery of each package being the same in
all directions. But for historical topographical reasons the ideal
never exists. Some peripheral communities can feel alienated from
others within their own 'hexagon' and more attracted to those in
neighbouring 'hexagons' - the seeds of political instability. To
minimize problems of this sort it is essential to take note of :

(i)   the size of a political division (municipality)
(ii)   its shape (concentration)
(iii) access to its 'centre'
(iv)   the distribution of the population within the political unit
       (problems of peripheral communities vis a vis those near the
       centre).

    The larger the political unit the greater the need for deleg-
ation and a degree of autonomy.

    Because of a variety of natural causes (population densities)
and historical reasons, administrative divisions differ in size,
shape and content. This simple fact is not always taken into account.
Bureaucracy can become top-heavy and uneconomic. These problems must
be avoided and several devices can be used to tackle them, for
instance, changes in the boundaries of political divisions so as to
move the focal points for decision-making. Several factors must be
taken into account -

(i)   if an area is too small - its administration will be uneconomic
       and out of balance,
(ii)   if an area is too large, peripheral populations will feel
       alienated by remoteness,
(iii) if the area is long and narrow, it will lack cohesiveness,
(iv)   the creation of two areas joined by a narrow neck is also an
       undesirable situation - it causes bipolarisation and less
       cohesiveness.

    A compact cohesive area with ease of access to its centre is
preferred. In existing situations, areas not close to the main
administrative centre must be examined with a view to associating

them with an existing closer centre by improving accessibility, thus,
if a fraction is physically separated it is better that it should
seek allegiance with its immediate neighbours. If the main focus of
a political unit is at its periphery (as so often happens) problems
of affiliation will arise. Who identifies with whom ?

Figure 4: The location of Vizela in northern Portugal. The shaded
           area has been enlarged in Figure 5.

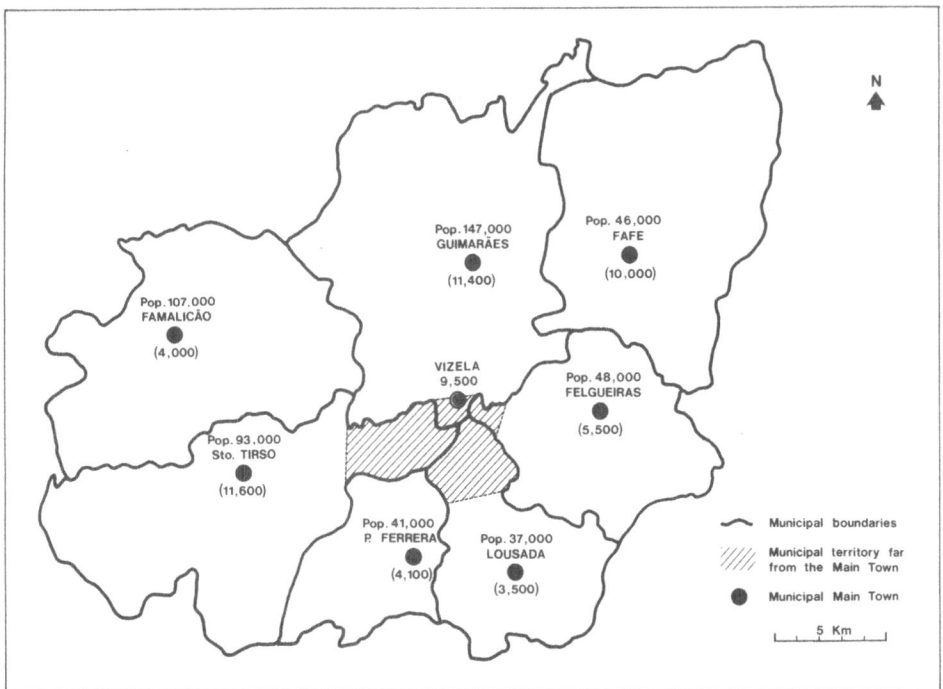

Figure 5:   The position of Vizela, in northern Portugal, relative
            to the nearest main administrative centres (municipal
            main towns). The figures represent the total population
            of each municipality with the town population provided
            in brackets. The population of Vizela and its surround-
            ing area would prefer to create a separate municipality
            with Vizela as its main town.

        Recently a problem of affiliation arose in a large inland munic-
ipality in northern Portugal (Guimaraes) where the inhabitants of
'freguesia' (an administrative subdivision of the municipality)
wanted to create a new municipality. The basis of the problem was
the long-standing rivalry between the main town and Vizela, the
town wanting autonomy, both important industrial centres (Figures 4
and 5). However, Vizela is also on the border of the municipality
and is adjacent to territory in three other municipalities which are
similarly far removed from their administrative centres. The people
in these areas wanted to have greater responsibility for their own
development. Traditionally the size of inland municipalities is
larger (average size 4827 ha) than those on the coast (average size
2204 ha) and opposition to this tradition had a bearing on the demand
to create a municipality of Vizela. It is clear that because of the

historic significance of Vizela and its surrounding territory this problem could have been anticipated and a Regional Plan implemented to avoid what has become a serious political issue.

History has been punctuated by periods of land subdivision followed by consolidation, of centralisation followed by decentralisation, of analysis followed by action but still the problems of land fragmentation remain. The ultimate cure must follow a consideration of cultural, ecological, economic, political and social factors which, together, each play a part in influencing the needs of a population. Having determined these needs, planners must provide a series of fully evaluated alternative proposals for scrutiny: we would be arrogant if we were to think that there was only one correct solution to each problem.

COMMENTARY : URBAN PLANNING

The topic of monitoring usually excites planners and scientists alike - it creates a great deal of concern. All types of monitoring, ranging from the collection of socio-economic data, through changes in land and climate to environmental pollution, are universally agreed to be important. However the value of the observations obtained is, in many instances, hotly contested. This is often because of the absence of a framework clearly setting out the rationale for monitoring in terms of decision-making objectives, frequently involving the need to integrate biological, physical, economic and sociological information. The need to present multi-faceted information, and the implications of this information in ways that are readily grasped by non-expert (in a technical sense) decision-makers often at the political level, is a major challenge that has not yet been satisfactorily resolved. It is necessary to have a framework with a structured hierarchy extending from local to national considerations, but most ecological and social information systems are site specific; they rarely advance beyond the regional level.

In many instances, public administrations do not perceive the need to monitor: they think that usually unstructured historical information, often the only indicators of past trends are sufficient, even though they may relate to completely different sets of circum-stances, and are less than relevant for the evaluation of the success or failure of new policy or administrative changes. It is easy to monitor if housing is expanding into permitted or prohibited areas, or if desirable landscapes are being protected, but monitoring the environmental impacts of pollution, or of infrastructural dev-elopments, is another matter. The environmental effect of building a road between two centres is not confined to the immediate distortion caused by its construction; this is insignificant compared with the impact of its existence and use as a vehicle for commun-ication, as a focus for housing and industrial developments etc. that would not have otherwise occurred.

Too frequently it is assumed that everything is known about the undisturbed physical environment of a site designated for develop-ment - planners would be wise to assume that this is not so. Canadian studies have shown that the extra costs of monitoring, often considered excessive, can usually be absorbed when it is realised

45

that monitoring can provide an early warning of something unexpected. The implementation of all planning decisions should be
monitored.

Planning at the urban - rural interface is a matter of great
concern, focusing on the 'conflict' between urban sprawl and the
loss of prime agricultural land. In many countries, the agricultural
lobbies effectively protect the financial interests of farmers while
being powerless to restrict the physical effects of urban encroachment. In its development plan, the expanding city of Marseille
ignored the rural hinterland; furthermore, its back-up studies
focused on the physical environment, while largely ignoring social,
and political, aspects. In Belgium the new town of Louvain-la-Neuve
was located on prime agricultural land; in this instance political
considerations were over-ridingly important. Changes in land use
as in parts of Portugal and Quebec Province, Canada, are often complicated by the fragmentation of land holdings by traditional methods
of inheritance so resulting in units of uneconomic size: so as to
ensure that the 'holdings' of all heirs retained some river frontage,
some of the land holdings in Quebec are unserviceably narrow.
Various methods have been devised to prevent sub-division into un-
manageable parcels such as the enforcement of size restrictions, as
in parts of South Africa. Illegal subdivision has been a particular
problem in some areas of Portugal, where developers often appear to
have been more sensitive and responsive, than administrators, to
the needs of society. In some parts of India, landowners seem to
be willing to subdivide their farms so as to provide building plots
for sale as 'pieds-a-terre' for individuals who have temporarily
moved to the Persian Gulf Oil States for work. Whatever the reasons,
uncontrolled land speculation can become damaging in the long term.
However the system of "land banking" - the consolidation, by government purchase, of land of widely dispersed ownership - which has been
extensively practised in Sweden, has become less attainable in the
current era of high interest rates.

The extremes of urban sprawl on the one hand, and aggregation
on the other, are represented by the cities of Los Angeles and New
York. Each, in its own way, is an object lesson for planners who
have a duty to analyse the implications of alternative scenarios.
Current approaches point to the distribution of populations so that
individuals have reasonable access to cultural amenities combined
with the provision of urban and social services without the creation
of environments that are superior for some people and inferior for
others. Decisions relating to the real conditions in which people
will live can only be made on a rational basis if there is a
suitably comprehensive planning framework for enabling analyses to
be made, for example, the Ranstad approach for the Rotterdam-
Amsterdam conurbation in the Netherlands. In the new town of
Louvain-la-Neuve it was decided to relocate the old university of
Louvain assuming a 'high density-low rise' approach to accommodation

and buildings with their own individual gardens providing a mosaic but using less land than the 0.1 ha. usually required in North America. This design recognised the need to view a development from the point of view of the inhabitant and not from that of the planner. The importance of 'defensible' private space, often in the form of small gardens or yards, has been recognised as a major element in family, and communal, life, even in poor conditions as in parts of India, Zambia and the U.K.

The solution to the problem of relocating Louvain University was announced before national resource and ecological considerations could be included in the decision process. Its relocation - the university is essentially an urban institution - on highly productive agricultural land, in a small municipality in a rural area posed many major problems for the planners. Interestingly, the new town of Louvain-la-Neuve is attracting an increasingly balanced population some of whom work in, and service, the University while others travel to Brussels using the convenient autoroute and rail connections, or work in the nearby industrial park.

Farmers who lose parts of their land for industrial/residential developments are often forced to fundamentally reconsider their approach to farming. Some may simply change the crops that they grow, while others may tackle the problem much more radically by switching from agriculture to forestry and vice versa. A third group have responded to pressure by intensifying fertilizer and pesticide usage with potentially damaging effects on the environment and thus contribute to the present debate about agriculture and environment which includes the rights and wrongs of culverting streams, removing hedgerows etc. in the name of efficiency. Each landscape tends to be unique, some being more valuable than others, a subjective judge-ment of the beholder. However, the time has come when their conser-vation may be regarded as more important than land improvement.

How will this be judged? How responsive should land-owners, however small, be to the needs of urban societies? How should urban societies attempt to seek a rapport with the owners of rural land recognising that self-interest should be directed to the long-term needs of everyone? Clearly, elements of planning are essential.

A

IDENTIFICATION OF ECOLOGICAL FACTORS CHARACTERISING THE RANGE OF *TERRESTRIAL* HABITATS

ii-Rural
a-Geology

4

GEOLOGY AND LAND-USE: THE PRODUCTION OF 'ENVIRONMENTALLY'

ORIENTATED MAPS FOR DECISION-MAKERS

E.F.P. Nickless

Natural Environment Research Council
North Star Avenue, Swindon
Wiltshire, SN2 1EU, England

INTRODUCTION

In his Presidential Address to the Geological Society of London on 20 February 1836 Charles Lyell reported that Henry de la Beche, the Society's Foreign Secretary, was appointed, the previous year, to organise and direct a geological examination of the English Counties concurrently with the geographical survey then in progress. Such was the beginning of the British Geological Survey. Lyell with Buckland and Sedgwick, the professors of geology at Oxford and Cambridge respectively, had been invited by the Master-General of the Board of Ordnance to state the case for geological work. They foresaw the advantages of such an undertaking as 'not only as calculated to promote geological science, which would alone be sufficient object, but also as a work of practical utility, bearing on agriculture, mining, road-making, the formation of canals and rail roads, and other branches of national industry' (Lyell, 1836).

Since its inception almost one hundred and fifty years ago the Geological Survey has been concerned with economic aspects of geology and the practical application of the science, as well as the maintenance of the geological map of Britain. Not surprisingly, the emphasis has changed from time to time, being directed to the mapping of mineralised areas such as Devon and Cornwall, the investigation of the coal-fields and ironstone-fields, and, at other times, to the survey of what might be termed the classic areas of geology including the north-west Highlands of Scotland, the Welsh Borderlands and many others. During times of war, attention was focussed on indigenous mineral and water resources, whereas in more tranquil periods mapping has been carried forward in order to complete the coverage of the United Kingdom.

Since the introduction of the Town and Country Planning Acts in
1947 there has been a steady increase in the control of land-use in
Great Britain. Public awareness of the need to develop a more
harmonious use of the environment, growing shortages of basic
raw materials, and changes in the funding of scientific research
have combined to bring the application of geology strongly to the
fore. The way in which the (now) Institute of Geological Sciences*
has responded to changing circumstances is perhaps best illustrated
by considering some of its recent maps likely to be of interest to
physical planners. Among the topics considered in this chapter are
(a) environmental geology which aims to portray in a simple and
readily understandable way those aspects of the geology of importance
in land-use planning; (b) industrial minerals surveys which outline
the distribution of materials and provide guidance on thickness,
quality or grade; (c) engineering geology which emphasises geotechnical properties, that is, the suitability of the ground for building
on; (d) hydrogeology which considers rock formations in terms of
their potential as aquifers, that is, their ability to store and
transmit water; and (e) geochemical surveys which quantify the
regional abundance of selected, mainly metallic, elements, of economic
or environmental significance. Thus it is possible here to refer only
briefly to what is an extensive series of maps and to make passing
reference only to the memoirs and reports of the Institute, but the
annual Ordnance Survey map catalogue (Ordnance Survey, 1981) and
Sectional List 45 (HMSO, 1981) summarize the range of publications.

THE TRADITIONAL GEOLOGICAL MAP

Apart from a few areas of 'classical' geology where the
1:25,000-scale is used, the standard published map of the Institute
of Geological Sciences was at 1:63,360 though this is being
progressively replaced, when reprinted, by sheets at 1:50,000 using
the same sheet outlines and, where possible, the latest topography.
In Scotland the 1:63,360 sheets are generally divided into two
1:50,000 sheets (western and eastern halves) when being reprinted.

Although the latest maps show more detailed information than
their predecessors - the additions being achieved, in the main, by
the use of improved cartographic and printing techniques - they do
not differ fundamentally from the maps produced at the beginning
of the century. The traditional map remains a regional summary of
diverse data sets. It embodies the results of primary and revision
field surveys, usually at 1:10,000 or 1:1,560, and incorporates
the findings of site investigations, including drilling, to delimit
the extent of solid rock and unconsolidated drift deposits. From
the surveys and a consideration of petrology and palaeontology
(which can assist the classification, description and interpretation
of age sequences), much can be learnt, and deduced, about the
formation of rocks and their subsequent history.

*From 1984 renamed the British Geological Survey

Limitations of Traditional Geological Maps

These maps are the geologist's primary tool both for summarising data and assisting investigations of geological history: in the main, they show the disposition of rocks by age (chrono-stratigraphy) rather than by rock type (lithostratigraphy), though in practice many maps combine both aspects; they rely on a high degree of geological understanding and awareness and, as such, are not easily understood by the layman, an unfortunate limitation at a time when public interest in environmental sciences is increasing. To the trained 'reader' geological maps 'comment' on the capacity of the ground to bear structures; they suggest the distribution, at the surface and at depth, of mineral resources including water, indicate

Table 1. Professional interrelationships during planning.

Legend:
- ■ Primary interrelationship
- ▨ Secondary interrelationship
- □ No interrelationship

Selected Professional Disciplines (columns, left to right): Geologists, Geographers, Civil Engineers, Sanitary Engineers, Landscape Architects, Architects, Recreation Planners, Planners, Conservationists, Public Administrators, Lawyers, Sociologists

Major Study Groups	Typical Study Topics
Regional economy	Economic base, resource potential
Regional population	Population studies, socio-economic studies
Transportation	Transport facilities, public transport, parking facilities
Natural environment and public utilities	Natural resources, hazards protection, land reclamation, public utilities
Community facilities	Schools, libraries, police and fire, parks, recreation
Land use	General plans, neighbourhood plans, commercial developments, industrial developments
Housing and public buildings	Private dwellings, public buildings
Aesthetics	History, cultural values, community improvement, legislative controls
Administration and legislation	Legislation, administration
Finance	Capital improvements, aid programmes
Other planning studies	Defining goals, urban renewal programmes, waste disposal, public health, civil defence

Source: Turner, A.K., and Coffman, D.M., 1973, Geology for planning: A Review of Environmental Geology, 2. Colorado Sch. Mines, 68/3.

Table 2. Geological conditions that
affect typical land-use.

Source: Turner, A.K., and Coffman, D.M., 1973, Geology for planning:
A Review of Environmental Geology,
2. Colorado Sch. Mines, 68/3.

likely areas of landslip or undermining, and much else besides.

The search for a higher standard of living, coupled with dramatic
increases in population over recent decades and the rapid growth of
urban settlements, have all intensified the use of land and the
Earth's resources. The geologist with experience in evaluating the
Earth's capacity to provide water, mineral resources, waste
disposal and building sites should be well placed to contribute
towards the solution of a variety of problems concerning the physical
environment. Unhappily, decisions affecting the use of land may be
made with an insufficient understanding of geological factors and
avoidable mistakes are common. Table 1 shows some of the possible
interrelationships between and among geology and other professional
disciplines during planning. It clearly indicates that the geologist
is an important member of a planning team and emphasises that unless
geological information is provided in an easily understandable form
to decision-makers, planners and lay readers with an interest in
land-use it is unlikely to be considered at a formative stage in
the decision-making process. Of necessity, the specific advice
which a geologist can offer will vary from area to area and project
to project, but the importance of some geological conditions for
typical land-uses are listed in Table 2. Failure of motorway
cuttings because the walls were cut at too sharp an angle (a
reflection of the nature of the rock itself), collapse of buildings
sited on abandoned underground mine workings, and the unwitting
sterilisation of resources by ill-conceived siting of pipelines are

some of the more obvious examples of the failure to take sufficient
account of geology. But that is not to say that improved geological
knowledge would cure all: given sufficient money it is technically
possible to build anything, anywhere; and it is perhaps inevitable
that pipelines and other route-ways will sterilise resources.
Rather, the matter is one of choice, and it would be better for
those making decisions about the use of land to have a fuller
appreciation of the principal environmental factors involved, be
they agricultural or scenic land-values, amenity considerations or
geology.

ENVIRONMENTAL GEOLOGY MAPS

    What is the role of the traditional geological map and the
geologist? If the traditional map is too complex to be easily
appreciated by the non-geologist then a simpler product is needed.
Since the early 1970s considerations of this sort have given rise
in West Germany, North America and subsequently elsewhere, to a new
style of geological mapping, often referred to as environmental
geology, whose aim is the display of geological data in a form
readily understandable to, and usable by, non-technical readers.
An example is the report on the urban geology of the Edmonton area
published by the Alberta Research Council (Kathol and McPherson,
1975). In the Edmonton area the most important natural resources
are groundwater, and sand and gravel: other resources, of lower
potential, include oil and gas, coal, marl, ceramic and brick clay,
silica sand, peat and salt. The report details the engineering
properties of the deposits which together with information about
the nature and distribution of the sediments, enabled the
preparation of a series of land-use maps relating to mineral
resources and water, general construction conditions, suitability
for solid waste disposal, susceptibility to erosion, slope
stability and deep sewer construction. In total the report has
forty-five figures, sixteen of which are reproduced as multicoloured
maps at 1:50,000. As has already been mentioned, similar work is
being done elsewhere, for example Kansas, but usually on a more
restricted scale. In West Germany, an attempt is made, by overlaying different components of the geology, to recognise areas of
potential conflict and, having done so, to resolve the issue by
making specific land-use recommendations based entirely on
geological considerations (Lüttig, 1975).

    With hindsight, it is possible to discern changes and
developments in presentational techniques used in the United
Kingdom (by the Institute of Geological Sciences and its predecessor,
the Geological Survey of Great Britain) which may now be regarded
as conforming with the emergence of environmental geology. Over the
last twenty or so years, traditional memoirs or pamphlets,
particularly related to mineral resources, have been supplemented

by specialist maps dealing with engineering geology, hydrogeology, mineral resources, geophysics and geochemistry. Usually each of these topics has been pursued and presented in isolation - quite simply, until recently no coordinated attempt has been made in the United Kingdom to present a wide range of simple interpretative geological information. In 1980, recognising this deficiency, the limitations of traditional geological maps and mindful of the developments in this field being made overseas, the Department of the Environment and the Scottish Development Department, in consultation with the Institute of Geological Sciences, selected a pilot study area where environmental geology mapping techniques might be assessed. A single 1:25 000 sheet, NO 20 based on Glenrothes, Fife Region was selected because it was representative of a large part of central Scotland with regard to geology and the availability of geological information; it was also chosen because, at the time, the use of the land encompassed by the map was not a contentious issue (Nickless, 1982).

The aim of the pilot study was to collate and interpret extant geological data and present each element of the interpretation in a simple cartographic form, the different aspects being presented separately on element, or basic, maps. The concept, therefore, is to simplify the traditional geological map by isolating and giving equal treatment to the significant geological components. This procedure does not necessarily involve the acquisition of additional information, but the levels of confidence which could be attached to the element maps, reflect the sophistication, completeness and accuracy of the basic data.

For Glenrothes, the basic information has been displayed in eighteen element, or monothematic, maps (Table 3), so demonstrating the versatility of the technique and showing how information can be presented in a variety of ways (in this instance one map has been broken into eighteen components). However, the nature and number of element maps are likely to vary for different locations in response to the data available. By combining information from two or more element maps, four derived maps were produced itemising underground storage potential within 100 m of the ground surface, sand and gravel potential, foundation conditions and groundwater resources.

The derived maps demonstrated four of many hypothetical planning options in the Glenrothes area. Except for the sand and gravel potential map, which also involves quasi-economic considerations, only geological factors have been considered and, in the main, the necessary background information is contained in the element maps. The procedures involved in preparing a derived map may be illustrated with reference to that for underground storage potential. This map indicates the relative suitability of the various geological formations in the area for the construction

Table 3.   Environmental geology maps relating to Ordnance Survey
           sheet NO 20 (based on Glenrothes in the Fife Region
           of Scotland).

Element maps (basic data)

    1   bore sites
    2   unconsolidated deposits
    3   lithology of the unconsolidated deposits
    4   engineering properties of the unconsolidated deposits
    5   thickness of the unconsolidated deposits
    6   depth to water in the unconsolidated deposits
    7   sand and gravel thickness
    8   bedrock geology
    9   bedrock lithology
    10   rockhead contours
    11   shallow undermining
    12   natural landslip potential
    13   opencast workings
    14   hardrock aggregate resources
    15   limestone resources
    16   brick and tile clay
    17   mudstone for brick making
    18   hydrogeology

Derived maps (combining two or more elements)

    19   underground storage potential within 100m of the surface
    20   sand and gravel potential
    21   foundation conditions
    22   groundwater resources

Environmental Potential (summary maps based on the element and derived
maps)

    A   development potential
    B   priority areas for on-site investigation
    C_1 resources at or near the surface which might be won by opencast
           working
    C_2 buried resources which might be won by opencast working
    C_3 buried resources which might be won by pumping or mining

of large underground caverns. It is based on consideration of rock
strength (element map 9, bedrock lithology), physical homogeneity
(element map 8, bedrock geology), permeability (element maps 8,
bedrock geology and 18, hydrogeology) and mechanical discontinuities
(element map 8, bedrock geology).

At Glenrothes three major aspects were singled out for special
attention; they are the subject of Environmental Potential, or
summary, maps and were prepared by synthesizing data from different
element and derived maps:

    a. development potential and,
    b. conversely, priority areas for on-site investigation;
    c. resources, including water

The development potential map shows the suitability of the
ground to support lightweight structures, such as two-storey housing
and single-storey factory development which form the majority of any
new urban building. The siting of heavy structures was not included
as generally many factors in addition to ground conditions, as such,
influence site selection.

It is recognised that, given the need, any structure can be
built anywhere - but at a cost. The aim of the map is to distinguish
those areas where, subject to normal site investigation practice,
lightweight structures might be built at average cost, from areas

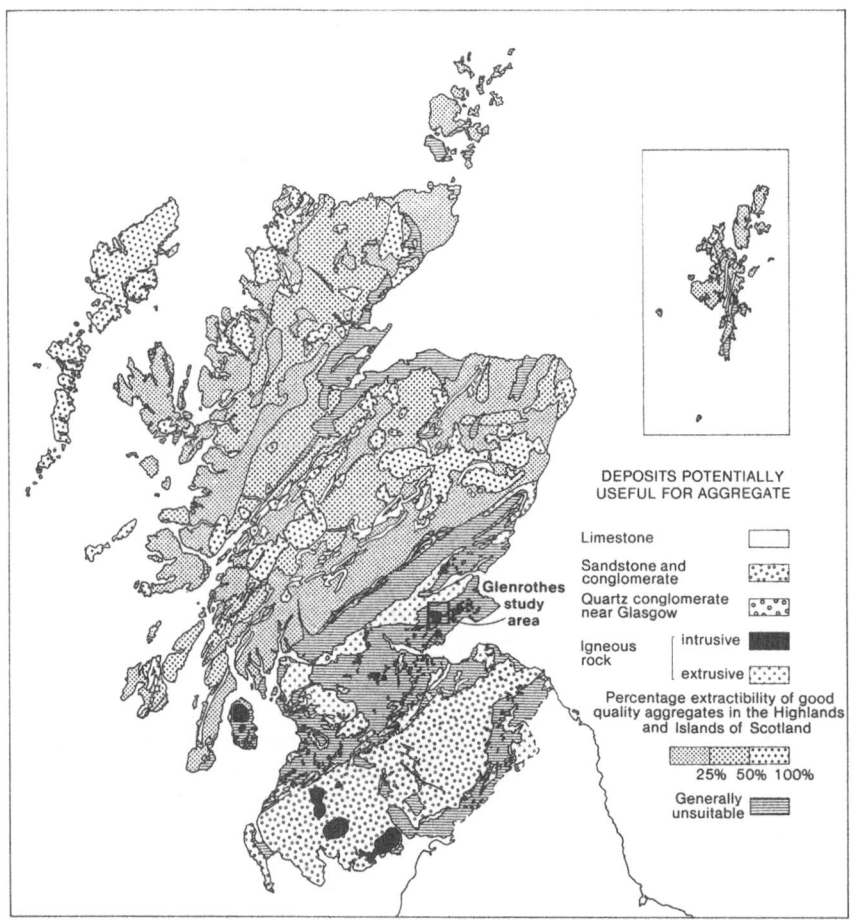

DEPOSITS POTENTIALLY
USEFUL FOR AGGREGATE

Limestone

Sandstone and
conglomerate

Quartz conglomerate
near Glasgow

Igneous   intrusive
rock   extrusive

Percentage extractibility of good
quality aggregates in the Highlands
and Islands of Scotland

25% 50% 100%

Generally
unsuitable

Glenrothes
study
area

Source: "Aggregates : the Way Ahead. Report of the Advisory Committee on Aggregates,"
Her Majesty's Stationery Office, London.

Figure 1.   Resources of hard rock aggregates in Scotland.
            Source: "Aggregates: the Way Ahead, Report of the
            Advisory Committee on Aggregates", Her Majesty's
            Stationery Office, London.

where it can be recognised now that development will require prior
detailed site investigation and possibly some form of ground treatment.

The second summary map, priority areas for on-site investigation,
highlights items from element maps of natural landslip potential,
shallow undermining, unconsolidated deposits and opencast workings.
It shows the geological factor which on the development potential
map causes areas to be singled out for detailed site investigation
prior to development. The nature of the geology will suggest the

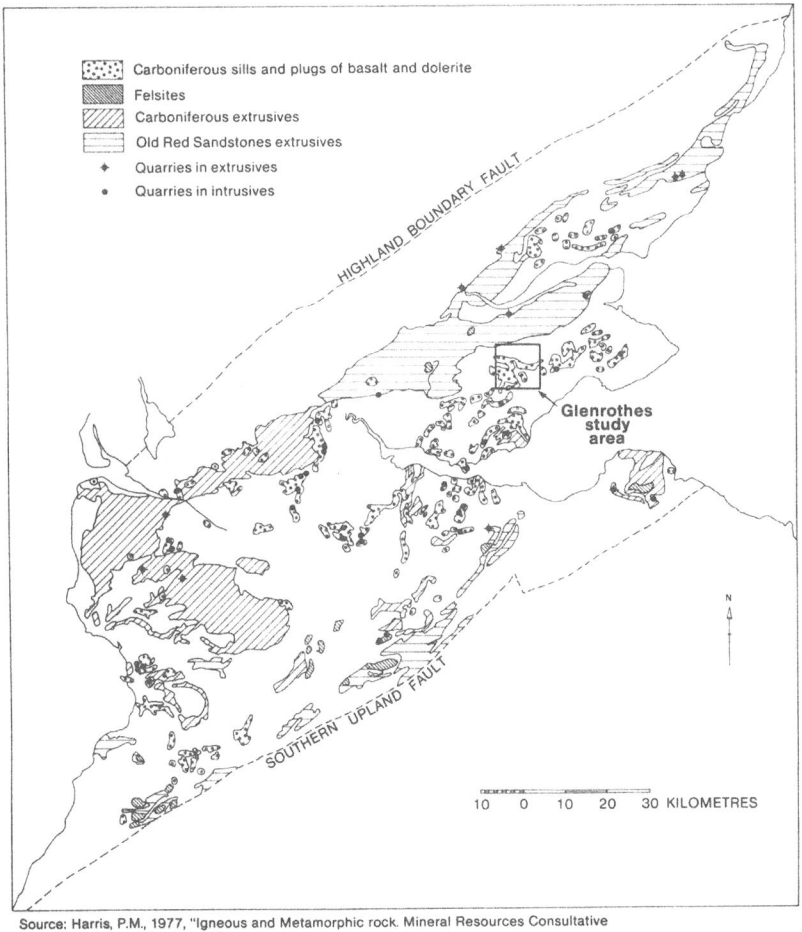

Source: Harris, P.M., 1977, "Igneous and Metamorphic rock. Mineral Resources Consultative
Committee, Mineral Dossier 19,"
Her Majesty's Stationery Office, London.

Figure 2.   Central Scotland: important quarries and igneous
            formations. Source: Harris, P.M., 1977, "Igneous
            and Metamorphic rock. Mineral Resources Consult-
            ative Committee, Mineral Dossier 19". Her
            Majesty's Stationery Office, London.

appropriate style of survey needed to investigate the site properly.

Resources form the topic of the third summary map. In the event, the presentation of resources took three maps separating surface or near surface resources, which might be 'won' by open-cast working, and buried resources which might be won by either open-cast working, or pumping or mining. The resources considered, hard rock aggregate, limestone, brick and tile clay, mudstone for brickmaking, and water, are each the subject of an element map which gives supporting detail. Although coal is included on the summary map no element map was prepared.

It is unlikely that, as at Glenrothes, environmental geology maps would always be at 1:25,000. There is no philosophical difficulty in presenting the maps at any scale providing that for large scales supporting data are available, and for smaller scales that the amount of generalisation, of both the cartography and content, is not such that the result is meaningless. Work at 1:25,000 is possibly only necessary for areas of particular concern; elsewhere 1:50,000 or a smaller scale may be more appropriate in providing a broader view. Whatever scale is chosen, unless there is a reasonably extensive series, there will always be the wish to look beyond the margins of a particular map to assess the local, regional and national significance of selected geological factors as they affect a variety of planning options.

Although it is not possible to present British examples of environmental geology maps, as such, at various scales, an idea of the way in which such maps might differ can be gained by considering three studies of hardrock aggregate. Figure 1 summarises the resources of all potential hard rock aggregates in Scotland, and, for the Highlands and Islands allows the reader to identify the percentage extractability of good quality aggregate. Where there is a diversity the user may select that rock which will best meet his requirements. Because of the scale at which the work was done the map is, of necessity, highly generalised. Nevertheless, it serves as a useful guide providing that the reader is able to judge the relative merits and uses of the rocks shown. Arguably, in lowland Scotland, igneous rocks will meet most user specifications and will command most interest. Figure 2 delineates and subdivides the igneous formations of the Central Belt. As the various rock types identified have different mechanical and physical properties the map will serve to identify broad areas where material for specific uses, such as road dressing or concrete aggregate, may be found. However, the map uses geological terms and fails to rank or identify the resource by end-use. Figure 3 is the environmental geology hard rock aggregate resource map for Glenrothes. This element map is a detailed analysis of a small area. Consideration of factors including thickness, depth of burial and grain-size of the rock may allow a site with commercial opportunity to be

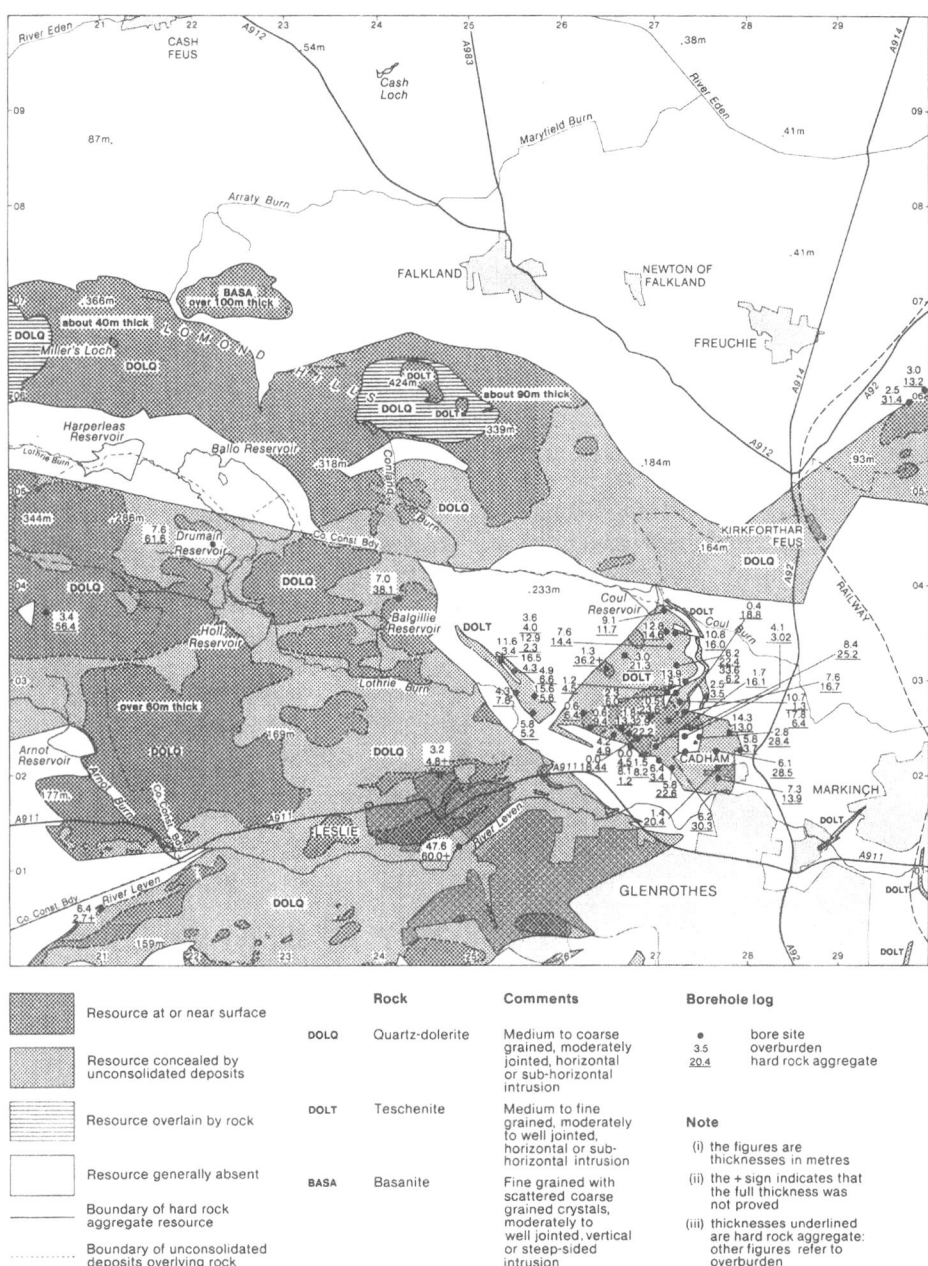

Figure 3. Environmental geology maps (Glenrothes): hard rock aggregate resources.

selected. Figure 1 would enable such a site selection process to
be placed within a regional context.

Since August 1980 when the pilot Glenrothes project was
completed, discussions with central government, regional planning
authorities and other interested parties have shown that there is
a significant interest in the interpretative approach to displaying
geological information. Environmental geology maps are seen as
overcoming many of the drawbacks of the traditional geological map
with the thematic approach giving decision-makers, at a formative
stage of the planning process, an improved appreciation of the
geology itself and the consequences of it. Unlike the traditional
product, the environmental geology maps display data with specific
end-uses in mind.

The Department of the Environment with the Scottish Development
Department has recognised the need to test the environmental geology
map-concept in a variety of terrains; work is currently in progress
in estuary, urban fringe and inner city areas. Although it is
unlikely that the environmental geology survey of these areas will
depart radically from the philosophy already outlined, geological
differences and user requirements will undoubtedly necessitate
additional solutions to the display of data: further developments
may be expected.

INDUSTRIAL MINERALS SURVEYS

The Industrial Minerals Assessment Unit (formerly the Mineral
Assessment Unit) of the Institute of Geological Sciences was
established in 1967 to undertake regional resource assessment
surveys of industrial minerals, primarily those used in construction.
The main effort has been focussed on sand and gravel (Laxton and
Ross, 1981), first in East Anglia and Essex, and progressively
through the valleys of the Thames, Trent and Nene, and then into
northern England, Scotland and the Welsh Borders. In addition, the
limestones of the Peak District (Harrison, 1981), the Craven
Lowlands and north Pennines, the Sherwood Sandstone Group (Bunter)
of the West Midlands (Rogers and others, 1981) and celestite in
Avon (Nickless and others, 1976) have received attention.

The environmental impact of open-cast mineral working is plain
for all to see. It is against a background of almost universal
opposition to the extension of such activity, increasing shortages
of material (particularly in the south-east of England), and an
interest in better resource management and husbandry that the
assessment surveys are made. They aim to provide the necessary
background data on quality and quantity of the different resources
so as to help formulate coherent land-use and mineral planning
policies.

The starting point, of these resource assessment surveys, is
the geological map, now available at a scale of 1:10,560 or 1:10,000
for much of the United Kingdom, which outlines the distribution of
materials but usually provides minimal guidance on the thickness of
deposits, their quality or grade. To correct these deficiencies
series of specifically-designed programmes of drilling, sampling
and testing are being undertaken. The surveys are made at a
regional scale and they are concerned with the appraisal of
resources but specifically not with reserves (reserves comprise
those parts of the resource which are proved locally to be
economically viable in current or immediately foreseeable market
conditions). In resource surveys the deposits are only outlined
and not proved in detail. Nevertheless, it is necessary to make
value judgements about the potential usefulness of deposits. To
do this, arbitrary physical criteria have to be established and
deposits generally satisfying these conditions are called 'mineral'.

The resource investigations made by the Industrial Minerals
Assessment Unit are each summarised on a 1:25,000 map with a
supporting explanatory soft-covered book. The resource maps are
of three kinds: those dealing with sand and gravel and
conglomerate (Bunter) show the surface and buried distribution of
the mineral; the limestone map indicates purity, whereas the
celestite map shows the degree of certainty with which the resource
has been identified.

The aims and limitations of the resource surveys are summarised
in the introduction to each report. A brief account of the geology
of the area under investigation is supported by more detailed
information on the composition of the mineral. The assumptions and
methods used in the assessment are stated. As appropriate, data on
the mechanical, physical and chemical properties may be given,
together with an estimate of volume, usually at the 95 per cent
probability level. The report also includes the results of detailed
tests, for example, the sieve-grading data of samples taken for
every one metre thickness when drilling sand and gravel. When there
is reason to be suspicious of the mechanical or physical properties
of the aggregate a limited number of tests, in addition to grading,
may be made.

To date (June 1983), the results from one hundred and twenty
eight study areas have been published: seven reports deal with
limestone, another with celestite, two with conglomerate, two with
conglomerate resources in the Sherwood Sandstone Group and the
remainder with sand and gravel. Of necessity, the detail at which
resource assessment surveys are made is a compromise in terms of
cost and time. They are done to substantiate and extend the scope
of the traditional geological maps by adding qualitative and
quantitative data. They will not supplant the need for the research
and development side of the extractive industry to do its own

customary detailed site analyses; however, they will assist the
selection of areas for further investigation, and enable the
planning authorities to consider applications to win minerals in a
broader, regional context.

Sand and Gravel Surveys

Because transport may account for a significant part of the
cost to the consumer, there is pressure to work sand and gravel in
close proximity to centres of demand. Inevitably, therefore, the
siting of mineral workings close to urban areas has to be
reconciled with other competing claims (agricultural and amenity)
for the use of the same and adjoining land.

Prior to the assessment survey in the Newport-on-Tay area,
Fife Region (Laxton and Ross, 1981), the principal source of
available geological information about the sand and gravel in that
area was the traditional geological map and the results of (a)
sixteen boreholes sunk, over a period of years, for a variety of
purposes and (b) six exposures mainly in working or defunct gravel
pits. To learn more about the thickness and quality of the sand
and gravel deposits a further 54 boreholes, totalling 933 metres,
were sunk, 34 shallow pits dug, and 590 sieve-gradings made. As a
result it has been possible to prove and extend the traditional
geological map to provide a resource map with areas of potentially
workable sand and gravel shown in shades of red, a dark tone
identifying areas where there is generally less than one metre of
overburden. The inferred distribution of buried deposits is
illustrated in cross-sections.

Two geological factors are important when considering the
usefulness of deposits, (a) the ratio of sand to pebble-size
material and (b) the depth below surface at which free water is
found. The latter is particularly important in Scotland where the
wet working of aggregate is the exception rather than the rule -
there is a reluctance on the part of planning authorities to allow
it and the industry generally seeks to avoid the additional
capitalisation (necessary for plant and equipment) to work wet
deposits. The borehole data from the assessment survey enable
areas of gravel to be distinguished from sandy deposits; similarly
the depth below surface to water may be contoured. By integrating
these sets of data, areas can be identified where gravel or sand
deposits of a determined thickness (say 5 or 10 m) might be worked
dry. So far two of a wide range of geological factors have been
emphasized but others, including the mechanical and physical
properties of the aggregate, the presence of deleterious material
such as shell debris, might also be considered. By preparing these

target maps, however, it was possible to see, at a glance, that
not all parts of the resource in the Newport-on-Tay district are
of equal interest, unlike traditional geological maps which only
depict the existence of a resource.

On geological reasoning the areas of greatest potential can
be readily identified, and when planning constraints such as
agricultural and scenic land value, pipeline corridors and similar
factors are considered, the choice of safeguarding deposits or
'sterilising' them by allowing building or other surface
developments may be made against a background of knowledge rather
than one of ignorance.

Celestite

The Central Unit for Environmental Planning (initially within
the Department of Economic Affairs but latterly within the Ministry
of Housing and Local Government) was commissioned in 1966 to
investigate the feasibility of large-scale urban developments on
both the Welsh and English shores of the Severn Estuary: its
report Severnside: A Feasibility Study, published in 1971,
concluded that the region had considerable development potential.

But the only commercially important deposits in Britain of
celestite (strontium sulphate), the main ore of strontium, occur
in the Northavon District of Avon, which borders the eastern shore
of the Severn Estuary. Until 1968 these deposits had supplied
fifty to seventy per cent of the world's annual production since
before 1875. Anticipating the Severnside Report, a study was made
of celestite resources in Avon with a view to (a) providing data
about the origin and distribution of the mineral, (b) discovering
areas likely to contain workable deposits and (c) establishing
sampling techniques which might be applied regionally and locally.
After fieldwork between 1970 and 1974, the findings were presented
in the form of a report and a 1:25,000 map showing areas of
probable, possible and hypothetical resources (Nickless et al.,
1976). The improved understanding of the origin and distribution
of the celestite and, in particular, the confirmation that the
mineral only occurs in the Avon area close to limestone deposits
with which it is genetically related - a coincidence also reported
elsewhere by West (1973) - suggests that it is most unlikely that
extensive deposits of the mineral occur outside the traditional
working area between Cromhall (ST 693 904) and Yate (ST 714 828),
the locale strongly favoured in the Severnside Report for large-scale
urban development. The results of the resource assessment survey
can be exploited in many ways - they certainly indicate those parts
of the celestite resource which are of greatest risk and which
should therefore be mined in advance of other forms of development.

ENGINEERING GEOLOGY MAPS

Engineering geology maps, like those for mineral resources, have
developed from traditional geological maps with an emphasis on
geotechnical properties.

Belfast

The Special Engineering Geology Sheet for Belfast and District
(1971) shows the solid and drift geology at a scale of 1:21,120
using conventional geological symbols and colours. On the reverse of
the map, each type of rock is described by thickness-range, lithology,
structure and occurrence; the major geotechnical properties are
listed and economic materials identified, together with comments on
underground water and the type of soils. The geology of the area
is briefly described and abbreviated logs given of strata proved in
selected boreholes. The Belfast map is a compilation of data and
indicates the normal range of various geotechnical properties:
although criticised by some who would have preferred to have the raw
data, the map has found favour with many.

Peterborough

A traditional geological map at 1:25,000 is accompanied by
structure contour maps on the Blisworth Limestone and Cornbrash.
The distribution of the Blisworth Limestone is of significance because
it underlies the greater part of the old city of Peterborough and
many of the 'high-rise' developments in this area will have pile
foundations taken down into the top of the formation; the Cornbrash
because, although many structures have been successfully built on
it, problems arise where the formation is thin or has been subjected
to superficial disturbance or deep weathering. In such cases,
structures should be founded in the underlying strata. A table to
the side of the map shows the main features of the geological
formations by distribution and landform, soil, lithological character,
thickness, engineering geology, economic potential and geological
history. A report (Horton et al, 1974a) accompanying the map
describes the geology of the district, its engineering geology,
economic products and water supplies. The major geotechnical
properties are described, based on an analysis of detailed data
held by the Institute, and are presented as a guide in the planning
of individual site investigations.

Milton Keynes

After the geological survey of Milton Keynes, the Engineering
Geology Unit of the Institute of Geological Sciences made extensive
geotechnical studies, compiling the results in the form of a
geotechnical map and report (Horton et al, 1974b).

Maps and reports, where appropriate, for Belfast, Peterborough
and Milton Keynes are essentially traditional geological maps
with supporting engineering geology commentaries; they are not
engineering geology maps in the strict sense. The latter have been
or are being prepared following the South Essex Geological and
Geotechnical Survey (for a possible third London airport at
Foulness), and studies of the engineering geology of (a) the Upper
Forth Estuary (Gostelow and Browne, in press) and (b) the north
side of the Cromarty Firth (Gostelow and Tindale, 1980).

The Forth Estuary study aimed to provide guidance on ground
conditions as they affect planning for industry, particularly the
siting of heavy structures; it was done between 1977 and 1981. The
study area within the Forth Estuary is characterised by very soft
Carse Clay which, together with the effects of past mining, poses
major problems for the preparation of foundations. Geological and
geotechnical data, in addition to the usual criteria for site
selection at a sub-regional level, have been used to zone the area
in terms of suitability for foundations. A study was made of basic
geotechnical and stratigraphic units and their sequences, their
lithological characteristics, geotechnical properties and lateral
variations; possible constraints from other geological hazards
(particularly mining) and foundation requirements. Records of
6700 boreholes in an area of 450 km^2 were analysed. By including
an interpretation of solid and drift deposits a statement of
engineering geology has been presented including a series of maps
and supporting specialist sections intended for geologists, civil
engineers and planners. The first map classifies solid deposits by
lithology rather than by age as in a traditional geological map:
it includes a geotechnical assessment of the major rock groups as
well as showing data about abandoned mine-workings and the areas where
rocks approach within 3m of the ground. The second map has contours
showing the thickness of the 'drift' (that is, depth to rock);
another indicates the depth of the upper surface of glacial deposits
(that is, the depth to which foundations of heavy structures must
be taken). The distribution of known mine workings and shafts is
the theme of a fourth map: the fifth is essentially a traditional
drift geology map with the lithology of deposits being emphasised
rather than their age. The engineering classification of the surface
sediments forms sheet 6 which is in essence a lithological map
interpreted in engineering terms as defined in CP 2004 (British
Standards Institution, 1972), the soils being described using an
alphabetic notation after Dumbleton (1968). Sheet 7 demonstrates
the three-dimensional relationships between deposits using geotechnical cross-sections and uses a simple graphic device to
illustrate the results of Dutch Cone soundings. The planning
implications for the location of heavy structures are shown on
sheet 8 which summarises information contained on earlier maps and
gives an indication of the reliability of the data used in the
assessment.

HYDROGEOLOGY MAPS

Over the last twenty years, great advances have been made in
the preparation and publication of hydrogeological maps. This type
of map disregards conventional geological boundaries and depicts
rock formations only in terms of their potential as aquifers, that
is, their ability to store and transmit water. The maps of greatest
value to the specialist are probably those at 1:100,000 (for example,
the Hydrogeological Map of Southern East Anglia, 1981) or larger,
on which a wealth of hydrogeological detail can be shown. However,
smaller scale maps at 1:500,000 or 1:1,500,000 (for example, the
International Hydrogeological Map of Europe, 1980) are sufficient
to show the non-specialist where underground water resources either
need to be conserved or are available for 'development'. The
recently produced 1:100,000 of the EEC and the smaller scale maps
of the whole of continental Europe are proving their worth in the
rational consideration of groundwater resources.

GEOCHEMICAL MAPS

In the United Kingdom, the Institute of Geological Sciences
began preparing geochemical maps in the late 1960s, with the principal
objective of supplementing available geological information and
providing geochemical data on a regional basis for studies of the
evolution of the Earth's crust and ore-forming processes. Additionally, by quantifying the natural abundance of a wide range of elements
it may be possible to predict the chances of contamination as well
as the role played between the amounts of trace-elements and the
health and disease of assemblages of plants and animals. Geochemical
data may also help define lithological, compositional and structural
variations in bedrock not identified by visual geological mapping
techniques thus pinpointing new occurrences of metalliferous
mineralization.

The distributions of up to thirty elements of economic or
environmental significance, including copper, lead, zinc, cadmium,
mercury and manganese, have been plotted on geochemical atlases now
becoming available for the north of Britain at a scale of 1:250,000
(Institute of Geological Sciences, 1978).

CONCLUSIONS

Society of the 1980s is radically different from that of 1835
when the British Geological Survey was founded and this is reflected
in the demands being made on that organisation. Increasingly, the
geologist is having to provide data in special forms to meet
particular customer requirements. This new role and the greater
involvement in decision-making is to be welcomed as many would

consider that for too long, geologists have kept themselves aloof
through pursuing scientific points of little concern to society.
But criticism of that sort is not justified because without the
thorough basic knowledge and extensive data base, many of the
questions posed of the Institute today would be unanswerable: it
is the accummulated information gained by basic research and routine
observations over the past 150 years which gives Britain an
unrivalled, if still imperfect, background of applicable geological
knowledge.

From its earliest years the Geological Survey has been intimately
concerned with the practical application of geology: the demands
of society have altered and the role has changed but the need for
geological advice continues. Where there was once an overwhelming
interest in coal (an interest which must revive as other non-renewable
energy resources are exhausted), the search for hydrocarbons, the
location of uranium deposits and the potential for the burial of
high-level radio-active wastes have recently commanded attention.

Everywhere there is an increasing demand for geological data to
be presented in a readily understandable form, perhaps the inevitable
consequence of the increasingly rapid accummulation of geological
data and their summary in ever more complex maps. It is of interest
that although the emphasis may vary, the developments in the maps of
the Institute in recent years have parallels with work in Europe and
North America. The environmental geology, resource and engineering
maps all seek to portray those aspects of data thought to be of
particular concern to broad areas of interest. The techniques
employed utilise the separation of the various geological data sets
as elements and their recombination to show various practical
consequences of the geology of an area. Inevitably the scope of the
work is limited by the interests and needs perceived by those
responsible for the production of maps.

Clearly the way forward is through the establishment of auto-
mated cartographic methods and geological data bases. A start has
been made in Britain and within the near future it should be possible
for a user to select the geological features of interest and 'overlay' them with whatever constraints he or she chooses. There is a
room for a lot of development, while always retaining flexibility.
But what of the data themselves? It would be unreasonable to
suppose that the results of routine geological surveys conducted
over many years will, if encoded, suddenly be able to service all
questions. Undoubtedly gaps in knowledge will appear; equally
certainly a new type of geological survey will be mounted. Within
five years it is not unreasonable to suppose that the establishment
of geological data bases and developments in computer technology
will literally allow a user to generate his own map, specifying
his particular requirements and knowing that it is based on the
most up-to-date information available.

ACKNOWLEDGEMENTS

The author wishes to thank his colleagues who have critically read the script and in particular, Dr I B Harrison and Mr D Long for their advice on hydrogeology and engineering maps respectively.

This paper is published with the approval of the Director, Institute of Geological Sciences (NERC).

REFERENCES

British Standards Institution, 1972, "Code of Practice for Foundations, CP 2004", British Standards Institution, London.

Central Unit for Environmental Planning, 1971, "Severnside. A Feasibility Study", Her Majesty's Stationery Office, London.

Dumbleton, M. J., 1968, The classification and description of soils for engineering purposes; a suggested revision of the British system, Rep. Trans. Road. Res. Lab., LR 182.

Gostelow, T. P., and Browne, M. A. E., in press, Engineering Geology of the Upper Forth Estuary, Rep. Inst. Geol. Sci., 81/13.

Gostelow, T. P., and Tindale, K., 1980, Engineering Geological investigations into the siting of heavy industry on the East coast of Scotland. The north side of the Cromarty estuary, Open file Rep. Inst. Geol. Sci., EG/80/S1.

Harrison, D. J., 1981, The limestone and dolomite resources of the country around Buxton, Derbyshire: description of 1:25,000 sheet SK 07 and parts of SK 06 and 08, Miner. Assess. Rep. Inst. Geol. Sci., 77.

Her Majesty's Stationery Office, 1981, "Government publications, Sectional list 45, Institute of Geological Sciences", Her Majestry's Stationery Office, London.

Horton, A., Lake, R. D., Bisson, G., and Coppack, B. C., 1974(a), The geology of Peterborough, Rep. Inst. Geol. Sci., 73/12.

Horton, A., Shephard-Thorn, E. R., and Thurrell, R. G., 1974(b), The geology of the new town of Milton Keynes: explanation of 1:25,000 Special Geological Sheet SP 83 with parts of SP 73, 74, 84, 93 and 94, Rep. Inst. Geol. Sci., 74/16.

Hydrogeological Map of Southern East Anglia, 1981, 1:25,000, Institute of Geological Sciences, London.

Institute of Geological Sciences, 1978, "Geochemical atlas of Great Britain: Shetland Islands", Inst. Geol. Sci., London.

International Hydrogeological Map of Europe, 1980, 1:1,500,000 Sheet B3, Edinburgh, Bundesanstalt für Geowissenschaften und Rohstoffe, Hanover and UNESCO, Paris.

Kathol, C. P., and McPherson, R. A., 1975, Urban Geology of Edmonton, Bull. Alberta Res. Council, 32.

Laxton, J. L., and Ross, D. L., 1981, The sand and gravel resources of the country around Newport-on-Tay, Fife Region: description of 1:25,000 sheet NO 42 and parts of NO 32 and 52, Miner.

Assess. Rep. Inst. Geol. Sci., 89.

Lüttig, G., 1975, Geoscience and the potential of the natural environment 'in: "Geoscientific Studies of the Potential of the Natural Environment", Deutsche UNESCO - Kommission, Koln.

Lyell, C., 1836, Address to the Geological Society delivered at the Anniversary on the 19th February, 1836, Proc. Geol. Soc. London, 2: 397.

Nickless, E. F. P., 1982, Environmental geology of the Glenrothes district: description of 1:25,000 sheet NO 20, Rep. Inst. Geol. Sci., 82/15.

Nickless, E. F. P., Booth, S. J., and Mosley, P. N., 1976, The celestite resources of the area north-east of Bristol with notes on occurrences north and south of the Mendip Hills and in the Vale of Glamorgan: description of 1:25,000 resource sheet ST 68 and parts of ST 59, 69, 79, 58, 78, 67 and 77, Miner. Assess. Rep. Inst. Geol. Sci., 25.

Ordnance Survey, 1981, "Map Catalogue", Ordnance Survey, Southampton.

Rogers, P. J., Piper, D. P., and Charsley, T. J., 1981, The conglomerate resources of the Sherwood Sandstone Group of the country around Cheadle, Staffordshire: description of part of 1:25,000 sheet SK 04, Miner. Assess. Rep. Inst. Geol. Sci., 57.

Special Engineering Geology Sheet for Belfast and District, 1971, 1:21,120, Geological Survey of Northern Ireland, Belfast.

West, I. M., 1973, Vanished evaporites - significance of strontium mineral, J. Sediment. Petrol., 43: 278.

COMMENTARY : GEOLOGY

Although the paper given by Nickless focussed on the activities
of the British Geological Survey, he referred to comparable activities
in Canada, the Federal Republic of Germany and the United States of
America; other countries were mentioned in discussion. In many
instances, the different geological institutions have, for sometime,
moved beyond the relatively straightforward records of the disposition
of rocks by age and/or rock type. Progress has been made in two
directions (a) the organisation of computer-stored geological data-
bases enabling ready access and interrogation and (b) the development
of interpretative approaches. Little more need be said about the
form, other than to stress that computer-oriented data-bases are a
'means to an end' with one of the requirements being compatibility.
To enable cross-linkages to be made between geological, climatological,
vegetational .... data, it is essential that the data bases are
compatible; they must therefore be structured in similar ways.

For many years there has been a strong interest in the applied,
interpretive aspects of geology. Earlier interest in the development
of interpretive maps showing, with a high degree of probability, the
occurence of minerals, has been augmented in recent years with an
ever increasing array of environmental geology maps concerned with
sand and gravel thickness, hydrogeology, depth of water in unconsolidated layers... leading to the preparation of "derived maps"
showing, for example, the conditions for the foundations of buildings,
the extent of water resources etc. Ultimately, maps concerned with
"environmental potential" were developed which indicated, for example,
the ability of sites to carry light- or heavy-weight buildings, or
identified areas which must be surveyed in detail before siting a
structure because of the perceived potential of landslips, the
occurence of shallow undermining, the susceptibility to erosion etc.
At present, interpretive geological maps are in their infancy. Their
potential is enormous. There is a need to encourage their development in geological institutions, recognising that data-portrayal has
not received the attention that it deserves: it should become an
established part of a geologists training. Similarly it is essential
to develop a sympathetic approach in decision makers to the exploitation of interpreted data, accepting that the level of detail for
decisions related to strategy is coarser than that required for
tactical implementation (small versus large scale maps).

   Times have changed, and with these changes there is evidence
that geologists have perceived the needs of Society.   In addition
to traditional maps, they have been preparing interpretative maps
highlighting the practical implications of basic data in a way that
is usable by the non-specialist planner/decision-maker.

A

IDENTIFICATION OF ECOLOGICAL FACTORS
CHARACTERISING THE RANGE
OF *ECOLOGICAL* HABITATS

ii-Rural
b-Soil Resources

5

THE FORMATION AND PROPERTIES OF SOILS

R. Mayer

Department of Urban and Regional Planning
University of Kassel (GhK)
Postfach 101380, D-3500 Kassel
Fed. Rep. Germany

INTRODUCTION

Before dealing with soils as a resource to be used by man, it is desirable to recall what these uses might be :-

- media for growing plants to produce food, energy and a variety of raw materials including timber, cotton, rubber
- substrates upon which to construct buildings
- sources of raw materials (minerals, water, sand, gravel, topsoil)
- repositories for waste materials which may accumulate or be transformed/degraded

and in an integrated sense - a key facet of terrestrial ecosystems.

Most of us tend to take soils for granted, not recognising that they have an important role to play whenever ecosystems are altered or used. In acting as repositories for pollutants they can minimize the damage done to water by absorbing and 'binding' toxic elements some of which in other circumstances may have been discharged to atmosphere; soils help sustain the quality of our environment. Our general understanding of what is meant by "soil" differs to some extent by the type of use one has in mind.

Scientifically, soil is defined as that part of the earth's surface where rocks are altered by substances and energy derived from the atmosphere and/or released by living organisms.

WHAT IS A SOIL ?
It is a mixed system of :-

(i)   altered rocks or mineral grains (absent from peat soils)
(ii)   organic matter, both living and dead
(iii) spaces (soil pores) between particles of mineral and organic
       matter, which are occupied by air or water (soil solution) in
       contact, and exchanging matter and energy, with the atmosphere
       in the case of terrestrial soils and with water in the case of
       subhydric soils.

Soils are found on all land surfaces except where unaltered rock,
devoid of vegetation even lichens, is exposed.

SOIL FORMATION
A number of factors act simultaneously or in succession to determine
the types of soil that will be produced. These include :-

       - bedrock (parent material)   )
       - relief (topography)   )
       - water regime   )   ABIOTIC FACTORS
       - temperature regime   )

       - flora   )
       - fauna   )   BIOTIC FACTORS
       - activities of man   )

       - time   TIME

    Under the influence of <u>water</u> and friction attributable to wind
and/or frost and ice (<u>climatic factors</u>) the surface of a solid <u>rock</u>
may disintegrate without necessarily a change in chemical composition (<u>physical weathering</u>). For example, a granite may be split into
grains of its component minerals - quartz, feldspars, and mica while
a sandstone may disintegrate into identifiable sand grains. These
particles can then be subject to a variety of processes, collectively
known as chemical weathering and including :-

       - dissolution by water
       - hydrolysis
       - oxidation reactions
       - reduction reactions
       - reactions with organic compounds (organic acids)
       - surface reactions (adsorption)

All of these processes are influenced by the amounts of oxygen in
soil pores which, in turn, reflect to a large degree the type of
<u>water regime</u>. They are affected by <u>temperature</u> which strongly
influences the rates of the different processes of chemical weathering, and thirdly by the activities of microbes, soil invertebrates,
higher plants and animals including man. As a result of physical and
chemical weathering mineral particles are corroded, become smaller
and are dissolved sometimes with the formation of secondary minerals
which mostly occur in very small particles.

As rocks are weathered, the surface layer of weathered material will, in most conditions, be invaded by living organisms. These live partly, or in total, in, or on, the weathering layer which has become a 'soil'. As the different microbes and invertebrates die, their debris contributes, mainly by decomposition, to soil specific organic compounds (humic substances); on the other hand part may be mineralised to simple inorganic compounds. Dead organic debris and newly formed organic matter often accumulates in surface layers while at other times it mixes with mineral grains in the weathering layer (A-horizon). In this horizon the individual mineral grains are coated with thin layers of humic substances, sometimes with the formation of organo-mineral complexes with chemical links (bonds) between the crystal lattices of the minerals and the molecules of the organic material. These coated mineral grains tend to stick together to form aggregates of greater or less stability, a process which is helped by the mixing of mineral particles and organic matter brought about by the activities of soil animals notably earthworms.

Soil forming processes are <u>time</u> dependent. Thus, a 'surface' exposed to soil forming processes for a few decades will have a soil at a different stage of development to that on a surface exposed for several thousand years. Soils reflect not only the influence of present day soil forming factors but also those of previous times, when they would certainly not have been the same. Vegetation would have been at a different successional stage; climate and rates of erosion or accumulation would have been different. Most soil forming factors are interrelated - assemblages of vegetation are influenced by human activity, the supply of water regime,the nature of the bedrock; the availability of water is dependent upon topography and temperature to mention only two factors.

In short - soils are constantly changing, the rates of change varying in time and with the strength of each factor or blend of factors.

At this stage, you may ask why knowledge of soils and soil prop-erties should be a prerequisite for planners ? The answer is simple - as a result of different combinations of factors a variety of soils exists, each soil having its own distinctive combination of physical and chemical properties which suit it for different uses.

SOIL COMPONENTS

About 50 % of the volume of soil consists of solid particles, the remainder, the soil pores, being 'occupied' by air and/or water, the proportion depending upon amount of soil moisture. Amounts of dead organic matter (humus) in subsoil are relatively small, 1 % or less by volume, while in surface horizons they may exceed 10 %.

In soils in which it is slowly mineralised, organic debris tends
to accumulate above the mineral soil. In these conditions the
organic matter, if neither removed nor mixed with mineral soil e.g.
by ploughing, forms a predominantly organic surface horizon which
gives rise to peat when there is an excess of water.

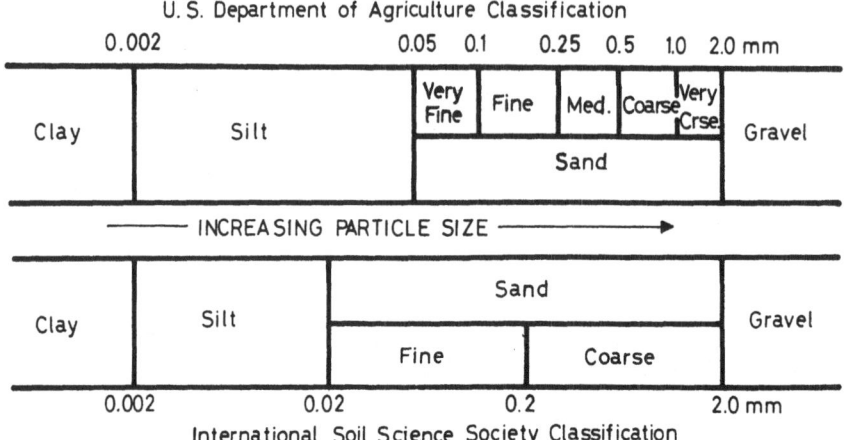

Figure 1: Two classifications of soil particles by diameter (note
that the scale is logarithmic).

SIZE OF SOIL PARTICLES

The physical and chemical properties of soils are closely
related to the sizes of their particulate matter and these in turn
influence the size and distribution of soil pores. By convention,
mineral particles are subdivided by diameter classes (Figure 1).

Most soils usually contain a blend of different diameter classes,
the blend of sand, silt and clay determining texture (Figure 2).
Thus a soil with 30 % silt, 40 % sand and 30 % clay is a clay loam
whereas another with 30 % silt, 60 % sand and 10 % clay is a sandy
loam. In addition to referring to particle size, the terms silt,
clay and sand imply mineralogical differences with the sand and
silt fractions being composed predominantly of mineral particles
derived from primary rock, rock fragments, or from sedimentary
rocks for example quartz, feldspars and mica (dark-coloured silicates like olivine, pyroxenes etc. are less abundant in most soils
because they are more readily weathered). In contrast clay
fractions are derived from minerals formed during the weathering of
rocks: clay minerals (kaolinite, illite, montmorillonite, vermi-
culite) and hydrated ferric and aluminum oxides.

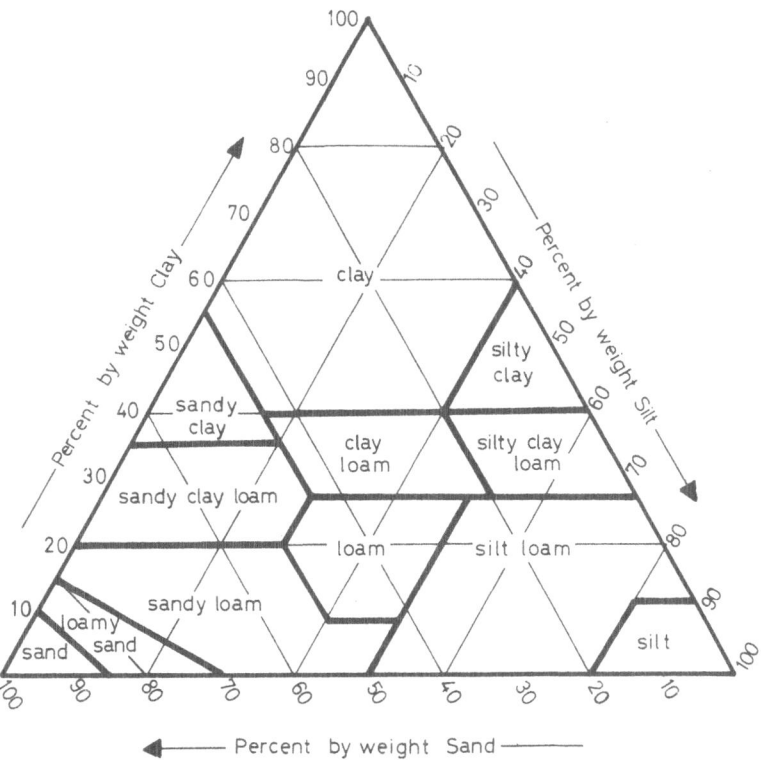

Figure 2: Triangle showing the proportions, by weight, of clay, silt
and sand fractions in a variety of soils with different
textures.

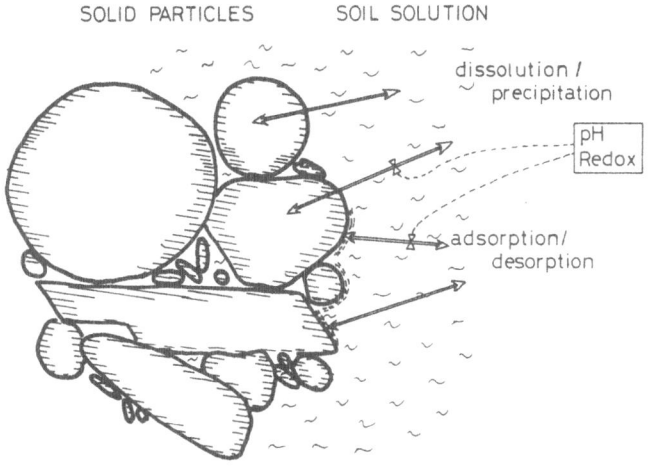

Figure 3: Schematic representation of soil chemical processes.

CHEMICAL PROPERTIES

A soil can be conceived as a complex ('multidisperse') physico-chemical system, with large numbers of organic and inorganic sub-stances in gaseous, liquid, solid and adsorbed states, constantly exchanging matter and energy with its environment, the exchanges being related to dissolution/precipitation and adsorption/desorption (Figure 3).

Dissolution/precipitation:

Solid matter (mineral grains) may be dissolved, the mineral constituents entering the liquid phase mostly in ionic form, a feature of importance as it influences nutrient uptake by plants and also the movement of nutrients out of reach of roots. On the other hand the solid phase may be increased by the precipitation of insol-uble compounds (when the solubility product of constituents in solution is exceeded).

Adsorption/desorption:

Very important processes for plant nutrition, because they maintain nutrients in a form available for root uptake while at the same time preventing the nutrients from being removed by seepage water. Of the many types of adsorption that occur in soils, ion exchange is the most important. Most organic and inorganic (mineral) particles have excesses of negative charges which are 'matched' by cations (such as Ca^{2+}, K^+, Mg^{2+}, Al^{3+}, H^+) that become, on substi-tution, available for root uptake.

Dissolution/precipitation and adsorption/desorption are strongly dependent upon soil acidity, pH, and redox potential. The soil pH is defined as the hydrogen-ion concentration (expressed on a nega-tive logarithmic scale) in its liquid phase, when that liquid phase is in equilibrium with the solid phase of the same soil. When hydrogen ions are added to the soils, some will be 'consumed' by chemical reactions, for example, the dissolution of limestone (cal-cite $CaCO_3$) and aluminum hydroxide (AlOOH):

$$CaCO_3 + H^+ \longrightarrow Ca^{2+} + HCO_3^-$$

$$AlOOH + 3\ H^+ \longrightarrow Al^{3+} + 2\ H_2O$$

As a result soil pH is lowered to less than might have been expected had these reactions not occurred - in short, soil has a built in buffering system. The redox potential of a soil is a measure of its ability to oxidize or reduce soil constituents. Water-logged soils, lacking oxygen, have small redox-potentials whereas well-aerated soils have large redox-potentials.

Dissolution and precipitation usually occur relatively slowly but nonetheless are important in the geological time scale of soil formation. On the contrary, adsorption/desorption occur instantaneously; they are important for the short term availability of plant nutrients and the immediate retention of many pollutants reaching the soil.

PHYSICAL PROPERTIES - SOIL WATER

The movement of water and the volume of soil air are closely linked to grain size distributions (texture) and to soil aggregation, the size of grains and their packing densities determining pore sizes. Water is held in soil pores by the surface properties (capillary forces, charged surfaces, water dipoles) of the particles. The closer a water molecule (dipole) is to a charged surface the tighter it is bound to that surface (Figure 4) and therefore the larger the gravitational, osmotic or suction forces needed to effect the release of water molecules.

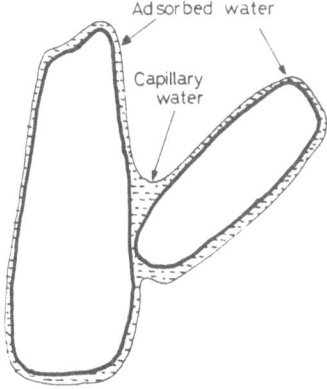

Figure 4:   Water film surrounding soil particles.
            Water molecules covering the solid phase are held very
            tightly (absorbed water); water in the space between
            particles (capillary water) is held more tightly and
            closer to the surface in small spaces than in large spaces.

From the concept of water binding in soils, it follows that soil texture and aggregation also strongly influence the air in soil as well as water transport through soil. Pores which are not filled with water must be filled with air. Soils with large pores retain water very loosely: it can therefore be readily removed by gravitational forces with consequent aeration. On the other hand, water is strongly bound in soils with small pores. The pores of these soils will be filled with water over extended periods of

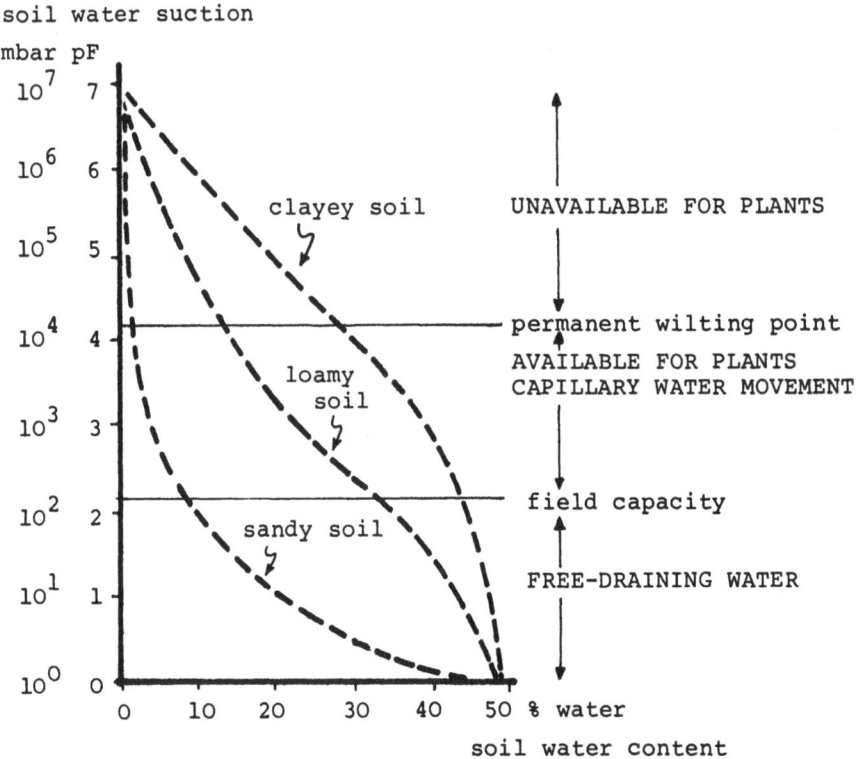

Figure 5: The relationship between soil water content and soil-
water suction in three different types of soil.

time. As a result soils with small pores are likely to lack aeration with potentially detrimental effects on plant growth. The
relationships between amounts of soil water and soil-water suction
(expressed on a logarithmic scale in mbar or pF value) are shown
in Figure 5 for three types of soils - clayey, loamy and sandy.
Critically there is a suction value beyond which plants are unable
to extract water from soils - $10^{4.2}$ mbar (=pF 4.2), the "permanent
wilting point". Below a soil-water suction of pF 2 ('field capacity')
water is bound very loosely and therefore drains under gravitational
forces. This difference between field capacity and permanent wilting
point corresponds to the amount of water available for plants. It
is less than 10 % of total volume in sandy soils, around 15 % in
clay soils and more than 20 % in loamy soils.

BIOLOGICAL PROPERTIES

    The dead organic matter in soil strongly influences soil physico-
chemical properties. It is also a source of plant nutrients, especially nitrogen and phosphorus. The decomposition, mineralization

and transformation,into humic substances, of plant and animal debris
is mainly done by microbes. Large animals, and the mesofauna, are
mainly responsible for the disintegration, comminuition and mixing
of organic debris. In general most soil organisms are active in
aerated, moist conditions, the rates of decomposition being faster
at high than at low temperatures and in neutral, rather than acid,
conditions. When conditions do not favour soil organisms dead organic
matter tends to accumulate.

SOIL FUNCTIONS

    Soils have many functions. As planners we are concerned to
consider these functions with specific objectives in mind - two will
be discussed in detail (a) soils as media for the growth of plants
and (b) as repositories for pollutants.

Table 1: Plant nutrients - forms, sources and functions.

	FORM	SOURCE	FUNCTION
C	CO_2	Air	Main constituents of
H	H_2O	Water	Organic molecules
O	O_2, H_2O	Air, Water	
N	NO_3^-, NH_4^+, N_2	Soil Water (Air), organic matter	Proteins and other Organic molecules
S	SO_4^{2-}, SO_2	Soil Water (Air), organic matter	Enzymes
P	Phosphates	Soil Water (organic matter, minerals)	Storage and transport of Energy in plant cells
K	K^+	Soil Water/Minerals	Activation of Enzymes
Ca	Ca^{2+}	Soil Water/Minerals	Cell Wall and Membrane
Mg	Mg^{2+}	Soil Water/Minerals	Photosynthesis (Chlorophyll)
Fe	Fe^{2+}	Soil Water/Minerals	Constituents of Enzyme
Mn	Mn^{2+}	Soil Water/Minerals	Systems
Cu	Cu^{2+}	Soil Water/Minerals	
Zn	An^{2+}	Soil Water/Minerals	

Soils as media for the growth of plants

    What does a plant need for good growth ?
(i)   Rooting space - soils in which roots may develop freely and
       without being limited by impenetrable rock, impermeable
       layers or water-logging.
(ii)   Mineral nutrients (Table 1) - these must be available close
       to roots and in a form that can be readily taken-up by roots,

namely the ionic form. However it is important to remember
that nutrients in this form are mobile and therefore can be
lost by leaching. The best supply of plant nutrient is provided by soils in which :-
- the parent material consists of easily weathered rocks
  containing minerals with large amounts of potassium, calcium
  (lime), magnesium, trace elements and phosphate.
- the ionic nutrients are bound loosely to prevent leaching
  but which are readily released (desorbed) into soil solution
  when concentrations drop below the demands of plants. These
  conditions are met by near neutral or slightly acid (pH-
  range 5.5 to 7.0) soils rich in both organic matter and
  clay minerals.
- nitrogen is readily mineralised by microorganisms, conditions
  are optimal in neutral or slightly acid soils rich in organic
  matter.
(iii) Water and air - Most plants need a continuous supply of water
together with well aerated soil pore spaces to facilitate
root growth, exchange gases and take up water and nutrients.
These requirements are provided by soils with a wide spread
of grain sizes (Figure 1). Whereas water is retained in
medium sized pores while still being available for root uptake,
the presence of large pores, from which water drains quickly,
ensures a replenished supply of air. Amounts of water available for plants and supplies of air are optimal in loamy soils
(Figure 5). Sandy soils do not retain water whereas clayey
soils, by retaining water too tight, restrict its withdrawal
by roots.

Soils as repositories for pollutants

Many pollutants are naturally or intentionally deposited on
soil
- atmospheric pollutants (acid rain, heavy metals, organic
  compounds)
- solid and liquid wastes (industrial, domestic, agricultural).
Non decomposable (persistent) pollutants can be stored and often
accumulated indefinitely; on the other hand they may pass through
the soil without being retained. Decomposable (non-persistent)
pollutants may pass through the soil, or be retained, stored and
decomposed.

The ability to store is generally found in :-
- fine textured clayey soils (*ipso facto* with large internal
  surface areas), where chemical binding and adsorption may
  take place. These soils would not be suitable for pollutants
  in solution as their clay contents impede percolation.
- neutral soils in which many compounds would be precipitated.
- strongly weathered, deep soils with large capacities for
  storing water.

Table 2:   Examples of the way soil characteristics appropriate for
           particular land-uses can be evaluated and how experts
           from different disciplines contribute to the planning
           process.

Land-use	Specific need or function	Soil characteristic relevant to specific need or function	Soil parameter characterising specific function
Growing an agricultural crop.	Air/water supply	Available water capacity Air capacity Permeability Infiltration	pF curve texture Kf value Infiltration rate.
	Nutrient supply	Exchangeable cations N-, P- content Organic matter	Cation inventory, soil pH. N,P, concentrations. Organic matter content.
	Rooting space	Depth of root-permeable soil. Water table Stone content	Depth to hard rock or layer with high bulk density. Depth to water table. Stone content.

Further examples of land-use, relevant needs or function and soil characteristics
(all in alphabetical order).

Campsites Forestry Heavy Industry Light Industry Residential area Roads Sports fields Waste disposal	Bearing value Corrosion potential Flooding hazard Shrink-swell potential Slope Susceptibility to erosion Trafficability	Clay content Exchange capacity Horizon sequence Hydromorphy Mineral composition Root density Salt content

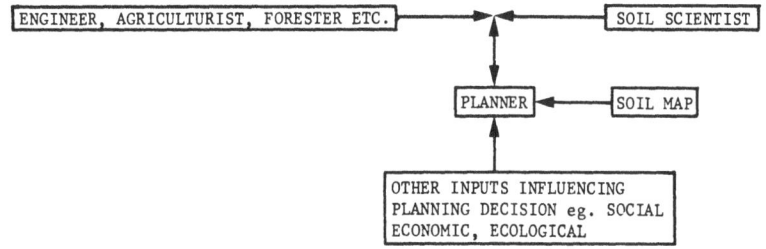

- soils or soil horizons with relatively large amounts of
  organic matter for adsorbing and chemically binding pollutants
  and complexing metals.

Many organic pollutants can be degraded by naturally occurring
microbes whose activities are usually intensive and include many
different groups of organisms, which is maximal in well aerated
neutral or slightly acid soils rich in organic matter.

MAN'S INFLUENCE UPON SOIL PROPERTIES AND FUNCTIONS

It goes without saying that soils are a priceless resource. But
how should they be maintained to ensure our future well-being ? On
the other hand it can be asked, how can soil conditions be improved ?
There are many techniques, whose use depends upon objectives, e.g.
the improvement of supplies of plant nutrients by applying fertilizers, improvement of water availability by irrigation, the prevention of excess water by drainage, the enhancement of rooting space
by deep ploughing. But, the good done by these techniques often
incurs a penalty, e.g. the pollution of groundwater by fertilizers
not taken-up by plants, salinisation by inadequate irrigation, the
increased loss of nutrients attributable to ploughing etc.

EVALUATION OF SOIL RESOURCES FOR PLANNING

An attempt has been made to show how decision makers could
rationally approach the selection of soils for maximal plant growth
and for disposing of wastes. But which attributes should be considered if he/she were seeking the best soil to grow irrigated rice,
or upon which to construct a tennis court - to name two contrasting
land uses. As Table 2 indicates there are many more soil characteristics that could, with advantage, be considered. But, a text
dealing with soils would be incomplete without a reference to soil
maps and mention of soil maps inevitably leads to a discussion of
soil classification. As is shown in Chapter 5, there are many
systems of soil classification some oriented to traditional soil
science concerned with the evolution of soils while others are
more readily interpreted in terms of utility. Soil maps outlining
the distribution of soils capable of carrying productive crops
are now commonplace. Soon we will need maps to show bearing
capacity (= clay content) for the construction industry; ability
to withstand erosion .... the possibilities are infinite.

6

CLASSIFICATION AND MAPS OF SOILS AND THEIR INTERPRETATION FOR

PLANNERS

C. P. Burnham

Wye College, University of London
Ashford
Kent TN25 5AH, England

INTRODUCTION

Land has many properties relevant to its use: some derive from
the natural environment, some from human activities. Even among the
natural factors relevant to agriculture and forestry, climate and
topography are of comparable importance to soil. Nevertheless if
land potential is being appraised in an integrated manner soil is
likely to be one of the most difficult factors to assess for, like
geology, most of the evidence about soil is concealed and generalisations must inevitably be made from a relatively small sample.
Indeed, even if the whole soil body could be investigated it would
be greatly altered in the process! Soils are very complex bodies,
and realistic models of their nature and distribution will, *ipso
facto*, be complex. A clear understanding of the problems involved
is therefore essential if expenditure on soil investigations including
mapping is to be used cost-effectively. For these reasons three
facets of the problem will be examined:

    (i)   soil classification
    (ii)   soil variability, and
    (iii)   soil mapping and the use of soil maps.

SOIL CLASSIFICATION

1. <u>'Soil' and 'not-soil'</u>

The first problem is the boundary of the 'entity' being class-

ified. Some examples of difficult categories are bare rock, the
surface layer of a spoil heap, bottom deposits under standing water,
dust on the lunar surface. If soil is defined as material at the
solid Earth-atmosphere interface which supports, or might become
capable of supporting, plant life, then the last two and possibly the
first are inadmissible, but the spoil heap certainly is 'soil'. The
planner is interested in the nature of any land surface, and therefore the definition of soil should be wide.

2.   The soil individual (Simonson, 1968)

        What are we classifying ? If we are classifying animals we know
what one lion is, but what is one soil ? Some soil properties will
change between two points however close to one another, but the soil
at a point cannot be sampled for examination and analysis. One might
classify bags of soil representing a mixture of that present over
some defined area and depth. This is done for agricultural advisory
purposes, generally on 'topsoil' only. However, Plot (1686) recognised
that soil tends to be layered 'like the skins of an onion', and the
soil profile, the assemblage of layers from the surface to more or
less unaltered geological material, is now accepted as the proper unit
of study. This implies that differences between layers and their
interrelations should be considered, and not merely the properties of
each layer in isolation.

        Some classification systems begin by giving names or indices to
separate layers or 'soil horizons'. Fitzpatrick (1980) puts most
emphasis on naming the individual soil horizons, a feature also
strongly influencing the United States Department of Agriculture
(1975) 'Soil Taxonomy', and other classifications including Avery's
(1980) used in England and Wales. Soils are a continuum vertically
as well as horizontally, and in many instances horizons are distinguished rather arbitrarily because of differences between properties
at different depths within the profile - soil profiles should be
studied as entities. They may be examined more or less at a point
using an auger, but then it is not easy to assess the relationships
between subsoil layers or to take uncontaminated samples. The definition of an individual soil or pedon specifies a body of soil
extending from the surface to more or less unaltered geological
material (the soil profile) and about 1 m^2 in area, i.e. just big
enough to facilitate a pit for examination and sampling (Simonson
and Gardiner, 1960; USDA, 1975). A pedon already includes some
spatial variation. The different sides of a small hole sometimes
show obvious soil profile differences and detailed analyses may
reveal differences when none can be seen. These differences may or
may not be significant in the classification adopted.

3.   The structure of systems of soil classification (see Schelling,
     1970; McRae and Burnham, 1976; Fitzpatrick, 1980 and Clayden,
     1982).

Single factor classifications. Great simplification can be
achieved if only one property is used as in classifications restricted
to a small area or devised for a limited purpose. The most likely
single property to be useful is particle size, usually known as soil
texture, which is well correlated with a number of other properties,
such as permeability, workability etc. It is usual to recognise 11 to
21 texture classes, based primarily on the proportions of sand, silt
and clay in the 'fine earth' (< 2 mm), as found by particle size
analysis in the laboratory. An experienced observer can place most
samples in their correct texture class by handling moist soil. One
trial involving several hundred samples gave precise agreement between
laboratory and hand assessments in two thirds, and at least a close
approximation in nine-tenths, of samples, so indicating that this
arbiter can be used in the field. The one tenth of discrepant samples
mostly contained large amounts of organic matter. Some would argue,
even in these instances, that 'field texture' is better assessed by
'feel' than particle size analysis as feel simulates workability.

If 21 texture classes are too many, then simpler systems are
available, e.g. with four divisions: sandy, coarse loamy, fine loamy
and clayey (Avery, 1980). This enables changes down the profile to
be recognised without the system becoming unwieldly, e.g. if the
predominant texture is assessed in two 'control sections', say 0-40
cm and 40-80 cm, there are 16 possible 'texture profiles': such as
'sandy throughout', 'coarse loamy over sandy' etc.

Other 'single property' classifications may be appropriate, e.g.
soil wetness classes (Jarvis and Mackney, 1979), depth to water table
(van Hessen, 1970) etc.

Coordinate classifications (Avery,1968; Webster,1977). It is
possible to combine two or more classifications based on single properties or criteria, without putting the properties or criteria in any
order of importance. In this instance the total possible number of
groupings is the product of the number of classes of each property.
For example, 15 texture classes, 6 drainage classes and presence/
absence of calcium carbonate (2 classes) might be combined in a co-
ordinate classification with 15 x 6 x 2 = 180 groupings in all. If
there are no more than 3 properties, combined names are possible,
e.g. 'calcareous, well drained, sandy loam'. With 4 or more properties, or if the soil is being considered as having more than one
layer, a symbol consisting of numbers and/or letters is likely to be
used. A good example is the Belgian system (Tavernier and Marechal,
1962).

It is not necessary for 'sub-classifications' in a coordinate
system to be simple gradings of one property. In the Belgian system
organic matter content is considered together with texture. In the
USSR a very elaborate 'ecological-genetic' sub-classification is
combined with 4 moisture regimes and 5 orders based on soil chemistry.

In these instances it is commonly found that some of the theoretical combinations do not exist in real life.

Coordinate classifications with two sub-classifications are easy to display in tables. Three sub-classifications make display a cumbersome problem, and with four or more a computer, able to cope with unimaginable 'multi-dimensional hyperspace', is essential. Computers can help to optimize a complex coordinate system by making the best use of the compartment 'boxes'; they can attempt to ensure that data from the pedons to be classified form clusters within the 'boxes' with comparatively few near the boundaries of the 'boxes', but this can be problematical as soils are part of a continuum (Webster, 1977).

TABLE 1 :   Structure of USDA Soil Taxonomy (1975), with details at each level of a soil pedon in the author's garden at Wye, Kent, U.K.

Name of category at the different levels of the hierarchy, also numbers of classes in that category in the U.S.A.	Name applied to the pedon, in the garden at Wye at the different levels	Summary of information implied (and required) by the name in the middle column
1. ORDER (10)	Alfisol	Relatively base-rich soil, not in arid climate, with clay enriched subsoil
2. SUB-ORDER (47)	Udalf	Moderately warm to hot, constantly moist climate
3. GREAT GROUP (185)	Hapludalf	Temperate climate, without a very compact, brittle, subsoil layer
4. SUBGROUP (970)	Typic hapludalf	Reasonably deep, loamy or clayey soil, with no unusual features
5. FAMILY (4,500)	Fine silty, mixed, mesic, typic hapludalf	Specified texture, mineralogy and temperature regime
6. SERIES (c. 10,000)	Hamble (a locality name near where soil was first described)	Specified profile features and parent material (brickearth)

Hierarchical classifications. Most soil classifications are hierarchical, i.e. divisions are made at several levels, each time subdividing the units at the level above but never recombining them. The name comes from the levels of jurisdiction of priests within certain church organisations. In the Church of England territorial authority has six levels: sovereign, archbishops, bishops, arch-

deacons, rural deans and parish priests. At each level the territory
may be subdivided, but a division once made persists to the bottom of
the system.

The U.S.D.A. Soil Taxonomy is a good example of a hierarchical
soil classification (Table 1), with six levels. Logic might suggest
that the criteria for the divisions at each level should be consistent.
But this is only nearly so at the fifth (family) level: all other
criteria used to define taxa are selected empirically, although the
presence of diagnostic horizons is the basis of most orders, and soil
moisture regime the principal arbiter of differentiation at the suborder level.

4. Choice of properties when classifying

Soil forming factors. Only a tiny proportion of the body of
soil to be classified can be examined, but some of the formative
factors can be examined more easily, or extrapolated more confidently,
than the soils themselves. From about 1870 to 1950 the 'zonal'
approach to soil classification was generally accepted, particularly
in Russia and the U.S.A. Soils characteristic of broad climatic and
natural vegetation zones were regarded as 'zonal', e.g. podzols in
the Northern Coniferous Forest, and chernozems in the Tall Grass
Steppe. Soils whose characteristics were dominated by an unusual
parent material, e.g. rendzinas, or by topography, e.g., gley soils,
were 'intrazonal' and little altered geological materials, e.g. raw
alluvium, were 'azonal' soils. Many of these group names have out-
lasted the framework in which they arose.

In Western Europe a geological approach to soil classification
was common, and emphasis on parent material is still strong in the
systems used in both West and East Germany and in the Netherlands.
A report on the soils of South East England published in 1911 had no
soil map, for it was assumed that geological maps would suffice.
One can still detect vestiges of the use of climate, geology,
topography and vegetation/land use in soil classification. Areas
covered by geological, but not by soil, maps are still found and
here useful statements can be made about 'granite soils', 'chalk
soils', 'alluvial soils' etc. In soil mapping great use is made of
topography, and terms like 'plateau soils', 'slope soils', 'floodplain soils' may form chapter headings in a soil survey report.
Conversely, while both natural vegetation and the activities of man
exert considerable effects on soils, there is a strong tendency in
some countries to avoid separating cultivated and uncultivated
examples of otherwise similar soils (USDA, 1975 p.8-9).

In recent years most scientists concerned with soil classification, excepting those in the USSR, have focussed on actual soil
properties instead of related factors and, from the planner's viewpoint, this is probably desirable. Thus he will seek information

about the soil itself from soil surveys, and prefer to consult geologists, meteorologists, ecologists etc. about their different spec-
ialities.

TABLE 2:   The Unified Soil classification system for civil engineers
           (United States Department of Defense, 1968).

Description   Code

Highly organic soils (Peat and Muck)   Pt
Coarse-grained soils (over 50% coarser than 0.074 mm)
  Gravel (% particles greater than 5 mm exceeds % particles between 5 mm
  and 0.074 mm)
    Less than 5% finer than 0.074 mm
      Well graded   GW
      Poorly graded   GP
    More than 5% finer than 0.074 mm
      Fines mostly silt   GM
      Fines mostly clay   GC
  Sand (% particles between 5 mm and 0.074 mm exceeds % particles greater
  than 5 mm)
    Less than 5% finer than 0.074 mm
      Well graded   SW
      Poorly graded   SP
    More than 5% finer than 0.074 mm
      Fines mostly silt   SM
      Fines mostly clay   SC
Fine-grained soils (over 50% finer than 0.074 mm)
  Low-organic silts (liquid limit is greater than (20 + 4/3 Plasticity Index):
    Low plasticity (liquid limit below 50% moisture by weight)   ML
    High plasticity (liquid limit above 50% moisture by weight)   MH
  Low-organic clays (liquid limit is less than (20 + 4/3 Plasticity Index):
    Low plasticity (liquid limit below 50% moisture by weight)   CL
    High plasticity (liquid limit above 50% moisture by weight)   CM
  Organic-rich silts and clays
    Silts and clays of low plasticity (liquid limit below 50% moisture
    by weight)   OL
    Clays of higher plasticity (liquid limit above 50% moisture by weight)   OH

Notes:

1. 'Well graded' means a wide distribution of particle sizes with no 'gap' or excess
    in any grade.
2. Plasticity Index = liquid limit - plastic limit.
3. The determination of liquid limit and plasticity index is described in standard
    textbooks of soil mechanics (eg. Jumikis 1967) and by Olson (1973).
4. A CL-ML intergrade is recognized for soils with a liquid limit below (20 + 4/3
    Plasticity Index) but above 10% and Plasticity Index between 4 and 7.

A Use-oriented choice of soil properties.   If it were possible
to identify one use, or a coherent range of uses, relevant soil
properties could be used exclusively in the classification. Good
examples are the soil classifications used by civil engineers, e.g.
the 'Unified' system of the U.S. Department of Defense (1968) and
the similar British Soil Classification system for Engineering
(British Standards Institute, 1976; Hartnup and Jarvis, 1979). In
the Unified system (Table 2) only particle size analyses using 4
mesh (5 mm) and 200 mesh (0.075 mm) sieves, measurements of liquid
and plastic limits and a broad assessment of organic content, are

needed. This classification is very useful for a wide range of
civil engineering uses (Holtz, 1969; USDA, 1971), but it would be
of little value to the agriculturist. It is usually agreed to be a
false economy to use a specialised classification for surveys over
large areas.

Genesis-oriented choice of soil properties. Most modern soil
classifications are based on intrinsic soil properties chosen to
reveal the way in which soils have developed; they highlight so-
called soil forming processes, for example the effects of water-
logging (see Clayden, 1982).

The preoccupation with soil processes has made life hard for
soil surveyors, and is also bad news for planners. For example,
there is great interest in whether a natural process called 'clay
eluviation' has moved fine mineral particles from higher to lower
horizons in a soil profile. This produces profiles with a subsoil
horizon containing more clay than horizons above, but such profiles
can also (a) be produced by the dissolution of clay under very acid
weathering conditions or (b) be a relic of less and more clay rich
layers in the parent material, e.g. coarser (sandy or silty) particles blown in by wind to bury an original clayey soil. To identify
this difference in origin, it is necessary to resort to the use of a
microscope. Soil surveyors must wait months or years before they
can place their soils in the classification, and then they can only
do so on the basis of a ludicrously small sample of supposedly
'representative' profiles. In the meantime the obvious property, a
relatively clayey subsoil, is made secondary even though, by rest-
ricting water movement, it has an important practical effect irrespective of the origin of the clay. An emphasis on genesis in the
higher levels of soil classification in effect means that only the
lowest level, often called the soil series, is of major importance
to the land planner (Simonson, 1964; USDA, 1969; USDA, 1978).

5. Numbers of categories

Many land use decisions really involve a binary classification,
i.e., either 'use for the specified purpose' or 'do not use'. The
soil investigator is often asked for a corresponding decision:
'cultivable soils' and 'soils not cultivable' etc. But even more
frequently it is convenient to consider a number of optional uses or
a number of levels of suitability for one use. Nevertheless
'suitability' classifications will normally have only a few divisions. General soil classifications, on the other hand, are normally
made to service a wide range of derived suitability classifications,
including those needed in the future, whose requirements cannot now
be precisely foreseen. This suggests that a general soil classification should not be over-simplified. When interpreting data in
terms of a complex classification system for a planning purpose, it
is readily possible to group together a number of the taxonomic

units, but not easy to make new divisions cutting across these units.
Also, when the system is used as a language summarising the nature of
soils, a reasonable amount of information should be recoverable from
the taxonomic name. This necessitates a system with many divisions
whose units are precisely defined with a large information content as
in the U.S.D.A. and Belgian systems. However, systems with many
precisely defined units are time consuming to apply, and cannot
readily be used by non-experts - a balance must be reached. The
U.S.D.A. (1975) remarks that 'The mind readily grasps 5 or 10 items,
but it cannot deal simultaneously with 100 or 1 000 items without
some ordering principle'. Experience suggests that laymen will grasp
a 'working vocabulary' of six or seven soil taxa, agricultural
advisors - 20 - 25, specialist University students - 35, professional
soil scientists - 50 - 60. However 'ordered', a system with more
than 50 taxa demands a specialist with a manual, and only a computer
can cope with more than 1 000.

But would a classification simple enough to be acceptable to a
'user' be capable of doing the tasks required of it ? If not, a
consultant must always be available to provide a digest of the data.

Fortunately the structure of a hierarchical classification offers
the possibility of selecting an appropriate 'level' with about the
right number of units. For example, using the U.S.D.A. classification (Table 1), the 'suborder' level would be appropriate for a
legend for a soil map of a continent or as a basis for a course in
world soil geography. The 'great group' or 'subgroup' level is
appropriate to an expert using the keys provided (U.S.D.A.,1975) to
characterize pedons. On a national scale, the soil series is only
suited to a data bank operated through a large computer. With a
coordinate classification the number of sub-classifications may be
reduced, or fewer divisions used in each sub-classification. Another
way of reducing complexity is to limit the geographical area, and
only concern oneself with the types of soil units that occur within
the limited area. The use of soil series with their high information
content may well be desired. The number of series, with a significant degree of representation, in a small area varies with the type
of terrain, but at least 20 may be expected in 100 sq. km and about
50 in 500 sq. km. In the USA there are about 9,500 series (USDA,
1972).

In the end, however, a <u>soil class</u> should enshrine as much useful information as possible. This generally means using low level
units, the soil series or a close equivalent, of which there will
inevitably be a large number. Simplification should be achieved
by grouping the series for the required purpose: e.g. good,
indifferent or unsuitable. Thus the kind of classification of
importance to a planner is that required to support soil mapping
and soil survey interpretation.

SOIL VARIABILITY (see Beckett and Webster, 1971; Webster, 1977;
                  Wilding and Drees, 1978).

1. Introduction

Lateral soil variation can usefully be divided, arbitrarily,
into three components: short, medium and long range.

2. Short range variation

This cannot be shown even on a detailed map. I recently examined
and sampled four pedons in coppice woodland with reasonably uniform
(i) slope, (ii) underlying geology and (iii) vegetation. The whole
area would be placed in one soil taxonomic unit, the Batcombe
series. Yet differences existed. Some of the pedons were moister
than others, reflected in more or less mottling at a depth between
40 and 70 cm. The topsoil varied appreciably in both colour and
thickness, apparently due to (i) slight sheet erosion after timber
extraction and (ii) a tendency for litter to accumulate in hollows
or in the lee of stumps. The most notable variation found was in
amounts of 'exchangeable cations' in topsoil, especially potassium,
calcium and magnesium, the largest amounts being more than twice the
smallest. The large amounts seemed to be associated with the occurrence of charcoal in the topsoil, i.e. with the sites of fires in
which branches and other waste were burned when the coppice was
harvested. Ball and Williams (1968) and Moormann and Kang (1978)
have noted analogous variability.

Short range variations of this sort are appropriately recognised
by characterising mapping units by means and standard deviations
following random sampling. While physical properties, such as particle size categories, are subject to only moderate short range
variation, planners should be warned that chemical 'fertility' is
often so greatly affected by past practices, including the application of fertilizers, that predictions can only be vague generalisations unless supported by the results of analyses.

3. Medium-range variation

When soil properties are measured at close intervals along transects it is commonly found that the rate of change is decisively
greater at some points than at others (Webster and Cuanalo, 1975;
Webster, 1978), for example at a boundary on a detailed soil map.
Usually the boundary represents a change in parent material, relief
(especially as it affects water in the soil) or in history of use
(e.g. from cultivated to uncultivated land). However, it seems
unlikely that subdivisions of landscape can account for more than
about 50 % of the total variation. Nevertheless, short range variation itself varies on a medium scale, so that some soil mapping
units are inherently more variable than others (Nortcliffe, 1978).

Sometimes this variation has a geological basis, e.g. closely inter-
bedded sandstone and clay, sometimes it is associated with micro-
topography, e.g. hummocky moraine or a 'meander scroll' on a flood
plain. In a crude way, the more variable areas have been recognised
for a long time on detailed soil maps by denominating them 'soil
complexes' instead of naming them after the dominant soil series.

4. Long range variation

When short and medium distance variations have been identified,
there remains an additional component of variation namely regional
differences which in part are attributable to climatic gradients,
past and present.

5. Implications for site investigations

Small sites. A soil map should not be taken as authoritative
for any area smaller than the planning unit indicated in Table 3
for that scale of map. Smaller areas are liable to be predominantly
occupied by 'inclusions' atypical of the mapping unit as a whole.
The only use of soil maps appropriate to small sites is for a desk
study to reduce a large number of possible sites to a smaller number
prior to 'on the spot' investigation. However small the site, a
field check for apparent uniformity should be made. If no differences
of significance are noted in soil or relevant external characters and
a sample is required, about 20 equal subsamples from the same depth
range can be well mixed to form a composite sample. If the area is
larger than, say, 5 ha, or if there are noticeable differences in
soil or external characters within the site, several separate samples
should be taken.   Either take at least 4 samples at random from
within each division of the area or take a composite sample from
within each division. If separate samples are taken, ensure that they
can be relocated in case the results suggest that the area should be
further subdivided or reclassified.

Larger sites. If a sufficiently detailed and reliable soil
survey exists, the information collected on site and the samples
taken can be assigned to its mapping units. If no suitable soil
survey exists, a soil map of the site, and perhaps of any other
areas affected by proposed changes, should be made. But it
should never be assumed that the mapping units are uniform. The
proportions of taxonomic groups represented by the pedons within
each unit should be determined, preferably using a different
population of samples from those used to plot the boundaries. If
it is desirable to take note of particular characteristics then
these should be estimated with statistical precision (Webster, 1977),
recognising that the required number of samples is likely to be
large. Care should be taken to avoid extrapolations from very
precise data obtained from an inadequate number of samples.
However, for many purposes approximations are sufficient: they can
always help in structuring more rigorous analyses

TABLE 3: Soil maps: Factors related to scale.

Scale	Designation of map	Typical Mapping units	Approx. no. of ground observations km^{-2}	Approx. Smallest area shown on map	Planning Unit
1:5,000	Very detailed	Subdivisions of series (variants, phases)	1,000	0.04 ha	PLOT (0.2 ha)
1:25,000	Detailed	Series and slope phases of series	100	1 ha	FIELD (4 ha)
1:50,000	Semi-detailed	Series and complexes	25 or 4, using interpretation of aerial photographs	4 ha	SMALL FARM (16 ha)
1:200,000	Reconnaissance	Associations of series	2 or 0.5, using interpretation of aerial photographs	64 ha	LARGE FARM or PARISH (250 ha)
1:1,000,000	National	Associations of major groups	Usually compiled	20 km^2	DISTRICT (80 km^2)

6. <u>Soil variability and soil maps</u>

    Soil variability is of the utmost relevance to soil mapping.
Even when mapped in detail only about one tenmillionth of the soil
body is actually examined. A perfect soil map is impossible. When
mapping soils in complex areas with a mixture of taxonomic units
(i.e. associations and complexes), it is reasonable to expect that
the taxonomic units encompassing more than 10 % of the pedons will
be mentioned in the map key and report - these can be expected to
account for 70 % to 95 % of the area being mapped. Sometimes soil
series are used both as mapping units and as low level taxonomic
units - it is important to remember the distinction. Many soil
series, used as mapping units, will be made up of 80 - 90 % of that
series in the taxonomic sense, with 10 - 20 % of other series (as
'inclusions' or 'impurities'). However, some mapping units, called
'soil series', may contain no more than 50 - 65 % of that series
(Ragg and Henderson, 1980),but this is not the catastrophe that it
may seem because much of the balance may represent very similar
soils only just outside the defined limits of the series. But,
the soil series naming a mapping unit is sometimes no more than the
commonest series within the mapping unit. In this instance the
mapping unit should be called a soil complex with every component
occupying 10 % or more of the area of the unit being described.
As customers, planners should insist on proper specifications and
quality control for soil maps (Bie and Beckett, 1971; Western, 1978).

SOIL SURVEYS

1. <u>Introduction</u> (Bridges, 1982; Burrough and Beckett, 1971; Dent
                and Young, 1981; Western, 1978; Vink, 1963).

    Soil surveys are expensive, so their objectives need careful
consideration. If the question to be answered concerns the percentage of specified soils in a given area, e.g. the proportion of
Staffordshire soils that would benefit from artificial drainage,
then assessments at a sufficient number of random sites will provide
an answer. However, in real situations we are likely to be interested also in where, and in what geological and topographical context,
the ill-drained soils occur. And while travelling round the area it
would be more economical in the long run to make a 'general purpose'
soil survey to enable additional future questions to be answered.

    In the late 17th Century, Plot, while grappling with the problem
of representing topography on maps, also saw that patterns of
geology and soils were related to topography, a thought which has
enabled small scale soil maps of extensive undeveloped areas to be
made fairly cheaply through the interpretation of air photographs
and satellite imagery in the last few years. For larger scale soil
maps many more ground observations are made; these can be obtained

and used much more efficiently when differences in topography, geology and vegetation are available to guide the insertion of boundaries.

2. Soil maps : factors related to scale

The nature of soil maps, the procedure best suited to making them, and their usefulness to the planner are all closely related to the scale of the map (Table 3). The expense of making soil maps increases rapidly as the scale becomes larger, hence the smallest scale that will serve the expected use, or uses, is the most appropriate.

It is necessary to encourage the 'customer' to think clearly about the size of the units to which land management decisions apply. For the farmer it is the field or equivalent management unit. In mechanised agriculture areas of less than 2 ha would seldom be managed separately, and parcels of less than 4 ha are undesirable. If we decide that the object of the survey is to know the nature of the predominant soil taxonomic unit in each area of 4 ha on the map, conveniently called the planning unit, 1:25,000 is the smallest suitable scale. If it is essentially a ground survey with no help from other sources of information it would be desirable to make at least one field inspection per hectare. At the other extreme, a survey of the West Midlands of England was made in 1962 to locate possible sites for 'new towns'. Among other requirements, these sites were to be on soils of below average quality for agriculture and 8 to 20 km^2 in area. A soil map at 1:650 000 was made (Mackney and Burnham, 1964), largely compiled from geological data and pre-existing soil surveys, supported by field observations where soil information was lacking or where the area seemed likely to be of particular interest. In this instance at least one inspection was made for every 2 km^2.

3. Purposes served by soil maps at different scales

International maps at small scales. Publication is now complete of the FAO/UNESCO Soil map of the World at a scale of 1:5 000 000 (FAO-UNESCO, 10 vols., 1971-80). Throughout the world this now forms a base line of readily available soil information. It has been compiled from previous surveys, in fact more detailed maps exist for most countries. These, however, are based on very diverse soil classifications, which have been 'translated' in terms of a standard newly-developed classification (FAO-UNESCO, 1974). These maps, in conjunction with 'agro-ecological' assessments including climate (FAO, 1978), are useful for international planning, e.g. in locating and roughly quantifying land available for development (Dudal, 1978; FAO, 1980). A 1:1 000 000 soil map of Europe has also been prepared.

National maps at scales smaller than 1:500 000. Most countries now have soil maps at scales around 1:1 000 000. In practice their value is greatly enhanced if complementary data on climate are

available. For example, for England and Wales a 1:1 000 000 soil map
(Avery, Findlay and Mackney, 1975) has been followed by maps of (i)
Winter Rainfall Acceptance Potential (Farquharson, Mackney, Newson and
Thomasson, 1978), (ii) Land Capability for agriculture (Mackney, 1979)
and (iii) Grassland Suitability (Harrod and Thomasson, 1980).
Together these aid decision-making at the regional level e.g. by
enabling comparisons of several conurbations in terms of the quality
of adjoining land.

   Reconnaissance maps (1:120 000 - 1:500 000). 'Reconnaissance'
soil maps are widely available in developing countries, where they
have proved to be an admirable basis for broad-based planning
decisions. For example, in West Malaysia a 'schematic reconnaissance
soil map' at 1:500 000 was completed in 1968. This was used, with
other information, as a basis for a Land Evaluation Programme (Panton,
1970). The Programme located 17 500 km^2 of unused land suitable for
agriculture; it also showed that much of the land legally alienated
for agricultural use was unsuitable or only marginally suitable. No
attempt was being made to farm 8 500 km^2 of it. It suggested that
11 300 of 32 400 km^2 of forest reserve were suitable for agriculture.
However, 42 000 km^2 of productive forest were unsuited to agriculture,
and much of this lacked formal protection, as was the case with most
of the 22 000 km^2 of unproductive forest, nearly all of which occupied
land liable to damaging soil erosion if stripped of vegetation. 15
areas, each averaging about 2 000 km^2, of unused potential agricultural land, were located in West Malaysia. This information has been
invaluable in drawing up Five-Year Development Plans enabling new
roads, settlements, land development schemes and crop processing
facilities to be established at suitable locations. There are
broadly similar programmes in other countries with resources of under-
developed land, e.g. the Canada Land Inventory (1965). Most are
based on reconnaissance soil mapping.

   Even among developed countries, the largest complete national
soil map is likely to be in the 'reconnaissance' category. For the
Netherlands a 1:200 000 soil map was published in 1856 by Dr. W.C.M.
Staring. The modern map, also at 1:200 000, dates from 1960, and
has been used as a basis for a 'General Soil Suitability Map for
Arable Land and Grassland' (Vink and van Zuilen, 1967). This small-
scale map is used in land use planning and in deciding the main
lines of land development programmes, (Haans and Westerveld, 1970).

   In England and Wales a 1:250 000 soil map has just been published.
As is usually the case at such a scale, the mapping units are not
individual soil series but 'associations' of two to five named series.
A relatively unusual, but very useful, feature would be a statement
of the estimated percentage of the mapping unit occupied by each of
the named series. This means that approximate quantitative statements can be made about the soils of an area. For example, suppose
that Association A contains Hamble series 60 %, and Hook series

40 %, and Association B contains Windsor series 50 %, Wickham series
30 % and Oxpasture series 20 %; then a parish with 50 % of its area
in Association A and 50 % in Association B would be expected to have
approximately 30 % Hamble series, 25 % Windsor, 20 % Hook, 15 %
Wickham and 10 % Oxpasture. Having consulted the report or a card
index of the properties of each series, one could further say that
about 65 % of the parish has silty/fine loamy topsoils and about
35 % clayey topsoils. About 30 % of the soils are moderately well
drained, 30 % imperfectly drained and 40 % poorly drained. Use-
oriented statements could also be made. If it is known that all
slopes are gentle and the climate is that of southern England, about
50 % of the land is likely to be '1', about 10 % '2sw' and 40 %
'3sw' in Agricultural Land Capability subclass (Bibby and Mackney,
1969). Also 50 % of the area (that occupied by Association A) will
have excellent stability for the foundations of buildings, while
50 % will be likely to experience cracking in dry summers.

The main limitation of reconnaissance maps is the rather large
size of the appropriate 'planning unit'; about 2 km^2, a large farm,
e.g. a square mile 'section' in the USA, or the land associated
with a village (parish or commune) or the land to be lost to a large
urban or industrial development, e.g. an airport.

Semi-detailed maps (1:30 000 - 1:100,000). This is the smallest
scale on which individual soil series can be used as mapping units;
semi-detailed maps enable assessments to be made of the quality of
land lost in fairly small developments. Both planners and agricultural advisers find that areas with particular problems, e.g. a
nutrient deficiency or liability of subsoil to swelling and shrinking
on wetting and drying, can be identified with fair accuracy on a
1:63 360 or 1:50 000 map.

Most countries have some semi-detailed maps, for 'sample' areas
in support of reconnaissance mapping, or covering the areas associated with particular problems or projects. Scotland is one of the
few countries to have mapped its entire lowland agricultural area,
at a scale of 1:63 360. These maps are useful to planners who
should be regarded as major 'customers' (Hartnup, 1976). Soil maps
of the Netherlands, at 1:50 000 are nearly complete, some having
been commissioned by municipal authorities (Haans, 1980). A
1:20 000 map of Belgium is expected in 1983 (Hanotiaux, 1980).

Detailed maps (1:10 000 - 1:25 000). Detailed maps can show
fields and comparable parcels of land which are constantly used as
planning units. Agricultural advisers can use these maps to guide
the detailed planning at the farm level, and planners use them for
agricultural impact studies, even of quite small developments. In
development studies, detailed soil maps are made of project areas,
e.g. irrigation schemes. In developed countries, comparable maps
of the urban fringe, would be extremely useful to resolve land-use

conflicts between agricultural, forestry, conservation and sporting
interests (Bartelli, Klingebiel, Baird and Heddleson, 1966; Simonson,
1974; Hartnup and Jarvis, 1973; McRae and Burnham, 1981; Jarvis,
1982).

In England and Wales a considerable number of 1:25 000 soil maps,
most of them accompanied by Land Use Capability maps and/or maps of
soil drainage (Sturdy, 1971), have been published. In the USA detailed maps at 1:15 840 have been produced for county reports, these
maps often having an air photomosaic base, which gives a useful impression of land use. The county reports (e.g. Petro, Shumate and
Tabb, 1967) usually contain a considerable body of information useful
to planners, the information e.g. suitability as source of topsoil,
and for highway construction, septic tanks, sewage lagoons, housing,
lawns, golf courses, athletic fields, cemeteries etc....., being
arranged by soil series. This enables planners to find, for example,
that a housing development proposed for the Carlisle series would be
subject to 'severe limitations due to water table and soft, unstable
soils'. 1:10 000 scale soil maps have often been produced for proposed developments, e.g. the Third London Airport site at Stanstead.
In Germany they have been made to guide land reapportionment and
forest planting schemes (Heide and Muckenhausen, 1980).

Very detailed maps (larger than 1:10 000). These are restricted
to small areas needed for special purposes. They enable soil series
to be subdivided so that virtually all significant variations in the
more stable properties of soil can be comprehended - the planning
unit is smaller than a normal field. Civil engineers commonly make
very detailed soil maps which can also be justified when landscaping
and planting prestigious sites; greens on golf courses and First
Division football pitches would justify metre by metre investigations.
They are of value in assessing the resources of topsoil and useful
subsoil prior to stripping the overburden from gravel workings and
opencast (strip) coal mines (McRae, 1983).

In the Federal Republic of Germany there is a long tradition of
exceptionally detailed soil investigations, beginning in 1934 with
field by field assessments needed to facilitate land taxation. Soil
maps at 1:5 000 have been prepared for land reapportionment schemes
in North Rhine Westphalia covering 6 000 km^2, in the Ems Valley for
2 400 km^2 to guide land reclamation, and for large areas along the
North Sea coast for planning drainage (Heide and Muckenhausen, 1980).

In summary the planner is the soil surveyor's most important
customer (Hartnup, 1976; Hartnup and Jarvis, 1979). To be cost-
effective he needs to think carefully, in advance, about what it is
that he needs to know, and why. The planner's objectives will then
decide what is surveyed and how, a process demanding a continuing
dialogue between planner and soil scientist.

REFERENCES

Avery, B.W., 1968, General soil classification: Heirarchial and co-ordinate systems, Trans. 9th. Int. Congr. Soil Sci. (Adelaide) 4: 169.

Avery, B.W., 1980, "Soil classification for England and Wales (Higher Categories)", Soil Surv. Tech. Monogr. 14, Soil Survey of England and Wales, Harpenden.

Avery, B.W., Findlay, D.C. and Mackney, D., 1975, Down to earth map of England and Wales, Geographical Magazine, 47:514.

Ball, D.F. and Williams, W.M., 1968, Variability of soil chemical properties in two uncultivated brown earths, J. Soil Sci. 19:379.

Bartelli, L.J., Klingebiel, A.A., Baird, J.V. and Heddleson, M.R., 1966, "Soil Surveys and Land Use Planning", Soil Science of America and American Society of Agronomy, Madison, Wisc.

Beckett, P.H.T. and Webster, R., 1971, "Soil variability - a review", Soils Fertil., 34: 1.

Bibby, J.S. and Mackney, D., 1969, "Land Use Capability Classification", Soil Surv. Tech. Monogr. 1, Soil Survey of England and Wales, Harpenden.

Bie, S.W. and Beckett, P.H.T., 1971, Quality control in soil survey, J. Soil Sci. 22: 32 and 453.

Bridges, E.M., 1982, Techniques of Modern Soil Survey, in: "Principles and Applications of Soil Geography", E. M. Bridges and D. A. Davidson, eds., Longman, London.

British Standards Institution., 1976, "CP2001, Site Investigations", British Standards Institution, London.

Burrough, P.A. and Beckett, P.H.T., 1971, The relation between cost and utility in soil survey, J. Soil Sci., 22: 359.

Canada Land Inventory, 1965, "Objectives, scope and organisation", The Canada Land Inventory Report No. 1., Department of Forestry and Rural Development, Ottawa.

Clayden, B., 1982, Soil Classification, in: "Principles and Applications of Soil Geography", E.M. Bridges and D.A. Davidson, eds., Longman, London.

Dent, D. and Young, A., 1981, "Soil Survey and Land Evaluation", George Allen and Unwin, London.

Dudal, R., 1978., Land resources for agricultural development, Plenary Session Papers, 11th Int. Congr. Soil Sci. (Edmonton) 2: 314.

FAO-UNESCO, 1974, "Soil Map of the World Vol. 1", (Legend), UNESCO, Paris.

FAO-UNESCO, 1971-1980, "Soil Map of the World 1:5000000". (Vols. 1-10), UNESCO, Paris.

Farquharson, F.A.K., Mackney, D., Newson, M.D. and Thomasson, A.J., 1978, "Estimation of run-off potential of river catchments from soil surveys", Soil Surv. Gt. Btn. Spec. Surv. 11. Soil Survey of England and Wales, Harpenden.

Fitzpatrick, E.A., 1980, "Soils, their formation, classification and distribution", Longman, London.

FAO., 1978, "Report on the agro-ecological zones project, Vol. 1:
    methodology and results for Africa", World Soil Resourc. Rep.
    48, F.A.O., Rome.

FAO., 1980, "Report on the second FAO/UNFPA expert consultation on
    land resources for populations of the future", F.A.O., Rome.

Haans, J.C.F.M., 1980, Soil resource inventories and their applic-
    ation in the Netherlands, in: "Land resource evaluation",
    J. Lee and L. van der Plas, eds., Commission of the European
    Communities, Luxembourg.

Haans, J.C.F.M., and Westerveld, G.J.W., 1970, The application of
    soil survey in The Netherlands. Geoderma 4: 279.

Hanotiaux, G., 1980, Land evaluation in Belgium, in: "Land resource
    evaluation", J. Lee and L. van der Plas, eds., Commission of the
    European Communities, Luxembourg.

Harrod, T.R. and Thomasson, A.J., 1980, "Grassland Suitability map
    of England and Wales, 1:1 000 000", Ordnance Survey, Southampton.

Hartnup, R., 1976, Soil survey as a tool for planners, Rept. Welsh
    Soil Disc. Gp., 17: 135.

Hartnup, R. and Jarvis, M.G., 1973, "Soils of the Castleford area of
    Yorkshire", Soil Surv. Spec. Surv. 8, Soil Survey of England and
    Wales, Harpenden.

Hartnup, R. and Jarvis, M.G., 1979, Soils in civil engineering and
    planning, Soil Surv. Tech. Monogr. 13: 110.

van Heesen, H.C., 1970, Presentation of the seasonal fluctuation of
    the water table on soil maps. Geoderma 4: 257.

Heide, G. and Muckenhausen, E., 1980, Land resource evaluation in
    the Federal Republic of Germany, in: "Land resource evaluation",
    J. Lee and L. van der Plas eds., Commission of the European
    Communities, Luxembourg.

Holtz, W.G., 1969, "Soil as an engineering material", Wat. Resourc.
    Tech. Publs. Rep. 17, United States Dept. of the Interior,
    Washington, D.C.

Jarvis, M.G., 1982, Non-agricultural uses of soil surveys. in:
    "Principles and Applications of Soil Geography", E.M. Bridges,
    and D.A. Davidson, eds., Longman, London.

Jarvis, M.G. and Mackney, D., 1979, "Soil survey applications",
    Soil Surv. Tech. Monogr. 13, Soil Survey of England and Wales.
    Harpenden.

Mackney, D., 1979, "Land Use Capability map of England and Wales,
    1:1 000 000", Ordnance Survey, Southampton.

Mackney, D. and Burnham, C.P., 1964, "Soils of the West Midlands",
    Bull. Soil Surv. Gt. Btn. 2, Soil Survey of England and Wales,
    Harpenden.

McRae, S.G., 1983, Soil survey and its role in the sand and gravel
    industry, Soil Survey and Land Evaluation, 3:5.

McRae, S.G. and Burnham, C.P., 1976, Soil classification.
    Classification Soc. Bull, 3: 56.

McRae, S.G. and Burnham, C.P., 1981, "Land Evaluation",
    Clarendon Press, Oxford.

Moormann, F.R. and Kang, B.T., 1978, Microvariability of Soils in
    the Tropics and its Agronomic Implications with Special
    Reference to West Africa, In: "Diversity of Soils in the Tropics",
    American Society of Agronomy Spec. Pub. 34, Madison, Wisc.
Nortcliff, S., 1978, Soil variability and reconnaissance soil
    mapping : a statistical study in Norfolk. J. Soil. Sci. 29: 404.
Panton, W.P., 1970, The application of land use and natural resource
    surveys to national planning. In: "New possibilities and
    techniques for Land Use and related surveys", I. H. Cox ed.,
    Geographical Publications, Berkhamsted.
Petro, J.H., Shumate, W.M. and Tabb, M.F., 1967, "Soil Survey of
    Ross County, Ohio", United States Dept. of Agriculture, Washington,
    D.C.
Plot, R., 1686, Natural History of Staffordshire, Oxford.
Ragg, J.M. and Henderson, R., 1980, A reappraisal of soil mapping
    in an area of southern Scotland, J. Soil Sci. 31: 559.
Schelling, J., 1970, Soil genesis, soil classification and soil
    survey, Geoderma 4: 165.
Simonson, R.W., 1964, The soil series as used in the U.S.A., Trans.
    8th Int. Congr. Soil Sci. (Bucharest) 5: 17.
Simonson, R.W., 1968, Concept of soil, Adv. Agron. 20: 1.
Simonson, R.W., 1974, "Non-agricultural Applications of Soil Survey",
    Elsevier, Amsterdam.
Simonson, R.W. and Gardiner, D.R., 1960, Concept and functions of
    the pedon, Trans. 7th Int. Congr. Soil Sci. (Madison) 4: 127.
Sturdy, R.G., 1971, "Soils in Essex I", Sheet TQ59 (Harold Hill),
    Soil Survey Record 7, Soil Survey of England and Wales,
    Harpenden.
Tavernier, R. and Marechal, R., 1962, Soil survey and soil classi-
    fication in Belgium, Trans. Int. Congr. Soil Sci. Wellington
    (New Zealand): p.298.
U.S. Dept. of Defense, 1968, "Unified soil classification system
    for roads, airfields, embankments and foundations (MIL-STD-
    619B)", U.S.Department of Defense, Washington, D.C.
USDA., 1969, "Soils interpretation for regional planning", United
    States Department of Agriculture, Washington, D.C.
USDA., 1971, "Guide for interpreting engineering uses of soils",
    United States Department of Agriculture, Washington, D.C.
USDA., 1972, "Soil series of the United States, Puerto Rico and
    the Virgin Islands : Their taxonomic classification", United
    States Department of Agriculture, Washington, D.C.
USDA., 1975, "Soil Taxonomy : a basic system of Soil Classification
    for making and interpreting soil surveys", Agric. Hbk. 436,
    United States Department of Agriculture, Washington, D.C.
USDA., 1978, "Application of soil survey information", National
    Soils Handbook Notice 24, United States Department of Agric-
    ulture, Washington, D.C.
Vink, A.P.A., 1963, "Planning of Soil Surveys in Land Development",
    Intern. Inst. for Land Reclamation and Improvement Pub. 10,
    Wageningen, Netherlands.

Vink, A.P.A. and van Zuilen, E.J., 1967, "De Geschiktheid van der Bodem van Nederland voor Akker - en Weidebouw: Kaart, schaal 1:200 000", Stichting voor Bodemkartering, Wageningen, Netherlands.

Webster, R., 1977, "Quantitative and numerical methods in soil classification and survey", Clarendon Press, Oxford.

Webster, R., 1978, Optimally partitioning soil transects. J. Soil Sci., 29: 388.

Webster, R. and Cuanalo de la C., H.E., 1975, Soil transect correlograms of North Oxfordshire and their interpretation, J. Soil Sci., 26: 176.

Western, S., 1978, "Soil survey contracts and quality control", Clarendon Press, Oxford.

Wilding, L.P. and Drees, L.R., 1978, Spatial variability: A pedologist's viewpoint. In: "Diversity of Soils in the Tropics", American Society of Agronomy Spec. Pub. 34, Madison, Wisc.

7

SOIL REQUIREMENTS FOR FORESTRY

E. L. Stone

Soil Science Department
University of Florida
Gainesville, Florida 32611
U.S.A.

INTRODUCTION

Non-urban land use planners in countries with an established history of land-related investigations face twin difficulties. These are, on the one hand, assembling, or even recognizing the existence of, a vast amount of detailed information and, on the other, appreciating their good fortune in having such substantial bases for their proposals and options. A visit to the library, access to agency files, a telephone call to colleague or consultant, likely will turn up more data on geology, climate, soil, natural vegetation, current agricultural and forestry practices, yields, etc., etc., than any non-specialist can easily assimilate.

The situation is otherwise in regions that have not yet accumulated such detailed information on land resources, environments, and management practices. Here, an inadequate factual basis may frustrate attempts at detailed planning.

I have written with a view to both situations. The conclusions can be stated at the outset: The kind of information needed for understanding and planning forest land use is more often not accessible in conventional, convenient and detailed form than not available at all. Even when the latter is the case, a sufficient first approximation of resource opportunities and limitations in a given area often may be obtained by rapid survey or sampling procedures, provided that these are conducted with requisite local knowledge, field skills, and objectivity.

Because published sources of data on forest resources are often
less widely based or more generalized than for agriculture, there is
a need to seek out and consult expert opinion before devising
specific plans, rather than relying on archives and official statements. Difficult as interdisciplinary planning may seem, its value
as a means of assembling and rationalizing diffuse or even contradictory information recommends its use in dealing with forest
resources.

Having reached these conclusions the following discussion
attempts to illustrate various kinds of soil-related issues and the
judgements that must be made in planning for forests and forestry.

As a first generalization, forest trees are subject to the same
kinds of environmental controls as agricultural and orchard crops.
In all, the species that prosper, potential growth rates, and realized productivity are determined by temperature, rainfall, soil
properties, pests and hazards, as well as human intervention. In
principle, then, a planner should have little difficulty in assessing
suitability of land for forests in those regions where highly detailed maps of climatic variables and soil are available, and the
ecological attributes of major trees or forest associations are well
understood.

Unfortunately, only a fraction of the world has such maps and
a yet smaller fraction has both maps and the requisite understanding.
Even where knowledge of tree or forest requirements is substantial,
it is seldom expressed in the simple codified forms desired by
planners. Among the several reasons for this is that residual
forests often occur on rough topography, where both soils and local
topography change rapidly within short distances, and species suitability is expressed in terms of "site", rather than soil and climate
independently. Further, some species that grow vigorously in native
forest develop reluctantly or not at all when planted on adjacent
cleared land. Such performance has no parallel in temperate zone
agriculture. A consequence is that large scale afforestation or
reforestation programmes have come to rely on relatively few species,
such as some conifers and eucalypts, that combine the capacity to
grow when planted on a range of soils with known wood values.

Planners or their advisors familiar only with the economic
culture of these few species may be bewildered by the great diversity in natural forests throughout the world. Likewise, the values
of these forests vary with locality, for example, from mangrove
forests that serve as marine nurseries and coastal protection as
well as for fuel and house posts, to subalpine forests critical for
slope protection as well as useful for wood.

How, then, in the face of all this does a planner go about
assembling sufficient information ? Yet an earlier question, how-

ever, is how much of what kind of information will be needed in any
specific instance. "Forests" and "Forestry" are neat collective
terms but their reality encompasses a wider spectrum of expectations,
activities and physical diversity than normally found in "agriculture". The kinds of information sought, and the levels of detail
and accuracy needed for good decisions are not the same throughout
this spectrum, as exemplified by Table 1.

Existing native forests obviously are adapted to the environments in which they occur. More specific information about species
requirements may be unnecessary, especially where such forests are
valued chiefly for protection of unstable slopes, for recreation,
or as wildlife refuges, with wood yields being of lesser consequence.
This may also be true where present forests simply represent residual land use after other demands for land have been largely met,
although here estimates of productivity under management may be
needed. In contrast, where "forests" and "forestry" imply rehab-
ilitating damaged lands, creating protection forests or shelterbelts, or replanting food and forage species in open savannas,
specific information about species adaptations and the measures
required for successful establishment become essential to detailed
planning.

Still further along the spectrum, the large investments in
creating forests for industrial wood production can be justified
only by exact information on the physical resource, hazards, and
growth rates of appropriate species, as well as required material
inputs such as fertilization or drainage.

Thus an initial analysis and definition of what constitutes
forestry in the region of concern may either simplify or complicate
obtaining adequate information for planning purposes.

SOIL REQUIREMENTS

"Soil requirements" for forestry begin with a concern for
regional climate, topographic effects on local climate, and soil-
geomorphology relationships, especially those influencing slope
stability and water movement. Although much the same might be
said about the requirements for agriculture, the frequent association of forests with steep mountainous land, shallow soils and
marginal climates places a special emphasis on the foregoing.

Regional temperature and growing season are the first determinants of species adaptation and potential productivity. Amount,
distribution, and year-to-year variability of precipitation directly
influence growth in obvious ways, and also determine which soil
properties must have first consideration. Those regions having
adequate long term meteorological records probably will have been

Table 1: How the detail and precision of information needed for forest land planning may vary with forest situation or intended purpose.

Forest Situation or Intended Purpose	Species Adaptations			Potential Productivity		
	Low	Med	High	Low	Med	High
Parks and refuges	X			X		
Protection forests	X			X		
Natural forest to be exploited	X				X	
Degraded woodlands		X			X	
Managed natural forest		X			X - - -	X
Rehabilitation of eroded lands		X - - -	X	X - - -	X	
Shelterbelts and strip plantings			X	X - - -	X	
Fuelwood plantations, intensive			X			X
Industrial wood plantations			X			X

Detail and Precision of Information Concerning:

classified climatically and the classes related to native vegetation.
Where this is not so, an initial classification must be developed,
using existing data and some bioclimatic or life zone scheme such as
that of Holdridge (1971). An alternative, recently employed in the
Jalapa region of Mexico, is to identify probable climatic zones
through tonal contrasts on LANDSAT images, and then characterize
these through cluster analyses of existing weather data (Marten,
1981).

     Such general stratifications are useful in planning forest
planting or rehabilitation schemes, as they are in agriculture.
Their importance for understanding the already existing natural
forest is usually less, inasmuch as the information needed to interpret the strata is already given by the forest itself. Exceptions
occur, however, when the growth or composition of the present forest
(or scrub) falls short of the inferred climatic potential. In such
instances there will be reason to suspect degradation by past treatment, or acute soil limitations such as inadequate nutrient supply
or moisture storage.

     Where detailed soil maps (i.e., 1:15 000 to 1:50 000) and
accompanying data exist, interpretations for forestry can be developed in a manner similar to those for agricultural purposes, (Bibby,
Chapter 9; Burnham, Chapter 6). Usually quantitative estimates of
growth rate are available for only a few species and so additional
information may be needed. Even then, soil classifications designed
primarily for agricultural purposes may give inadequate attention to
the deeper soil layers into which tree roots extend. The main
problem in many forest regions, however, is simply the lack of any
detailed soil maps. Even where agricultural lands have been satisfactorily mapped, the included or adjacent forest areas often are
mapped in lesser detail, or ignored.

     Lacking such soil maps or the prospect of having adequate
coverage in the near future, agencies in many countries have developed various systems of land capability, land classification, or
"biophysical" classification. Several are described by Carpenter
(1981). Most are based on land form and/or vegetational differences
recognizable on aerial photographs or LANDSAT images, plus any
existing maps. The scales vary from semi-detailed, useful for
generalized planning purposes, to small scale, providing only a
broad overview.

     In any case, where soil properties and patterns are not well
known, and existing knowledge of forest requirements is scanty or
not organized for use, some format is needed to facilitate the
accumulation and interpretation of soil-related data. The vast
array of interacting soil, topographic, and local climatic variables
can be reduced arbitrarily to a few clusters of "functional
properties" that express their net influence on tree species or

vegetation. Leaving aside low temperature, the important functional
groupings in most landscapes include :

(i)   Soil moisture and aeration
(ii) Anchorage
(iii) Nutrient supplies
(iv) Relative freedom from root damaging agents
(v)   Accessibility and trafficability
(vi) Soil erosion and watershed behaviour.

When so phrased, some properties of many soils are easily rated
for some forestry purposes, often through skilled observation alone.
Likewise, the reciprocal adaptation of tree species to major soil
features can be estimated. This allows the planner or his advisors
to concentrate attention on major unknowns in unfamiliar areas.
Inasmuch as each of the six aggregate properties has been subject to
book length treatment (although not in such context), the following
comments serve only as introduction.

Moisture and aeration

This is the most complex of the aggregate properties, and the
most difficult to generalize about over a range of climates. As
Mayer (Chapter 5) notes, there is a reciprocal relationship between
soil moisture and aeration in the ideal upland soils of humid
regions, and thus it is impossible to maximize both at the same
time. Over the great variety of soils on which forests grow,
however, either low moisture storage or poor aeration alone may
provide the major restriction to growth. Additionally, compact
subsoils and poor aeration may combine to limit rooting depth to
such an extent that trees also suffer from moisture stress during
subsequent dry periods.

Total water use by plant canopies - evapotranspiration - is
regulated first of all by available energy, hence climate, then by
availability of water within rooting depth and, finally, by the
particularities of plant species. Among the latter are season of
leaf, crown density and the ability of a species to retard loss
when evaporative stress is high. Most upland species outside rain-
forest climates tolerate some degree of drought, but sustained
growth requires continuous availability of water. For most upland
species, such availability depends not only on the amount and
distribution of precipitation, but on how completely it enters the
soil - infiltrates, rather than running off the surface, and how
much is retained within rooting depth. Even shallow soil depths
support continuous growth where rains are frequent or evapotranspiration is reduced by cloud and fog, as in rainforests.

In contrast, soil features affecting moisture storage and
rooting depth become overwhelmingly significant for productivity

in monsoon and Mediterranean climates, or wherever long rainless
periods occur. Growth rate or potential growth with added fertility
may be predictable simply from estimates of soil moisture storage.
The forestry literature of northern Europe and northern North America
tends to neglect the importance of deep rooting inasmuch as many
soils there restrict rooting to a metre or so. In contrast, *Pinus
taeda* in the southeastern United States has been found to extract
nearly all the available water to a depth of 4.9 m in a penetratable
soil (Patric et. al., 1965), and *Pinus patula* and *Eucalyptus saligna*
plantations in east Africa to greater than 6 m (Russell, 1973).
Their growth rates necessarily would have been much less in shallower soils.

High available storage capacity alone cannot compensate for
low average rainfall or prolonged droughts, although it may moderate
their effects. In environments subject to frequent drought, forests
survive, grow better, or show greatest diversity in topographic
positions that either provide additional water through subsurface
flow, or reduce evapotranspiration demand by shadowing or slope
inclination away from the sun. Such features are well treated in
the forestry literature (Carmean,1975); their importance for the
planner is that he not mistake forest productivity in such favoured
locations as indication of the potential elsewhere.

Likewise, a number of topographic situations present permanent
or temporary ground water within reach of deep-rooted trees and
shrubs, allowing them to be somewhat independent of local precipitation and soil storage. Sometimes such vegetation is readily
seen to be "phreatic" (reaching the water table), as along water
courses in arid climates. Elsewhere the relationship may not be
obvious without examination as, for example, a pine plantation
growing vigorously on a seemingly dry sand terrace that actually has
a water table within reach of deep roots.

Especially in humid regions, the rooting depth of trees is
often restricted by subsoil compaction, by poor aeration in such
layers, or both. Soil profile descriptions and soil maps reveal
these features (Bibby,Chapter 9; Troy,Chapter 8) and allow interpretation of the limitations that they impose.

In the absence of such maps, schematic diagrams can be developed to rationalize and apply the above generalizations to broad
soil physical groups. Figures 1 and 2 were devised for a humid
region having periods of excess and deficient rainfall. The centre
of the arc represents an idealized "moist, well-drained fertile
soil," well aerated and deeply penetrable by roots. Physical stress
increases in intensity outward along each radius, with the nature
of the stress depending on attributes of the particular soil group
and climate. Accordingly, productivity decreases outward and, in
natural forests, species and vegetational characteristics often

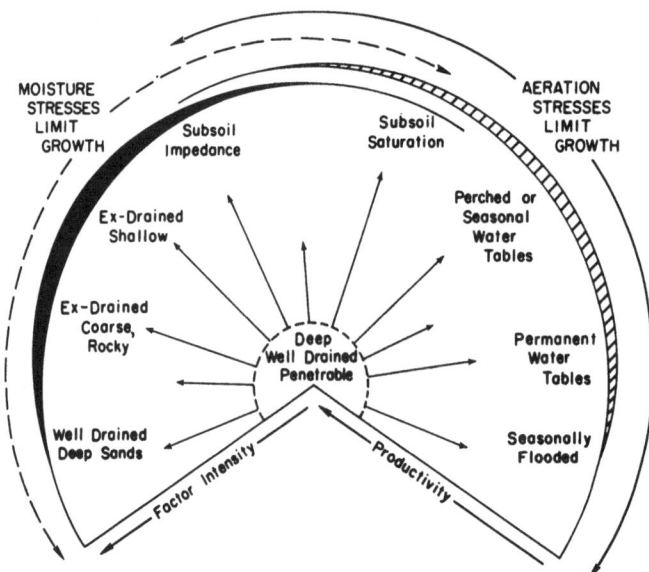

Figure 1: A spectrum of soil moisture regimes and productivity
           gradients in a humid climate with variable rainfall.
           The centre represents ideal conditions of moisture avail-
           ability and aeration; extreme deficiencies of both are
           arrayed around the periphery. Most actual soils are
           located somewhere along the radial gradients between the
           centre and some extreme. Note also :
           1) Sands that allow deep rooting impose less moisture
           stress on adapted species than do shallow soils;
           2) Riverine species adapted to seasonal flooding may be
           less affected by soil saturation than others on inter-
           mittently wet soils; and
           3) Some soils impose alternate moisture and aeration
           stresses as rainfall varies. (Stone, 1977).

shift also. Although such a simple display cannot cope with the
issues of nutrient supply, freedom from root damage, or soil stability discussed later, some of these concerns will be associated
with the same soil groupings.

     Information about growth limitations and damage in any part-
icular landscape also can be organized around such a scheme.
Figure 2 illustrates this for pine forests of the southeastern
United States. A similar display for plantations would emphasize
hazards to survival and early growth, and planting difficulties.

     Moisture deficiency is an absolute, in that growth of all
species is retarded, regardless of how well they survive drought.
In contrast, the action of poor aeration is relative. Most upland

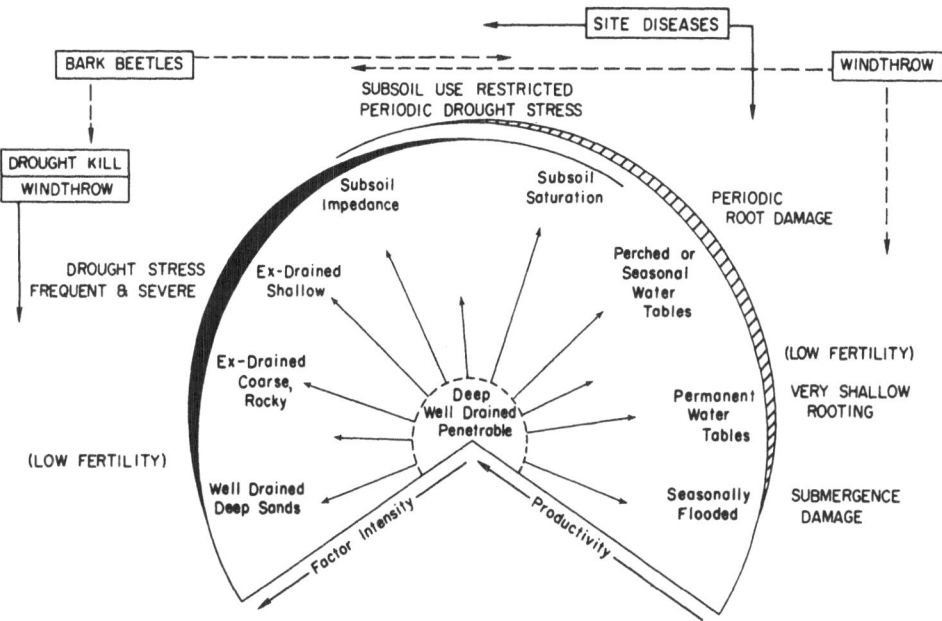

Figure 2:   Some growth limitations and hazards in established pine
            forests, associated with the extreme moisture regimes of
            Figure 1. Low fertility and poor anchorage often occur
            within certain extremes. (Stone, 1977).

species are affected to a greater or lesser extent, but some are
tolerant whereas others are damaged by even a single episode of
exceptional wetness. Actually, poor aeration occurs in a variety
of circumstances (Figure 1), on soils that are permanently moist,
or seasonally wet and dry, and over substrates that either prevent,
restrict, or allow free penetration of roots. The extent of
adverse soil chemical effects (e.g. soluble manganese, sulphide,
salt) associated with poor aeration also varies widely in different
landscapes. In plantation forestry, as in agriculture, the adverse
effects of poor aeration may be mitigated by ditching to carry off
excess surface water, or by planting on formed mounds, ridges or
beds.

On the other hand, a variety of "swamp species" and mangroves
throughout the world are adapted to saturated or frequently flooded
soils and may grow poorly elsewhere. Such forests are too diverse
for generalization, other than that they are sometimes an important
fraction of the residual native forests in some densely settled
districts, and that they may be vulnerable to sedimentation or
changes in water flow resulting from upland land use.

Anchorage

This is a rare concern in horticulture, apart from rubber, but a frequent one in forestry. European forestry literature is replete with accounts of large scale uprooting or stem breakage resulting from violent winds, and almost no forest in the world is immune from at least occasional damage. Stem breakage is actually a consequence of anchorage too great for rapid overturning by wind or ice loading, and more serious because it reduces the salvage value. Uprooting, however, is the far more common cause of loss. Factors of soil and topography, as well as species characteristics, size and silvicultural treatment, all influence vulnerability and can be used to assess the extent of hazard. Forestry Commission planting in windswept districts of western Scotland, for example is guided by a calculated "exposure factor".

With a few such exceptions, anchorage is seldom the decisive factor in selecting species or planning forest use; rather it is classed among the natural hazards affecting any long term crop. Nevertheless, prudence argues against concentrating industrial wood plantations in any one single locality with a high proportion of susceptible soils.

Nutrient Supplies

Trees require the same essential elements as agricultural crops, and all of the "deficiency diseases" known for crops have been found in one or another forest species in some part of the world. Likewise, the general effects of inadequate nutrient supply, ranging from retarded growth to characteristic symptoms and death, are similar in both kinds of plants, and many of the same diagnostic criteria and soil relationships apply.

Soil nutrient supply or "chemical fertility" varies enormously throughout the forests of the world. The sheer range in quantity of plant-available nutrients is far greater than the equivalent range in moisture storage capacity, for example. A result is that the acute nutritional problems of one location may be wholly unknown at another.

The native forests of an area have evolved in response to the chemical environment in which they grow. Not surprisingly, species that succeed on low fertility soils have low nutrient requirements and/or mechanisms for both absorbing and conserving meagre nutrient supplies. Although the productivity of these forests is sometimes very low, the trees rarely display visible symptoms of dysfunction. In consequence, the existence of severe nutrient limitations may be unsuspected until "nutritionally unadapted" - although useful - species are planted on the same soils. This has been a common event in Australia, where pine plantations replaced low value native eucalypts.

Native forest ecosystems and, to a variable extent, planted forests also use limited stores of nutrients with high efficiency, recirculating many elements internally as well as externally through litter fall and canopy washout followed by readsorption. Losses that occur through leaching and fire are compensated for in some degree by releases from soils and weathering rock within reach of roots, by atmospheric deposition, as of dust and ash, and by biological fixation of nitrogen. Conventional soil analyses provide useful information about the fertilizer requirements for growing agricultural crops on such soils, but very little insight into the rates of nutrient circulation and replacement under natural forest systems. Hence reliable information about the probable responses of such systems to intensified wood harvest, changes in management, or conversion to plantation species usually must come from local observation, other empirical evidence, and the results of systematic trials.

Attention to soil nutrient status is required wherever the tasks of forestry include planting eroded or otherwise degraded lands, or developing large scale plantations for fuel or industrial wood. The soils available for the latter are commonly low in fertility and necessitate use of relatively undemanding species such as pines and eucalypts. Although many highly successful plantings of these have been established without nutrient addition, elsewhere fertilization is often needed either for rapid early growth or to maintain high yields. The high productivity of introduced pines in Australia and New Zealand, for example, has been achieved only by systematic applications of phosphate on many soils, together with other nutrients as needed. On the other hand, recognizing favourable combinations of naturally high fertility and moisture may allow successful culture of a wider range of species, even without nutrient additions.

Freedom from Root Damaging Agents

The threads that link this diverse group are that all such agents act on or through the root system, their existence is commonly overlooked until widespread damage occurs, and at that point no remedy is possible. Large scale planting enterprises that place a few standard species on a variety of soils without prior surveys or experiment are especially vulnerable to such damage.

Although root pathogens are numerous, the great majority are local or regional in importance. A contrasting example is a "water mould", *Phytophthora cinnamomi*, which attacks roots of many horticultural crops and presumably by this avenue has spread into many parts of the world and many soils where it had not been known previously. It has killed susceptible native trees and shrubs, including some eucalypts (Newhook and Podger, 1972). It also causes death or slow decline of planted pines of several species, especially

on ill-drained or poorly aerated soils that favour its multiplic-
ation. In some instances the source of infection was the forest
nursery in which the seedlings were grown. Such infections can be
prevented by surveys and sanitation discipline. The situation is
otherwise when the fungus is already well established in soils to
be planted. Here the only feasible options are to avoid susceptible
species, or to avoid planting them on soils where damage is predictable.

Parasitic nematodes - microscopic root worms - are major pests
of many horticultural crops but as yet little is known about damage
to forest trees outside nurseries. It now appears that nematodes
may be responsible for the puzzling decline of slash pine, following
satisfactory initial growth, when planted on dry deep sands in the
south-eastern United States (Bengtson and Smart, 1981). Sand pine,
which is native to such soils, seems unaffected. Similar instances
doubtless will occur elsewhere, and emphasize the need for broadly
based field trials.

Several soil chemical features also act directly, with large
differences in susceptibility among species, or even varieties
within a species. High levels of soluble aluminum and/or manganese
cause failure or poor growth of several major crop species in
strongly acid soils. Trees native to such soils obviously are
tolerant, whereas those introduced from different habitats may not
be. A conspicious problem, though relatively small in area, arises
from toxic concentration of copper, zinc, lead, etc., in mine spoils
and tailings. Likewise, extreme acidity develops in certain coal
mine wastes, and sometimes in the vicinity of smelters. There is
a well-nigh irresistable temptation to recommend planting trees on
such highly adverse sites as the simplest means of stabilization,
concealment or restoring productivity. Although some species
display a degree of tolerance, successful afforestation is generally
either costly or unfeasible.

Many otherwise hardy species will not tolerate strongly calcareous or moderately alkaline soils. Such problems are most likely in
subhumid climates where marginal soils are planted without adequate
testing. A variety of other species well adapted to such soils is
known although their growth is often less than that achieved by
susceptible species growing under optimal conditions. Yet another
hazard sometimes encountered in subhumid regions, and also along
some low-lying coasts in any climate, is soil salinity (Bevege and
Simpson, 1980). This possibility should be recognized before any
existing salt-tolerant native vegetation is converted to unadapted
species. Likewise, salinity surveys should preceed any plans for
forest or windbreak plantings in such landscapes.

Accessibility and Trafficability

Several soil and topographic properties affect the feasibility of forest planting, management, and harvesting operations, rather than biological suitability. The need for roads (or water systems) to transport heavy loads of wood is not likely to be ignored in planning the exploitation of existing forests, but it is easy to overlook the difficulties in bringing wood to such roads over wet soils, steep or broken topography and long distances. Or, the cost effectiveness of heavy equipment operated without constraint may not be weighed against prospective watershed damage. Similar factors together with unplanned seasonal interruptions may act against the large scale development of remote lands for low value products, such as fuel or pulpwood.

Such obstacles also influence the efficiency of transporting nursery stock and labour for planting and tending to remote planting areas. Obviously, the importance of these limitations is greatly affected by social and economic factors, and their mention here is only to insure consideration.

Soil Erosion and Watershed Behaviour

Generally, forest covers are highly effective in preventing surface erosion, reducing the frequency of landslides and other forms of mass erosion, and lessening stream sediment loads. Replanting or restoring forest cover to barren lands usually re-establishes these protective functions, although sometimes incompletely or only at high cost. These costs, or the lower costs and political difficulties of avoiding forest destruction on erosion-prone lands, are often amply justified by the protection of downstream agricultural lands and water resources, including impoundment storage.

Conservative cutting and wood extraction practices tailored to the landscape will often be compatible with the protective functions even on steep terrain, as well demonstrated by Swiss experience. Conversely, heavy exploitation, or conversion of steepland forest to agriculture, or grazing without enforceable safeguards against erosion, can set the stage for disaster.

In general, total water yields will be less, not more, from forest or shrub cover than from grass, and less from evergreen as compared to deciduous canopies. These differences may be inconsequential on very shallow soils, or where rainfall is abundant. On the other hand, the greater consumption of precipitation or ground water by tree cover will sometimes be a cost that must be set against the expected benefits from forest planting.

The generalizations of the foregoing three paragraphs are useful first premises, but the ranges of soil erodability, infilt-

ration and storage capacity, and stability under use are very wide
and, of course, strongly affected by precipitation patterns. Thus
local data and observations rather than generalizations are
essential when forecasting the probable impacts of forestry activities and land use changes on erosion and watershed performance.
For example, the recent volcanic ash soils of Java are stable under
careful cultivation on slopes far steeper than would be tolerable
with less porous soils. Again, the common generalization that
forested soils do not generate surface runoff unless saturated
applies very well to the southern Appalachian Mountains of the
United States (Hewlett and Hibbert, 1967), but not to similarly
steep watersheds in Queensland (Cassells, et al.,1983). Conversion
of poor forest or brushland to grass may increase erosion only
slightly or not at all in some circumstances. Such conversion on
steep slopes in California, however, led to landslips that had
been restrained by the woody root systems and deep soil drying
beneath the previous shrub cover (Rice and Foggin, 1971). And,
finally, to deal with a long standing conflict in case history
folklore, forest planting may either dry up existing minor springs
and streamlets because of greater consumption of subsurface water,
or restore their flow because improved infiltration rate upslope
diminishes surface runoff and increases subsurface water.

SUMMARY

    The two common conditions faced by the planner are existing
natural (or regrowth) forests, and open land on which forest establishment is considered. Along with the latter may be low quality
native forest or brushland available for conversion to plantations.
In the first instance, the species are already in place and hence
their biological suitability is not at issue, although the economic
utility, other values, and appropriate management of those forests
may be. In the second and third instances the soils or "sites"
are the starting point, and both the kinds of trees to be planted
and established practices must be matched to them. In principle,
a very large number of combinations is possible. Unless adequate
knowledge about these is already available, the land use plan
should focus on the objectives sought and provide mechanisms for
obtaining the specific information needed to achieve them.

REFERENCES

Bevege, D.I. and Simpson, J.A., 1980, Salinity associated with
    exotic pine plantation establishment in coastal Queensland, in:
    "Proc. Symp. Salinity and Water Quality". Rixon, A.J. and
    Smith, R.J., eds., Darling Downs Inst. Adv. Educ., Toowoomba,
    Australia.

Bengtson, G.W. and Smart, G.C., 1981, Slash pine growth and response
    to fertilizer after application of pesticides to the planting
    site, Forest Sci, 27: 487.
Carmean, W.H., 1975, Forest site quality evaluation in the United
    States. Advances in Agronomy 27: 209.
Carpenter, R.A., 1981, "Assessing Tropical Forest Lands: Their
    Suitability for Sustainable Uses", Tycooly Internat. Publ.,
    Dublin.
Cassells, D.S., Gilmour, D.A., and Bonell, M., 1983, Watershed
    management in tropical forests - some experiences from north-
    eastern Australia, XV Pacific Science Congr., Dunedin, New
    Zealand.
Hewlett, J.D. and Hibbert, A.D., 1967, Factors affecting the
    response of small watersheds to precipitation in humid areas,
    in: "Forest Hydrology", Sopper, W.E., and Lull, H.W., eds.,
    Pergamon Press, New York.
Holdridge, L.R., 1971, "Forest environments in tropical life zones",
    Pergamon Press, Oxford.
Marten, G.G., 1981, A landscape approach to regional land use
    planning in Tropical Mexico, in: "Assessing Forest Lands: Their
    Suitability for Sustainable Uses", Carpenter, R.A. ed.,
    Tycooly Internat. Publ. Co., Dublin.
Newhook, F.J. and Podger, F.D., 1972, The role of *Phytophthora
    cinnamoni* in Australian and New Zealand forests, Annu. Rev.
    Phytopathology 10: 299.
Patric, J.H., Douglass, J.E., and Hewlett, J.D., 1965, Soil water
    adsorption by mountain and piedmont forests, Soil Science
    Society of America Proc., 29: 303.
Rice, R.M. and Foggin, G.R., 1971, Effect of high intensity storms
    on soil slippage on mountainous watersheds in southern
    California, Water Resour. Res., 7: 1485.
Russell, E.W., 1973, "Soil Conditions and Plant Growth", 10th ed.,
    Longman, London.
Stone, E.L., 1977, A critique of soil moisture-site productivity
    relationships, in: "Proc. Soil Moisture-Site Productivity
    Symp." Balmer, W.E., ed., U.S.Forest Service, Southeastern
    Area State and Private Forestry, Atlanta, Georgia.

SUGGESTED READING

Evans, 1982, "Plantation Forestry in the Tropics", Oxford Univ.,
    Press.
Morgan, W.P. and Moss, R.P., 1981, "Fuelwood and Rural Energy
    Production and Supply in the Humid Tropics", Tycooly Internat.
    Publ. Co., Dublin.
Thirgood, J.V., 1981, "Man and the Mediterranean Forest. A History
    of Resource Depletions", Academic Press, London.

8

TROPICAL FOREST SOILS

J. P. Troy

Ecole National du Genie Rural, des Eaux et des Forets
19 avenue de Maine
75015 - Paris, France

INTRODUCTION

The population explosion in most "third world" countries, the persisting economic crisis in the world, and their converging effects on the tragic impoverishment of many nations in the tropics have suddenly highlighted distinctive regional problems which are matters of general concern for everyone, wherever they live; these problems require sympathetic and in most instances new solutions. The target of regional self sufficiency for food and basic goods is a first priority; it necessitates a much improved understanding of biological production potentialities. In view of the elaboration of new methods of management which protect the future while meeting immediate needs, it is necessary to go beyond the compilation of a comprehensive inventory of existing resources. There is an urgent need to understand production processes so as to predict their possible responses to different forms of management. Although we have some knowledge of events in the tropics more remains to be discovered.

Tropical forests are of prime importance, yet they tend to disappear rapidly as a result of the activities of man especially as they provide goods, space and fertility, which can be used to accomodate man's expanding activities. The concept of fertility is commonly associated with forests in traditional tropical societies where systems of shifting cultivation have been extensively developed within forests. But the increasing pressures of population have resulted in shorter intervals between successive periods of cultivation and an increasingly settled population with, as a consequence, an appreciable drop in site fertility and the initiation of quasi-irreversible processes of degradation - although structur-

ally and floristically complex tropical forest ecosystems are fragile
and as a consequence of the specific properties of their soils they
must not be abused.

As noted by U.N.E.S.C.O. in 1978, most soil studies in the tropics
have been restricted to agricultural lands, few having been made of
forest soils. But because the agronomist's concept of soil differs
from that of the ecologist who includes genetic and evolutionary
processes, the latter can make only limited use of the experience
of agronomists. But what is the ecologists concept of 'soil' with
special reference to tropical systems ? How do the present tropical
landscapes reflect the genetic and historical links between soils
and forests ? From an examination of these links, what can be
learnt to provide guidelines for the future management of the
tropics ?

THE CONCEPT OF SOIL : A COMPREHENSIVE APPROACH

As stated by Cruickshank (1972), "for most people living now,
the idea of soil is little changed from that of prehistoric man -
simply a substance in which plants will grow, a plant growing medium".
Soil also provides a platform for structures (houses, roads, dams,
etc.) and a source of raw materials (peat and ores). Very few people
imagine that soil is a medium teeming with living organisms whose
biomass turn-over may exceed that of tops and roots of structured
terrestrial ecosystems.

Spatial variations in soil properties are commonly related to
conspicuous changes of colour, stoniness, drainage, etc., but evolutionary changes with time are rarely or poorly perceived except
when due to erosion or when a quick loss of fertility follows the
introduction of more intensive management.

For the scientist, 'soil' is a complex natural entity which
deserves methodical and detailed studies. But what is soil ?
Boulaine (1975) suggested that soil is a "4-dimensional structure"
(in euclidian space and time), developing itself and evolving in
the course of time at the interface between geological materials,
atmospheric agents and living systems, and permanently crossed by
variable flows of matter and energy, the greater part of which is
exchanged with the environment the rest remaining temporarily
included in the structure itself. But even an inclusive concept,
such as this, does not yield the limits of a soil : what is or is
not soil ? Should the concept include the organisms living within
the soil matrix remembering that most present-day soil scientists
think that their products, e.g. the products of litter decomposition,
are regarded as fractions of soil. These products and (i) the production of free organic molecules (acids, amino-acids, sugars,
quinonic products, enzymes, insoluble macromolecules of the humic

phase, etc), (ii) the formation of organo-mineral complexes, many
of which are pseudo-soluble in water, (iii) the changes induced in
the physico-chemical conditions of the matrix and (iv) the structural
rearrangements of particulate matter, determine changes in the
dynamics of flows across the soil mantle.

The soil mantle (or soil cover) must be considered as an intrinsic
component of the biosphere forming, locally, an essential part of
each and every terrestrial ecosystem. Its characteristics are obviously related to the present biocenose which in turn depends on prevailing environmental conditions (mainly climatic) but also reflect
the history of each ecosystem.

THE DEVELOPMENTAL LINKS BETWEEN SOILS AND FORESTS IN THE TROPICS

Soil Characteristics and the History of Soil Development

The extent to which the present characteristics of soil are
linked to existing communities of plants and animals is difficult to
determine. An approximation, although not causal, can be expected
from extensive surveys of the correlations between soil characteristics and the assemblages of plants and animals particularly the
former, the surveys being made on a cartographic basis or using more
recent tools of data handling and analysis. But which features, if
any, of soil, or communities of plants and animals, or other environmental factors could "explain" the present state of the observed
ecosystems ? In fact, the soil mantle and the communities of plants
and animals may have evolved simultaneously, vide the concept of co-
evaluation developed by eco-geneticists, the necessary flows of
energy and matter (mainly water) depending on the interplay of environment : climate, rock substratum and relief.

But the evolutionary processes influencing soil and living communities differ in two major respects (i) drastic changes in living
communities may occur within a short (or very short) time as the
consequences of fire, flood, storm, unprecedented drought periods,
or human interference whereas correlative effects on soils tend to
take longer to be expressed, and (ii) perhaps more importantly, quasi-
cyclic dynamic processes can affect living communities while changes
in most soil characteristics are unlikely to be reversed: the only
cyclic soil changes that can be conceived are those associated with
the exposure of rock after the total removal of soil by erosion or
burial beneath water-borne deposits of alluvium or volcanic emissions/
outflows.

Most continental regions in the tropics are formed of ancient,
crystalline rocks commonly covered by deep to very deep soils.
Coastal regions, river valleys, continental depressions, on the
other hand typically have sedimentary deposits whereas other areas

may be affected by recent and/or presently active volcano flows or
locally intense erosion following the uplifting of rock disturbed
by recent tectonic movements.

What is meant by "deep to very deep soils" ? - a thick mantle
several metres or tens of metres deep, of transformed mineral material
derived from underlying rock (parent material). Depths of this magnitude suggest that soil forming processes continue for very long
periods of time while parent rock is continuously weathered. Even
if temporary episodes of erosion could disturb, or partly remove,
the upper layers of the soil mantle at various epochs, the bulk
balance between soil formation (pedogenesis) on the one hand and
erosion on the other has remained in favour of the former. This is
different from the temperate regions (especially in the N. hemisph-
ere), where drastic climatic changes occurred throughout the quaternary era up to 12 000 years B.P. (with the glaciations inducing
widespread erosion in the mountains and hilly areas and the extensive deposition of silts in the plains); therefore the soils are
young and relatively simple. Thus the time scale of pedological
development differs considerably between tropical and temperate
zones. The pedological evolution in progress, or scarcely started,
in temperate soils may have been completed in tropical soils thereby making it more difficult to describe.

The Forests in the Tropics - Extent and Diversity - Past and Present

Apart from the humid tropics on either side of the equator, where
vast areas are still covered with natural forests (for instance : the
Congo basin in Africa, the Amazon basin in S. America, and the large
and sparsely populated islands of Borneo and Sumatra in the Sunda
archipelago), forests account for only a small part of the present
tropical landscape, being mostly represented by isolated patches of
different sizes amidst generally open areas, either cultivated or
left to grassland formations with sparsely distributed trees and/or
shrubs (the savannas). Yet, except in the deserts and sub-deserts
to the north and south of the tropics, and the mountains higher
than 4 000 m, forest relicts still persist in most of the major
tropical habitats.

These relicts suggest what existed in the recent past before
the upsurge in human activity: a diversity of conspicuously distinct
forest types associated with a spatially heterogeneous mixture of
different environments. At low and medium altitudes (below 1 300-
1 500 m in elevation) the major influence is rainfall with its two
main characteristics : abundance (total amount in the year) and
seasonal variation.

Dense evergreen rain forests are associated with the most humid
conditions (mean annual rainfall exceeding 2 000 mm, evenly distributed over time, and with small annual differences). They have
complex architecture, a high to very high upper canopy, and typically

'broadleaved' foliage which severely restricts the amount of solar
radiation directly reaching the ground at all periods of the year.

   Thorn forests or thickets, in contrast are associated with drier
(but not desertic) environments (mean annual rainfall between 250
and 600 mm concentrated within a short summer rainy season of one to
three months and with large differences from year to year). They are
short and open with relatively simple structure, the ground having a
variable cover of herbs. Between these two extremes, a number of
other distinctive forest types can be identified. At least one of
these deserves special mention, namely the economically important
dry deciduous forests found in regions with mean annual rainfall
between 800 and 1 200 mm, a distinct dry winter season (4 - 6 months)
and conspicuous differences between years. These ecologically intermediate deciduous forests are totally defoliated during the dry
season; they usually have 2 - 3 storeys with a sparse ground cover
of herbs. In Asia, their medium-sized upper storeys include valuable
timber species such as teak (*Tectona grandis*), and rosewood
(*Dalbergia* spp.). In wetter locations dry deciduous forests are
replaced by wet deciduous formations which include taller dominant
trees and are floristically distinct. These in turn give way to
semi-evergreen types, which architecturally are similar to the dense
evergreen rain forests but distinctively include gigantic deciduous
trees which tower over the usually closed upper canopy of evergreen
species. In drier areas the deciduous forests become less dense and
as the canopy cover is broken they are invaded by a variety of tall
grasses, which grow profusely during the rainy season. These forests
are particularly subject to fire during winter when the grasses are
desiccated; not surprisingly they are floristically dominated by
fire resistant species of trees. Over the years these open forests
have suffered severely as a result of man's activities, relatively
large areas having been degraded to savanna woodlands, tree
savannas, or even to grass savannas, in which species have been selected for their tolerance of fire and/or grazing by domestic animals
(e.g. thorn bushes). Other types of forests peculiar to tropical
environments include elfin forests (intricate mixtures of dwarf and
stunted trees abundantly colonized by lichens) characteristic of the
permanently cloudy mountain environments at medium elevations in the
humid tropics; the fresh water peat swamp forests of equatorial low
lying areas; the mangrove forests of estuarine and coastal environments; and the tropical pine forests of SE Asia and central
America.

   In addition, there are large areas in the tropics where primary
forests have been converted to secondary forests by man especially
as a result of shifting cultivation and the selective harvesting of
particularly valuable timber species. Secondary forests differ from
primary forests by the greater occurrence of fast growing light
demanding species which in primary forests occupy the niches created
when dominant trees die. Their wood is usually soft and only suitable for pulp and plywood.

Conclusions

Many examples of local, yet spectacular, soil degradation have
been recorded following deforestation in the tropics. In tropical
locations with a marked dry season degradation is associated with
intensified erosion either as a result of the swift notching of deep
gullies or the less spectacular, yet rapid, thinning of the mantle
of soil by diffuse erosion. In those parts of the humid tropics
without a marked dry season, degradation of another type may occur
namely the hardening of subsoil, into true laterite, after the
removal of the upper soil horizons by diffuse erosion with subsequent
replacement of luxuriant forests by sparse shrubs and grasses. Such
evidence supports the opinion that thick to very thick mantles of
soil could only develop and/or survive in the tropics when protected
by forests. But this suggestion is too simple. It is now known
that drastic climatic changes occurred simultaneously in most tropical, and many temperate, regions during the Quaternary. These
climatic changes induced large fluctuations in the distribution of
vegetation with concomitant alterations to upper soil horizons
induced by erosion.

AN OUTLINE OF TROPICAL PEDOGENESIS : THE MAJOR FACTS

Pedogenesis : Generalities

The word pedogenesis embraces the processes which transform
parent material (bedrock) to differentiated soils. These processes
can be divided into three major sets of processes, often interdependent:

Weathering processes, or the conversion of parent material by
percolating solutions resulting in the identification/separation of
four types of product - (i) unweathered, resistant primary minerals
(ii) soluble ions (iii) unstable mineral gels and (iv) amorphous or
crystalline secondary minerals. Together, the unstable mineral gels
and secondary minerals constitute what is known as weathering complexes which differ according to parent material, climate, topography
etc.

Incorporation of organic matter (humic impregnation) as a result
of the biological decomposition of litter and moribund roots mostly
in the upper level soil horizons.

Spatial redistribution of matter in suspension, or dissolved
in soil solution; movement being either vertical through soil
profiles or lateral through the soil mantle. In both instances part
of the matter is added to groundwater or drained to streams and
rivers.

The most important among the weather resisting primary minerals
is quartz. Most of it remains in soils in gravel, sand and silt
components but some may be slowly dissolved under aggressive tropical
environments. Most of the minerals of weathering complexes newly
formed from parent materials, are of two types: (i) iron and aluminium
hydroxides (sesquioxides) - the former being mostly in amorphous
forms (goethite, haematite) whereas the latter are generally crystalline (gibbsite) and (ii) the mineralogical clays with flaky
(phyllitic) microcrystalline structure which behave as electroneg-
ative colloids. They can be formed either as a result of the progressive transformation of primary phyllitic minerals like micas (or
even pre-existing primary clays) or more generally in the tropics
by the recombination of free silica and alumina liberated from
weathered primary silicates.

Mineralogical clays are classified by their chemical and crystal-
lographic characteristics. Clays with equal amounts of silica
and alumina belong to the kaolinite group: they are (i) physically
and chemically relatively inert with low cation exchange capacities
(C.E.C.) and (ii) characteristic of rather poor soils. On the other
hand silica-rich clays (2 or more Si: 1 Al) are considerably more
active: they include the very important montmorillonite group, characteristic of tropical and subtropical black soils (vertisol class)
containing significant amounts of calcium and magnesium.

The unstable mineral gels of the weathering complex are metal
hydrates of silicon, iron and aluminium. They are particularly
abundant in the andosol class of soils derived from recently weathered vitreous rocks (basalts and volcanic ashes).

The weathering processes in the tropics

Primary weathering in the tropics is characteristically greatly
influenced by (i) high and relatively constant temperatures at which
chemical reactions (and particularly those breaking down primary
minerals) proceed apace throughout the year; (ii) heavy and intensive
rainfall which flushes through soil and in so doing redistributes
suspended and dissolved substances and (iii) the considerable depth
of most tropical soil profiles in which the primary processes of
weathering at the base of the profile are spatially separated from
the inclusion of organic matter into the upper horizons. As a
direct consequence, the solutions that percolate down to the parent
material (bedrock) are quasi neutral, the predominant neutral hydrolysis of rock weathering in the tropics contrasting with the acid
hydrolysis (or even acidolysis) in cool temperate regions.

The water regime within the soil mantle, which is directly
dependent upon the climate and local topography (relief), influences
rates and amounts of transport and therefore affects the availability of ions for incorporation into secondary minerals. When the

water balance, expressed as rainfall minus potential evapotranspiration is low, the movement of water through the soil mantle is restricted with a consequent decrease in amounts of soluble elements.
However free silica remains in sufficient quantities to recombine
with aluminium to form kaolinite. In contrast, when there is a
large excess of rainfall over evapotranspiration, deficits of silica
develop and as a result gibbsite (alumina) accumulates. Where
drainage water tends to accumulate, as in low lying areas, the
increased concentration of free silica and cations facilitates the
formation of montmorillonite.

Basic rocks (i.e. those poor in combined silica) are particularly
liable to weathering with the liberation of large quantities of
sesquioxides and bases if circulating water is plentiful. Gibbsite
precipitates on the periphery of each particle/crystal in the
weathered matrix of the rocks, so fossilizing its granular structure
and adding to the porous nature of the derived material without
changes in apparent volume (isovolumetric weathering). In freely
drained locations, e.g. on slopes, sesquioxides of iron and aluminium
are the dominant fractions of the weathering complex, kaolinite is
rarely formed. The derived material is known as ferrallite (plinthite in U.S. terminology); it may harden into laterite when directly
exposed to the atmosphere. Acid rocks, (e.g. leucogranites, in
contrast, liberate smaller quantities of iron and bases and larger
amounts of silica with the consequent synthesis of kaolinite. Even
in wet climates the derived material is richer in clays and therefore
much less porous than would be the case with basic rocks.

It should not be assumed that weather alone controls the water
regime of soils - the role of vegetation can be decisive. Vegetation, by its utilization of water, can regulate the rates and extent
of drying and wetting in addition to physically minimizing erosion.

Broadly speaking plentiful rain in the tropics enables weathering to produce an infertile mineral soil with few bases and a poor
C.E.C. Most of the remaining minerals are stored in the unweathered
primary minerals and are therefore not readily available (to plants)
except in (i) low lying areas where rich montmorillonitic clays are
synthesized; (ii) areas with recent volcanic deposits whose plentiful supplies of bases are weathered into rich andosols with exceptionally large cation exchange capacities.

Tropical Forest Humus : its role in maintaining fertility

Humus (or humic system) has, in the broad sense, three distinct
components: i) litter, exclusively organic, comprising dead detached
plant fragments with leaves, branches, bark etc., being deposited on
the ground and roots being located *in situ*, (ii) exudates and
leachates, materials on the surfaces of leaves and branches which
amount to 10 - 15 % of the total organic inputs of tropical forest

soils, (iii) the soil horizons immediately below the litter layer in which organic material is intimately mixed and/or associated with the mineral phase. Fresh litter provides energy for microbes. As a result of microbial activity, litter is transitorily transformed to a range of more or less stable products (humidification) before ultimately being degraded into carbon dioxide, water and mineral forms of nitrogen (mineralization). At an early stage mineral ions are lost from decomposing litter to become available for uptake by plants and microbes.

Our understanding of organic matter and its breakdown in soils is relatively recent and still far from complete - it is mainly derived from studies of soils in temperate regions. Humic systems which are characterized by small accumulation of litter, and other free organic material, and which soon become intimately mixed in mineral soil are known as mull systems. They indicate a fast rate of turnover and typically reflect productive soils. These are extensively represented in the tropics, particularly in well drained locations in wet zones. Bearing in mind the dearth of bases in the mineral phases of most tropical soils, it is clear that tropical soils greatly depend for their fertility upon humus (the humic system) - the litter acting as a reservoir of nutrients which can be retained, after being released during decomposition, by adsorption by organic macromolecules which contribute to the high cation exchange capacity of the litter layer (humic phase).

An active humic system implies that humification and mineraliz-ation are proceeding rapidly. If the supply of fresh organic matter is drastically curtailed the humus resource will be rapidly depleted as when natural tropical forests are destroyed for intensive systems of cultivation with consequent deleterious effects on fertility. This loss of fertility was recognised by pastoral people, the problem being overcome by shifting cultivation.

CONCLUSIONS

Most tropical soils are deep having been derived by the weather-ing of bedrock over long periods of time: hundreds of thousands to millions of years. As a general rule it can be considered that they mostly developed as parts of forest ecosystems, upper layers have soil horizons often showing signs of major perturbations attributed to exceptional weather or the activities of man. The pedogenetic processes involved in their formation lead to the diff-erentiation of poor to extremely poor derived minerals, mostly sesquioxides and kaolinitic clays. For fertility tropical soils depend upon humus incorporated into the upper soil horizons; they are therefore at risk when the supply of litter is disrupted. Physical properties depend on the nature of the weathering complexes: the weathering of basic rocks leads to porous soils where the

derived material, ferrallite (= plinthite), is stable and resists
erosion: acid parent materials lead to more compact and imperfectly
drained soils which as a result are subject to erosion and landslips.
Soils in low lying areas tend to be more fertile than average but
their dominant montmorillonitic clays which swell and shrink to a
considerable extent, are an obstacle to the cultivation of trees and
the construction of buildings. These soils are sensitive/susceptible
to diffuse and linear erosion even in flat areas with slopes less
than 0.5 %.

The complexities of tropical soils and their very rapid responses
to environmental changes highlight the need for sympathetic attitudes
when planners consider the requirements of development programmes
particularly those affecting forested areas. A wide appreciation
of implications is necessary, bringing together decision-makers and
soil ecologists, regretably as yet a rare breed. Local and traditional practices also deserve careful attention, being put into the
context of the most recent scientific advances in soil science and
ecology. Such dialectic approaches may provide the key to successful management.

REFERENCES

Bocquier, G., 1973, "Genese et evolution de deux toposequences de
    sols tropicaux du Tchad. Interpretation biogeodynamique,"
    Office de la Recherche Scientifique et Technique Outre Mer
    (ORSTOM), Paris.
Boulaine, J., 1975, "Geographie des sols" Presses Universitaires
    de France (PUF), Paris.
Bridges, E.M., 1970, "World soils", Cambridge University Press,
    Cambridge.
Chatelin, Y., 1974, "Les sols ferrallitiques - L'alteration",
    Office de la Recherche Scientifique et Technique Outre Mer
    (ORSTOM), Paris.
Cruickshank, J.G., 1972, 1974, "Soil geography," David and Charles,
    Newton Abbot.
Duchaufour, Ph., 1977, "Pedologie - Pedogenese et classification",
    Masson, Paris.
Dudal, R., 1967, "Black clay soils of tropical and sub-tropical
    regions," F.A.O., Rome.
Erhart, H., 1967, "La genese des soils en tant que phenomene
    geologique," 2e ed., Masson, Paris.
F.A.O.-U.N.E.S.C.O., 1973, "Soil map of the world. The legend,"
    Rome, Paris.
Millot, G., 1964, "Geologie des argiles," Masson, Paris.
Ollier, C.D., 1969, "Weathering. Geomorphology," Oliver & Boyd,
    Edinburgh.
Schnell, R., 1970-1976, "Phytogeographie des pays tropicaux,"
    Gauthier Villars, Paris.

Segalen, P., 1973, "L'aluminium dans les sols," Office de la
    Recherche Scientifique et Technique Outre Mer (ORSTOM), Paris.
Troy, J.P., 1982, "Tropical forest humus", Biotrop, Bogor.
U.S.D.A., 1975, "Soil Taxonomy", U.S.D.A., Washington D.C.
U.N.E.S.C.O.-U.N.E.P.-F.A.O., 1978, "Tropical forest Ecosystems",
    U.N.E.S.C.O., Paris.
Whitmore, T.C., 1975, "Tropical rain forests of the far East,"
    Clarendon Press, Oxford.

9

PRINCIPAL FEATURES OF THE FORMATION OF HILL LAND SOILS, THEIR

MANAGEMENT AND CAPABILITY IN COOL, MOIST, TEMPERATE CLIMATES

J. S. Bibby

Department of Soil Survey
Macaulay Institute for Soil Research
Craigiebuckler, Aberdeen AB9 2QJ

INTRODUCTION

The hill lands considered in this chapter are confined to cool, moist, northern temperate climates found almost exclusively in Britain and, in particular, in Scotland: some offshore islands of British Columbia, and perhaps Tasmania have conditions approaching those in the United Kingdom. Most other regions experiencing even remotely similar conditions are either influenced by their proximity to continental land masses (e.g. Scandinavia); or by a warmer climate (e.g. New Zealand) which considerably affects their 'capability'.

More than one third of the total land area of the United Kingdom can be considered hill land. In the context it is used in this chapter hill land does not imply a specific altitudinal range, instead it denotes an environmental type.

HILL-LAND ENVIRONMENT

The soils of British hill lands are very strongly related to their environments, in particular to (i) the parent rocks from which they were formed and (ii) the many climate-dependent processes which have affected these rocks, and which are very difficult to separate. Perhaps the most effective way of describing the hill land relationship in Scotland between geology (topography and parent materials) and climate, is to examine a slightly exaggerated and adjusted east/west transect before using it as a baseline for assessing the variations that occur to the south and north. A line running north-east from the south-west corner of the Island of Mull (Lat. 56.19 N.Long. 6.25 W.) to the North Sea at about Rattray Head in the north-east corner of Scotland (Lat. 57.37 N.Long. 1.49 W) will serve the purpose (Figure 1).

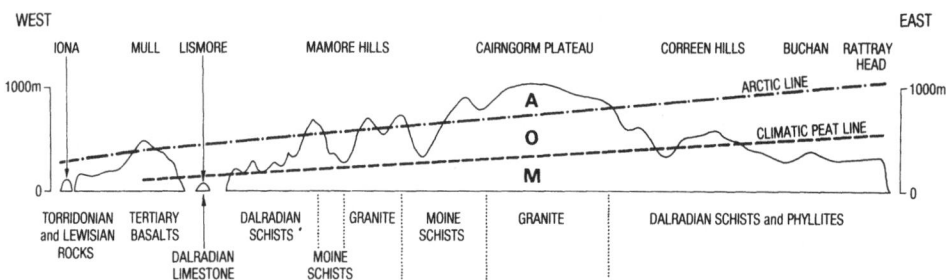

Figure 1 : Diagrammatic section of Scotland along a south-west/north-
east transect starting at Iona (Lat. 56° 19'N, Long. 6°
25'W) and stretching to Rattray Head (Lat. 57° 37'N, Long.
1° 49'W). The principal soil zones (A = Arctic soil zone,
O = Organic soil zone, M = Mineral soil zone) are temperature dependent and their altitude falls steadily westwards under the influence of increasing oceanicity,
despite changes in underlying geology.

The rocks along this transect are mostly Pre-Cambrian, those in
the west being Lewisian gneisses and Torridonian sandstones, some
of the oldest known. They comprise parts of the Island of Iona,
the Ross of Mull, the islands of Coll, and Tiree and are found on the
mainland of north-west Scotland in Wester Ross. Apart from the Ross
most of the Island of Mull is considerably younger with Tertiary
basalts lying above Mesozoic limestones, clays and sandstones, and
their associated soils contrasting with those found over the older
rocks. To the east of Mull but west of the Scottish mainland is the
limestone island of Lismore.

The Tertiary, an era of volcanic eruption in the west, was a
period of significant erosion on the mainland. A platform, created
at about 600 m along the west coast of Scotland was severely dissected by subsequent glaciations to form a series of ridges. The
platform cuts across various formations - Dalradian mica schists and
soft black graphitic slates in the south and acid gneisses and
granulites of Moinian age in the north with granites in between.
Inland the same rock types rise to over 1200 m in a series of jagged
peaks deeply dissected by glacial processes. The effects of glacial
erosion lessen in an eastward direction and the massive plateaux of
the Cairngorms and Monadhliath Mountains are notable features of
the eastern central Highlands. Towards the extreme north and east
of the section the mica schists and phyllites of Dalradian age once
more appear and are eroded into a series of low, steep sided rolling
hills, mostly below 200 m high.

Most, if not all, of the Scottish landscape was subjected to
glaciation and evidence of erosive processes is common in the hills,
with depositional features prominent at lower levels. The rocks of
the section are fairly typical of those of Scotland; they are sandy

and acid with some finer grained, though metamorphosed, siltstones and mudstones and, occasionally, rocks of more basic chemical composition like basalts and limestone. The soils of the Highlands were developed from the debris of these rocks, produced by glaciation.

Climate has strongly influenced the distribution of soils. Cool temperate areas characteristically have small ranges of annual temperature with either an even distribution of rain or a winter maximum, the highest being associated with the western coastal areas (Figure 2). These maxima decline steadily eastwards and on the east coast of Scotland and the western coasts of continental Europe tend to disappear. In British eyes, and from the point of view of soil development, the very large amounts of rain, low summer temperatures and mild winters (caused by the proximity of the Atlantic), distinguish the western, from the eastern, parts of the country. The effects of a large sea mass on wind speeds, and hence exposure, are also most marked on the western coasts.

The effects of weather can be illustrated by considering two climatic surfaces defined by the altitudes at which (i) an arctic climate influences vegetation and soils and (ii) the accumulation of surface organic matter becomes an important soil feature, the climatic peat line. The onset of arctic conditions on mountain summits can be detected along the line of section at approximately 400 m in the west, rising sharply inland at first and then more gradually to 650 - 700 m in the east (Figure 1). The climatic peat surface rises from below 100 m in the west to above 400 m in the east. The form of these surfaces, steadily declining westwards, is also affected by latitude. In mountainous areas to the south, Wales, the Pennines and the Southern Uplands there are larger accumulated temperatures, 875 - 1375 day ^{o}C, than in the Highlands, 550 - 1100 day ^{o}C. Accumulated temperature is a commonly used method of assessing seasonal warmth by adding the excess of temperature over a notional base (in this case 5.6 ^{o}C, the temperature at which major growth commences in most plants) for each day and averaging results from a number of years. Because of the influences of (i) the continent to the southeast of Britain and (ii) the Atlantic Ocean to north and west, warmth assessed by this method decreases from south-east to north-west Britain and as a result arctic conditions are virtually absent in the mountains of England and Wales, just clip the summits of the Southern Uplands and decline in altitude northwards appearing almost at sea level in the Shetlands. In parallel the climatic peat surface is found at higher elevations in England and Wales than in Scotland (Figure 1), where it is at sea level in much of the northwest Highlands and the Hebrides. These strong influences of climate on soils are important when considering their uses.

The arctic and climatic peat surfaces take the form, in general outline, of planes inclined from south-east to north-west with a sharp downwarp at their western edge. In detail, however, there

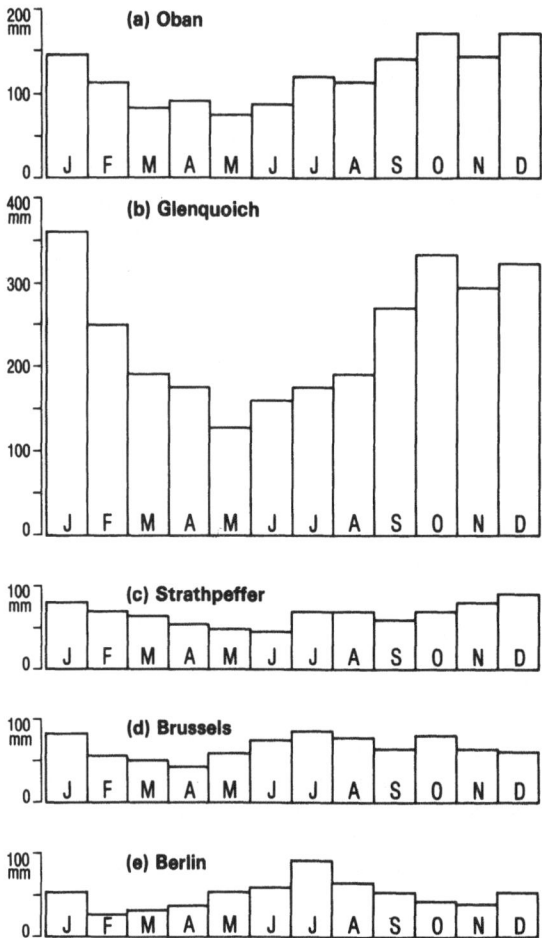

Figure 2 : Rainfall values for three Scottish locations, Oban (west
coast), Glenquoich (Highlands) and Strathpeffer (east
coast) and two continental locations, Brussels and Berlin,
showing decreasing amount and shift in maxima from winter
to summer in moving from maritime to continental temperate climates.

are local undulations, reflecting the ways in which different rocks
have responded to past, and are responding to current, climatic
influences. This can be illustrated by considering two west coast
islands, Mull and Skye, which are similar geologically, before comparing them with adjoining mainland areas. These islands are largely
composed of igneous rocks (extensive outpourings of Tertiary basalt).
Their soils are similar morphologically; they also have similar
chemical characteristics. On the Island of Mull the climatic peat
surface (on the basalts) is at an altitude of 200 m whereas on Skye it
is more usually at 100 m. Since the islands are physiographically
similar, this difference approximately halves the area of brown

mineral soil on the more northerly Skye. In the south of Skye (much of the Sleat penninsula) and the neighbouring mainland at Kyle of Lochalsh where Torridonian sandstones occur instead of igneous rocks peaty soils are found down to sea-level, a drop in the altitude at which peat starts to form because of the inert character (biologically) of the parent material. On the mainland to the east of the Island of Mull, the principal rock type is an andesite lava with soils similar to those overlying basalt. Where somewhat more acid phyllites of Dalradian age occur, to north and south, the peat line falls to 30 - 50 m altitude. In summary, rock type, in conjunction with climate, plays an important role in determining the resources of soil and hence the use of land, the basic rocks (basalt and limestone) producing greater extents of mineral soils than the acid rocks.

Figure 3 : Gently sloping hills of the Southern Uplands, considerable
           areas of which are covered by peat. Attempts at drainage
           have been made but reversion to rushes (foreground
           fields) is common.

   Topography interacts with climate and, particularly, variations in the drainage characteristics of slopes influence soil development. For example peat formation will proceed at low elevations when slopes are slight and drainage impeded (Figure 3). Such conditions

Figure 4 : Steeper hill slopes are frequently full of peat but often
          have extensive peaty podzols and peaty gleys (upper
          slopes). Their vegetation contrasts sharply with that of
          the lower hill slopes (humus-iron podzols and brown forest
          soils) and the wetter mineral soils (gleys) of the valley
          floors, which often overlie moraine.

are common in parts of northern and eastern Scotland, the Southern
Uplands, the Pennines, some areas of the Lake District and central
and Southern Wales. On steeper slopes with more rapid run-off the
land surface is drier for longer periods of time (Figure 4) and
peat formation is confined to lowland depressions and the ridge
crests of high hills. On the upper parts of the steep slopes a thin
layer of organic material (less than 50 cm thick) overlies leached
subsoils (peaty podzols) or even wet subsoils (peaty gleys), while
the lower slopes are occupied by slightly leached mineral soils
(humus-iron podzols and brown forest soils). The different soil
types support very different vegetation communities (Figure 4) and
are thus significant both to ecology and to land use. Extensively
in north-western Scotland, but also in some parts of the Southern
Uplands, the Lake District and Wales, glacial erosion has produced
rugged landscapes of arêtes, U-shaped valleys and rocky slopes where
soils are very shallow (Figure 5). The upper slopes and summit
ridges have a hemi-arctic or arctic climate. In the eastern High-

Figure 5 : Extremely rugged, dissected country in the western
           Highlands has rock outcrops, boulders and very shallow
           soils (mostly with a peaty surface horizon) overlying
           rock.

lands (Figure 6) the mountains are not so dissected and high level
plateaux occur, covered with more extensive arctic soils the surfaces
of which have a broken vegetation cover and frequent signs of active
frost action (soil polygons and stone stripes). The soils are
classified as arctic variants of podzols and peaty gleys.

THE SOILS

     In classifying soil the aim is to organize knowledge about soils
so that it can be recalled and communicated; so that relationships
may be seen among (i) the arrays of different soils and between
(ii) soil properties and environmental factors, and (iii) soil properties and the potentialities of soils for different uses.

     Too often, however, systems of soil classification become revered,
irrationally, by a few adherents; at best they fulfil their objectives
for the specialist, leaving the user of soil information lost in a
forest of verbiage. There are many soil classifications, but few

Figure 6 : Broad plateau summits are more common in the eastern
           Highlands which, at elevations above 750 m, have a
           mountain arctic environment. The soils are primarily
           weakly-developed podzols with some gleys, and the veg-
           etation cover frequently incomplete.

have been designed for non-specialist users. Hill soils in Britain
can be classified on the basis of four properties - wet or dry;
mineral or organic surface material - to give an easy to remember
array of seven types each having a combination of characteristics of
importance to land users. The seven types include 3 mineral, and
4 organic soils :

Soils with a Mineral Surface Horizon

1. Brown soils are characterized by free drainage and brown colours.
    Their surface horizons have larger amounts of organic debris
    than subsoil horizons but the content varies according to geo-
    graphical position and land use history. Because of the heavy
    rainfall in mountainous areas these soils develop on coarse
    textured parent materials which may, or may not, be calcareous,
    but in areas with lower rainfall the soils may contain a higher
    proportion of clay.

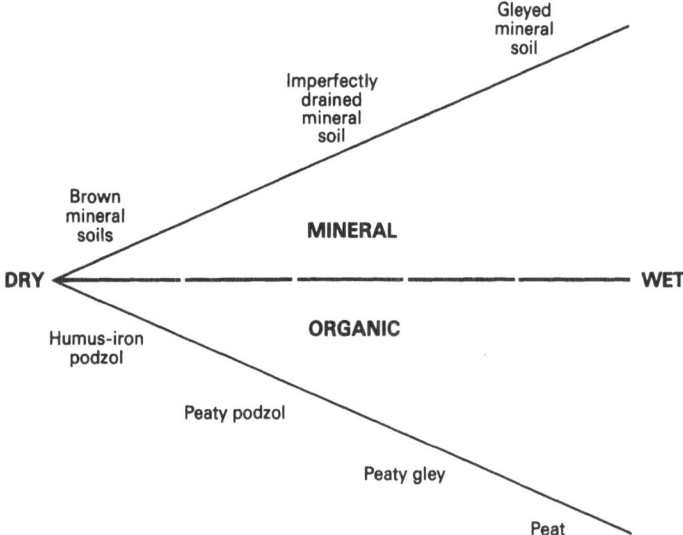

Figure 7 : A simple classification of the most common hill-land
           soils.

2. Imperfectly drained soils are closely related to brown mineral
    soils and occur, sometimes together, in areas with similar
    topography. As their name suggests they are not as well drained
    as brown mineral soils, with waterlogging persisting in the sub-
    soil during winter months as shown in the profile by mottling.
    It may be possible to alleviate this problem by drainage to
    produce valuable farmland.

3. Gleyed soils are soils which have developed in often persistent
    waterlogged conditions. In the wetter mountain zones the surface
    horizons of gleyed mineral soils accumulate considerable amounts
    of organic matter (and may intergrade to peaty gley) and in doing
    so develop a dark coloration. Their subsoils are drab in colour
    and often dominated by greenish and bluish weathered stones with
    occasional brighter mottles in a grey ground-mass. These soils
    are usually acidic; they develop in material which is either
    heavier in texture, i.e. contains more clay than 'silt' or has
    a horizon impeding water flow. It is usually possible to amel-
    iorate the wet conditions to some extent but these soils often
    continue to provide wetness problems for the user even after
    drainage.

Soils with an Organic Surface Horizon

4. Humus-iron podzols occupy, in many cool northern areas, the ecol-
    ogical niche in which brown mineral soils develop in warmer
    climates. It is often difficult to separate humus-iron podzols

from brown mineral soils when the surface horizons of the former
have been ploughed. They are more freely drained and as a result
are generally more leached than brown mineral soils hence the
reason for a brighter brown subsoil. The humus content of the
cultivated topsoil is sometimes a little greater than in natural
conditions. Humus-iron podzols are frequently found on coarse
textured parent materials or associated with steep slopes in wet
areas; they are the basis of considerable areas of arable land
in the north-east of Britain.

5. Peaty podzols have a distinct, black organic surface horizon that
    qualifies as peat. Peaty podzols are often underlain by a grey,
    leached extremely nutrient deficient mineral horizon, which in
    turn is underlain by an iron-pan. The latter impedes the flow
    of water and in consequence peaty podzols often have the appear-
    ance of being very wet. Because the soil below the iron-pan is
    freely draining it is often possible to change the properties
    of the surface horizons by breaking the iron-pan, for example
    by deep ploughing. Peaty podsols are commonplace in the mid-
    hills of north-eastern Scotland, in the Southern Uplands of
    Scotland, in Wales and in the Pennines of England: they are
    characteristically colonised by heather, whose colour gives rise
    to the colloquial term 'black hill' among the farming community.

6. The peaty gleys also have a distinct black organic surface
    horizon of peat which is usually wetter than that in podzols.
    Sometimes a grey, leached layer is present beneath the surface
    but the subsoils are extremely wet (in contrast to those of the
    podzols) resembling those of mineral gleys. Consequently, the
    properties and potential of the peaty gley are very different
    from the point of view of the user. Peaty gleys may have widely
    different textures, ranging from silts to clays, they are
    commonly associated with (a) broad rolling uplands overlying
    Carboniferous sandstones and shales, (b) depressional sites on
    coarse textured materials in the east and south; they are the
    predominant soil type in mountainous areas of north and west
    Scotland. They also occur on very steep slopes in these regions.

7. Peat is recognized when the organic surface horizon is more than
    500 mm thick. In hilly areas the organic horizon is usually
    acidic, sometimes extremely so. As with the organic surface
    horizon of other soil types, different kinds of peat occur with
    the types of colonising assemblages of natural/semi-natural
    vegetation being related to nutrient status. Peat in lowland
    areas is often a valuable agricultural asset because it maintains
    available water even in climatically dry conditions. However,
    in the hills this property may be a hindrance, the fine
    colloidal material holding so much water that effective drainage
    is exceptionally difficult. The reclamation of peat soils in
    areas with large (a) accumulated temperatures (>1375 day ^{o}C) and

(b) potential moisture deficits (>100 mm), is more successful than in cooler and wetter areas.

Two main variants of each of the seven soils already described, can be recognized (a) a shallow variant where rock is within 1 m of ground-surface and (b) an immature variant found in the arctic climate zone where cool temperatures greatly inhibit weathering and the soil types, although recognizable, are weakly developed.

ASSOCIATION	PARENT MATERIAL	SOIL GROUPS			
		Brown mineral soils	Imperfectly drained	Gley soil	etc.
INSCH	Drifts derived from gabbro	*		*	
STRICHEN	Drifts derived from schist			*	
	etc.		*		

* Named soil series separately coloured on soil maps

Figure 8 : A specimen two part array key to aid the identification of series map units shown on soil maps prepared by the Soil Survey Department, Macaulay Institute for Soil Research, Scotland.

In Chapter 6, Burnham outlined the role of Soil Surveys; in Scotland, soil surveys have been the responsibility of the Soil Survey Department of the Macaulay Institute since 1947. It has used a soil classification based upon the simple scheme already outlined, users having little difficulty in recognizing the soil types on the horizontal axis of the two part array (Figure 8) which is part of the key to a soil map. The vertical axis is taken up by groups of parent materials, so that gley soils on parent materials derived from gabbros for example are effectively delineated as map units from those derived from schists. This is the basis of the soil series described by Dr. Burnham and by which these groups of properties are displayed on the map.

Soils in hilly and mountainous terrain are intimately linked to landscape, geology and climate; they occur in complex patterns. For this reason, the recent soil mapping, done by the Soil Survey of Scotland, has focused on the delineation of soil complexes and a key of a different type has been used (Figure 13) which also includes information on land form and associated vegetation. In areas where soils have been managed, and used, intensively, these 'complex' units would not be of great value to users and planners. However, the areas of ground managed as 'units' are much larger in the moun-

Figure 9 : The character of land has a direct influence on the
            effectiveness with which it can be utilized. Even on
            flat, well-drained gravelly land (Corby series), stones
            hinder the emergence of crops and cause increased wear
            on tractors and implements.

tains than in the lowlands and because soils in an area tend to have
affinities (for reasons already explained) the scope and type of
opportunities for managing different soil types in a circumscribed
area are closely linked (Figures 9 - 12).

    When defining complex map units it is obviously much more useful
to users if the proportions of the different component soil types
are quoted. Some attempts have been made to do this based on subjective estimates. While these may be better than no information,
attempts are now being made to obtain accurate assessments when
mapping at 1:50,000 in some parts of Scotland. Initially it is
necessary to recognize the major complexes within the study area,
and as the recognition of patterns of natural objects can be highly
subjective, and land patterns particularly difficult in this respect,
a list of criteria (Table 1) assisted in defining mapping units used
in western Scotland. Delineation of these complexes is then
possible with the additional help of aerial photographs. This

Figure 10 : On wetter soils, for example peaty gleys, surface
horizons may be extensively damaged by cattle trampling.
Such damage frequently leads to infestation by rushes
and a decline in fertility and grazing quality. Wet
soils in areas of high rainfall require extremely careful management to avoid damage and to preserve soil
structure and the farmers investment in improvement.

having been achieved the complexes need to be described in terms of
soil types. On the Island of Mull 893 sites were visited on a 1 km
grid pattern where brown mineral soils accounted for 76, 3 and 0 %
of the Knockan, Mishnish and Cruachan complexes respectively whereas
peaty gleys in contrast accounted for 1, 16 and 25 % (Table 2). To
check the accuracy of air photograph interpretation vis-à-vis ground
survey all 1 km National Grid squares on the Island of Mull whose
ground area was more than 50 % occupied by the Knockan complex were
listed. Subsequently seven areas were selected randomly and twenty
points within each, again randomly distributed, were field-surveyed.
Although the proportions of brown mineral soils and shallow mineral
soils differed, the detailed survey in general confirmed the valid-
ity of the deductions made from aerial surveys (Table 3). Interestingly the balance of different soils in the Knockan complex on
nearby Morvern (on the mainland of Scotland) was the same as that on
the Island of Mull.

Figure 11 : Steep lands, even on dry soils, may cause problems for
            tractors and implements. At best they reduce the working
            efficiency of the machinery but serious injury may occur.
            Work at the Scottish Institute of Agricultural Engineer-
            ing has demonstrated that the angle of slope and the
            nature of the surface must both be taken into account.
            Long grass is particularly dangerous. Photo courtesy
            Scottish Institute of Agricultural Engineering, Penicuik.

The interpretation of aerial photographs and the application of
statistical techniques to map units, has shown the greater development of brown forest soils on basic parent material (basalt) whereas
podzolic soils and peaty gleys were more abundant over acidic
granitic parent material (Table 4). When comparing two essentially
peaty complexes with overlying acidic parent materials a larger
proportion of peats was detected over granite, whereas podzols were
more abundant over schist, differences reflecting the ridged nature
of the schist topography compared with a flatter landform resulting
from the weathering of granite.

The ranges of soil types are larger in the Knockan and Tearnait
than in the peaty Funtack and Assapol complexes. Where mineral
surface soils are common peaty soils develop in hollows and flush
channels in dips in the landscape, especially where drifts are
derived from acid rocks. However, mineral soils are not often
found even as minor components in the zones where peaty soil com-

Figure 12 : Hagging of peat, caused by changes of climate or
            injudicious management such as over-grazing or over-
            burning is a feature of many mountains. Extensive
            areas provide serious obstacles to mechanized vehicles
            and although grazing animals occasionally use the edges
            as shelter, few utilise the poor grazings in the centre.

SOIL ASSOCIATION	PARENT MATERIAL	MAP UNIT NO.	COMPONENT SOILS	LANDFORM	VEGETATION
INSCH	Drifts derived from gabbro	1	Brown forest soils; some non-calcareous gleys	Undulating lowland	Bent-fescue grassland; some arable
STRICHEN	Drifts derived from mica schist	2	Brown forest soils and non-calcareous gleys	Undulating lowland	Bent-fescue grassland and rush pasture; some arable

Figure 13 : A specimen key describing soil complexes shown on soil
            maps prepared by the Soil Survey Department, Macaulay
            Institute for Soil Research, Scotland.

plexes are dominant. For example in Table 4, peaty podzols, peaty
gleys, shallow peaty soils and peat all occur in the Knockan and
Tearnait complexes but there are no mineral soils (brown forest soils,
gleys or shallow mineral soils) in the Funtack and Assapol complexes.

In summary the soil mapping units overlying different rock types
depend on geological, geomorphological and climatic relationships,
also the occurrence of drift - inevitably strong and consistent
patterns can be identified; these are being predicted with increasing
accuracy.

LAND AND THE LAND USER

The land user both needs information about the distribution and
properties of different types of land with the combination of soil
properties, climate and landform determining the potential of land
for different uses. The distribution of areas with similar potential
often determines whether or not that potential can be realized recognising that social and economic considerations have their part to
play. While the latter will be described in Chapter 25 et seq.,
the rest of this Chapter concentrates on those properties which are
important to agriculture and forestry before concluding with two
examples of integrated agricultural and forestry planning.

TABLE 1 : The criteria used to assist in defining mapping units to
           enable recognition of the major soil complexes in western
           Scotland.

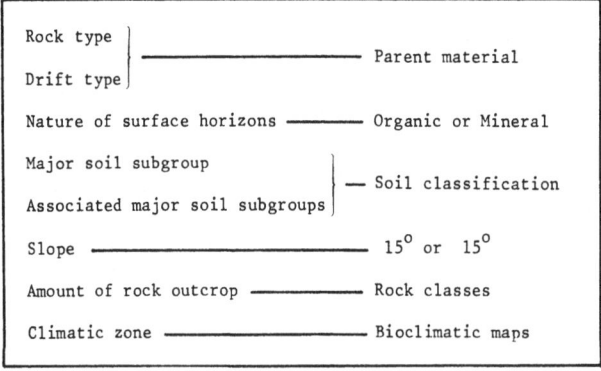

TABLE 2 : Proportion (%) of different soil types in three soil complexes overlying basalt on the Island of Mull, Scotland.

	Brown Mineral Soils Free and Imperfect drainage	Gleyed Mineral Soils	Podzols		Peaty Gley	Shallow Soils		Peat 45cm	Alluv.	Rock	Other	Number of sites examined
			Humus Iron	Peaty		Mineral	Peaty					
KNOCKAN COMPLEX	76	1	2	1	1	11	0	1	3	2	2	149
MISHNISH COMPLEX	3	2	39	16	16	3	7	2	2	3	2	61
CRUACHAN COMPLEX	0	1	0	14	25	1	7	45	0	7	0	167

TABLE 3 : Proportion of different soil types associated with the Knockan complex, determined by :

a) aerial photograph interpretation supported by grid sample survey on the Isle of Mull, Scotland.

b) detailed field assessment of 20 sites from each of 7 randomly selected locations on the Isle of Mull, Scotland.

c) detailed field assessment of 20 sites from each of 5 randomly selected locations on the Scottish mainland at Morvern.

Soil Subgroups / Location and type of survey	Brown Mineral Soils Free and Imperfect drainage	Gleyed Mineral Soils	Podzols		Peaty Gley	Shallow Soils		Peat 45cm	Alluv.	Rock	Other	Number of sites examined
			Humus Iron	Peaty		Mineral	Peaty					
Isle of Mull: a. Aerial photographs + grid sample survey	76	1	2	1	1	11	0	1	3	2	2	149
Isle of Mull: b. Detailed field assessment of 7 locations	65	1	0	2	2	20	2	2	3	3	0	140
Morvern: c. Detailed field assessment of 5 locations	63	3	5	0	3	17	5	1	1	1	1	100

TABLE 4 : Proportion (%) of different soil types in similar complexes overlying

a) basic (basalt) bedrock in Morvern and acid (granite) bedrock in Ardnamurchan, Scotland.

b) different acidic bedrocks (granite and schist) in Ardnamurchan, Scotland.

Parent rock : complex / Soil subgroups	Brown Forest Soils Free and Imperfect drainage	Gleyed Mineral Soils	Podzols		Peaty Gley	Shallow Soils		Peat 45cm	Alluv.	Rock	Other
			Humus Iron	Peaty Gley		Mineral	Peaty				
BASALT: KNOCKAN	63	3	5	0	3	17	5	1	1	1	1
a. GRANITE: TEARNAIT	23	4	23	5	14	8	7	3	7	6	0
GRANITE: FUNTACK	0	0	2	9	38	0	10	39	2	0	0
b. SCHIST: ASSAPOL	0	0	4	17	38	0	13	27	1	0	0

Agriculture

Agriculture in the hills and uplands is traditionally synonymous with stock rearing, usually either sheep or cattle but nowadays occasionally red deer. Stock farms are usually of 3 parts (a) arable land, used for the production of winter feed (hay/silage) or top quality grazing, (b) land which can be improved to provide supplementary areas of good quality grazing and (c) hill land of poor quality. Because the uplands are wet, their soils need to have good drainage (natural or man-made) and the ability to withstand the damage done by cattle and/or machinery if agricultural cropping is to be sustained (Figures 10 and 11). Access for machinery and stock will decrease with increasing slope.

In cold wet areas of northern and western Scotland, almost the only sites available for arable agriculture occur at low altitudes with shallow slopes; they occur on coarsely textured soils which drain rapidly. With the improvement in climate when moving southwards to the borders of Scotland, England and Wales it becomes possible to work some of the heavier land with impeded soil drainage. Thus arable agriculture spreads upwards from valley floors to valley sides. Although some very steep slopes are now being cultivated, slope remains a serious obstacle to cultivation. Because spring is often a dry period of the year cultivation is usually done at this time. Nonetheless, the slow transfer of material downslope, which can be seen in accumulation on the upslope sides of hedges and walls, is discernible on most farmed slopes in Britain. Although not serious in any one year the cumulative effects of soil erosion and the resultant thinning on crests should not be ignored. Management techniques should not accelerate soil erosion; they should, if possible, retard it.

The hill lands in Britain have relatively long growing seasons; the ample supplies of moisture enable the seasonal re-establishment of vegetation except on 'near-arctic' mountain summits unfavourable for agriculture. At these locations soils are put at risk by walkers and climbers, and in some instances by skiers but the damage by them can be dramatized - soil erosion is not the universal problem in these uplands that it is in arid lands.

Uplands are not prepared for grass as rigorously as land being used for arable crops. Nonetheless the land may be treated with herbicides to kill the extant sward before rotavation, discing and/or applying fertilizers while seeding. While peaty soils and peat may be successfully reseeded in north and west Scotland it is essential, because of severe problems of utilization and maintenance, to make regular inputs. In southern and eastern areas of Britain with a better climate and drier surface conditions, upland peats can be successfully reseeded with considerably less maintenance than in the north. In south-western England it is technically feasible to

TABLE 5 :   Land capability classification for agriculture developed
            for use in Scotland.

LAND SUITED TO ARABLE CROPPING

Class 1. Land capable of producing a very wide range of arable crops

          Cropping is highly flexible and includes the more exacting crops
          such as winter harvested vegetables. The levels of yield are
          consistently high.

Class 2. Land capable of producing a wide range of arable crops

          Cropping is very flexible and a wide range of crops may be grown
          but difficulties with winter vegetables may be encountered in
          some years. The level of yield is high but less consistently
          obtained than in Class 1.

Class 3. Land capable of producing a moderate range of crops

          Division 1. The land is capable of producing consistently high
          yields of a narrow range of crops (cereals and grass) or moderate
          yields of a wider range (potatoes, field beans and other veget-
          ables and root crops). Grass leys of short duration are common.

          Division 2. The land is capable of average production but high
          yields of grass, barley and oats are often obtained. Grass leys
          are common and longer than in division 1.

Class 4. Land capable of producing a narrow range of crops

          Division 1. Long ley grassland is commonly encountered but the
          land is capable of producing forage crops and cereals for stock.

          Division 2. The land is primarily grassland with some limited
          potential for other crops.

LAND SUITED ONLY TO IMPROVED GRASSLAND AND ROUGH GRAZINGS

Class 5. Land capable of use as improved grassland

          Division 1. Land well suited to reclamation and to use as
          improved grassland.

          Division 2. Land moderately suited to reclamation and to use as
          improved grassland.

          Division 3. Land marginally suited to reclamation and to use as
          improved grassland.

Class 6. Land capable only of use as rough grazing

          Division 1. Land with high grazing value.

          Division 2. Land with moderate grazing value.

          Division 3. Land with low grazing value.

Class 7. Land of very limited agricultural value.

reclaim practically all areas of moorland and as a result conservation interests may be at risk. Whether or not to encourage reclamation is a matter for consultation and planning, but detailed soil
surveys done in advance can enable the identification of areas where
there is likely to be a conflict of interest.

Conflicts of interest may be identified by overlaying maps of
land potential on those of current land use. Although the Department
of Agriculture for Scotland had an established land classification
for the low ground of Scotland in the 1940's there was not one for
the uplands. In 1966 the Soil Survey introduced a land capability
classification encompassing lowland and upland. Classes 1 to 4 are
considered as arable land, classes 5 to 7 as hill land, a classification which, after a trial period is gradually being phased-in as
the principal classification to be used in Scotland for planning
(Table 5). The classes involved in hill areas are Class 4 - marginal
arable land; Class 5 - reclaimable land, and Class 6 - hill land
which is not considered to be reclaimable; Class 7 refers to land
of no value to farmers.

These classes are defined by maximum potential soil moisture
deficits, accumulated temperatures and extent of exposure which give
rise to climate zones suitable for different types of enterprise:
within this climatic framework the effects of topography, soil,
wetness, erosion are considered with amendments being made where
appropriate. Within each class, land is further ranked by division,
based on criteria of importance to land users. For example Class
4 has two divisions based primarily on climate but with the interaction between soil texture and increasing rainfall in the hills
also coming into play. Within Class 5 three divisions have been
established based on soil wetness (particularly liability of land
to damage by stock trampling), soil pattern and slope, the latter
affecting the ease with which machinery can be used. Decisions
regarding Class 5 and Class 6 land will always be slightly contentious because they involve estimates, largely subjective, of whether
economic returns can be obtained from 'improvements'.   Inevitably
there are large areas of hill-land which cannot be 'reclaimed'.
Nonetheless they can be grazed, their value reflecting the quality
of their natural and semi-natural assemblages of plants. Although
still somewhat empirical, the subdivisions have been discussed by
several national committees including Department and College of
Agriculture officials and farmers, and accord with accepted farming
views. The subdivisions broadly reflect amounts of palatable grasses
and herbs in swards with good grazings found on brown mineral soils
- the higher the proportion of brown soils the better the grazing
potential of hill areas. Poor grazings are associated with bogland
communities on wet soils with organic surface horizons.

1:250,000 maps of Scotland showing these categories are now
available. They enable planners to make rational decisions about
land-use including the inter-relation between agriculture and forestry.

Forestry

Over a period of about 5,000 years, starting in the Neolithic
period, the hills of Britain have been progressively deforested.
Through the activities of both private and public sectors the trend
has been reversed in recent years with considerable areas of afforestation, primarily with rapid growing conifers, 'softwoods'.
Despite the existence of the Canadian land capability classification
for forestry and the United States Department of Agriculture Woodland Suitability Classification, a land capability system similar
to the one operated for agriculture has proved difficult to introduce into Scotland. A national map was proposed as part of the
recent 1:250,000 soil and agricultural capability scheme but was
discarded. The Canadian system is principally based on measurements of existing timber stands, but in Britain, there are as yet
few longterm stands of timber species. The British Forestry
Commission has a system of site classification linked to predicted
yield classes, but it is only applicable to land already afforested;
it is of little use to planning authorities wishing to know the
forestry value of different types of land which have not yet been
afforested.

Some of the properties of land important to foresters differ
from those required by agriculturalists. As with agriculture, it
is not so much the identification of important properties as the
establishment of limiting values within them which is difficult.
In terms of climate, moisture in British hill-land has not been
thought of as being limiting like temperature and exposure.
The vertical zonation of climate, described earlier, strongly
affects yield and hence the choice of tree species. Exposure, in
combination with soil depth 'shallowness', have been evaluated by
the Forestry Commission with the development of hazard classes. At
high altitudes, 'windthrow' is a serious limitation, curtailing
rotation lengths and strongly influencing management options.
Although farmers habitually apply fertilizers, foresters, because
of economic considerations, do not; they apply fertilisers to
assist trees at critical growth stages augmenting naturally available nutrients. In most instances the amounts to be applied can be
judged from the composition of swards of natural/semi-natural
vegetation that exists before afforestation. Ploughing, a ubiqui-
tous aspect of site preparation, suppresses competing weeds, provides a deeper and easier 'root-run' for young trees and, by
providing drainage, alleviates waterlogging. Topography may limit
mechanical operations, from ploughing to roading and eventually
harvesting, but it is not as important as in agriculture where
regular traverses are made annually. Because modern forest ploughs
have deep draughts, slopes as little as 5° can preclude two-way
ploughing, but this apart there is no particular impediment until
slopes are $30-35^{\circ}$ or more when the risks of accidents become acute.

Much is known therefore about the environmental characteristics of British hill-land; guidance is available but unfortunately, from the point of view of the planner, the requirements of agriculture and forestry, and their impacts on land capability, have not been brought together - a synthesis has not been prepared.

CASE STUDIES OF AGRICULTURE-FORESTRY INTEGRATION

In order to indicate how the type of information described in earlier sections is being applied to the use of hill lands, two case studies will be considered both relevant to the west coast of Scotland (i) an agricultural estate, Melfort Estate, some 20 kilometres south of Oban and (ii) the survey of the Island of Mull.

Melfort Estates, Kimelfort, Argyll

This estate is on the north shore of Loch Melfort, a sea-loch, and rises from sea-level to over 330 metres. With a rainfall varying from 1800 - 2500 millimetres per annum it can fairly be described as wet. The lower land on the estate is (i) underlain variously by epidiorite, phyllite and granite rocks and (ii) surrounded by a high escarpment (which is the southern edge of an extensive ancient volcanic plateau of andesite and tuff). Arable crops are grown on a gravelly raised beach of brown mineral soil. This is surrounded by an area with shallow, brown mineral soils, both freely and imperfectly drained, which, in parts, is strongly undulating, rocky and steeply sloping. The andesite scarp consists of very stony but freely drained mineral soils and the higher land a series of blocky irregular hills with peaty gley and peaty podzol soils with intervening upland basin peat (Figure 14).

Initially, the estate totalled 1205 hectares, but half was sold for forestry, principally the higher hill land. The areas retained included 557 hectares of hill land, 110 hectares of woodland and 48 hectares arable. The Hill Farming Research Organisation (HFRO) had developed over a number of years, a system of sheep management known as the two-pasture system. In co-operation with the Council of Scottish Agricultural Colleges, the HFRO was seeking to apply the system on commercial sheep farms. Having just acquired Melfort Estate, Mr. Clarke, agreed to implement the scheme which was carefully monitored by Mr A. MacLeod and his staff at the West of Scotland Agricultural College office in Oban. The principle of the two-pasture system is to improve a small area of hill pasture, grazing it intensively during lactation and early lamb growth (April-June) and prior to mating (August-November). The adoption of the system presupposes the existence of suitable areas of land for improvement (improvement procedures are expensive and their use is only justified if enhanced pasture, production and quality, are reflected in extra animal output and hence increased financial

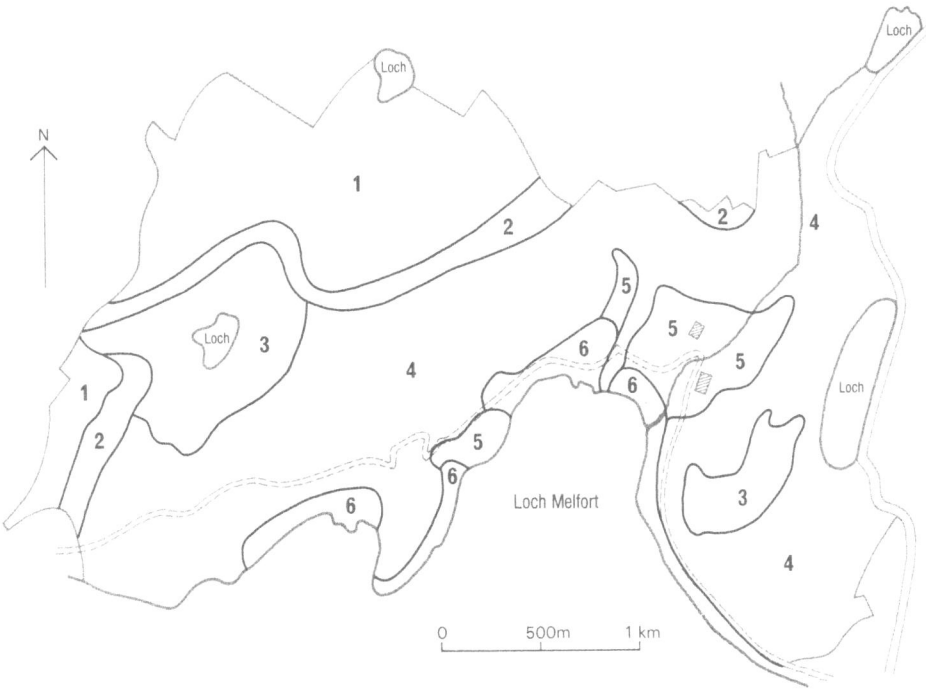

Figure 14 : Simplified soil map of Melfort Estate, Argyll, Scotland,
            showing distribution of different soil types :

    1. Peaty gley and peat on andesite
    2. Brown mineral soil on andesite
    3. Peaty gley and peat on schist and epidiorite
    4. Brown mineral soil, freely and imperfectly drained,
        on schist and epidiorite
    5. Brown mineral soil on raised beach gravel
    6. Peaty gley and peat on raised beach gravel

returns). As it happens improvable land was available at Melfort;
about 25 per cent of the hill land in the lower parts of the farm
and on the main face of the hill was improved including the provision
of 9,400 metres of road (Figure 15).

    Although ewe numbers have been very slightly increased, lambing
percentages at weaning have steadily increased from 83 % (1975/76)
to 145 % (1980/81) to give more production (in terms of number of
lambs) from 607 hectares than was originally obtained from 1,205
hectares (Table 6). The quality and weight of lambs increased from

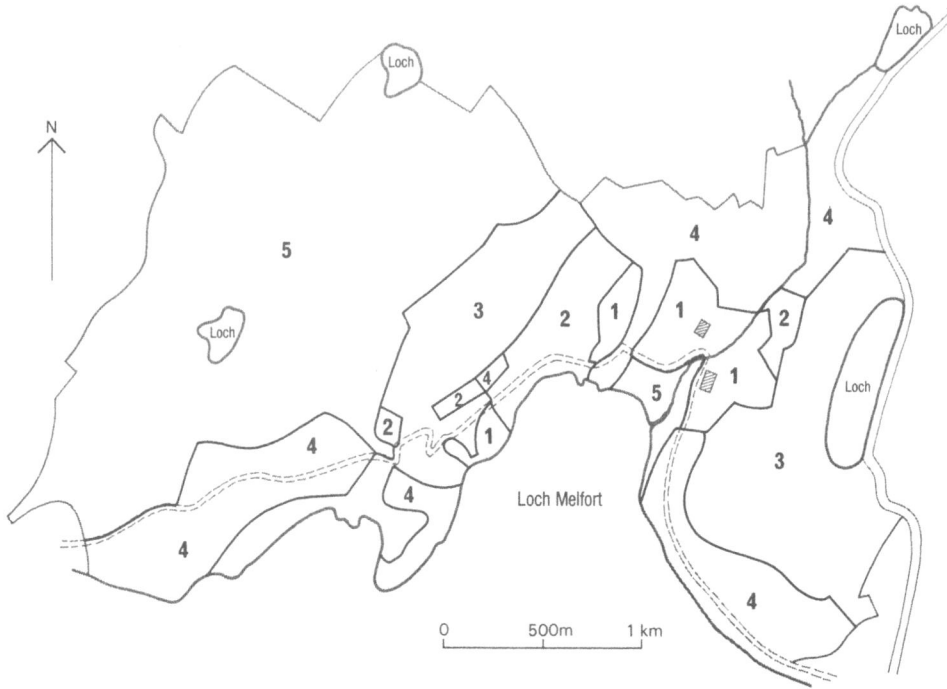

Figure 15 :   Map of land use of Melfort Estate, Argyll, Scotland,
              showing that area of old (2) and new (3) improved land,
              used to create a two pasture system of sheep farming
              since 1975 which has significantly increased sheep pro-
              duction (Table 6).

              Key to land use : 1 Arable
                                 2 Old improved land
                                 3 New improved land
                                 4 Woodland
                                 5 Open hill and foreshore
                      --------   Road

29.8 kilograms in 1975/76 to 32.5 kilograms in 1980/81 while almost
600 hectares were made available for forestry. The financial implications of these changes are given in Table 7. In 1976 under the
old system of management the gross margin for the farm was £8,482.
In 1980 this had increased to £19,535 against an estimated margin
of only £8,892 had the old management systems been continued. The
key to this increase in production was the adequate and efficient
utilization of the brown mineral soils in the farm unit. Soils of
this type are scarce in the Western Highlands: in a recent survey
it was calculated that mineral soils accounted for only 15 per cent

of the area while less than 10 per cent had combinations of other
land characteristics which made it suitable for improvement.
Despite superficial appearances, therefore, the scope for intensifying management in these uplands is limited, and thus considerable
care should be taken by planners when deciding the appropriate
course of action.

TABLE 6 :   A comparison of the total number of ewes and hoggs,
            their breeding success and the physical attributes of
            lambs in the flock at Melfort Estate, Argyll, Scotland
            since 1975-6 when a land improvement programme enabled
            a two-pasture system of sheep farming to be instituted.

	1975/76	1976/77	1977/78	1978/79	1979/80	1980/81
Ewe numbers	701	734	750	757	760	725
Hogg numbers	204	200	208	214	211	205
Weaning percentage	82.6	106.3	125.7	130.4	141	144.5
Average weaning weight (kg)	29.8	32.4	32.0	31.9	32.8	32.5
Total weight lamb weaned (kg)	17,254	25,272	30,176	31,485	35,292	34,060
Weight lamb weaned/ewe	24.6	34.4	39.9	41.6	46.4	46.9
/ha	28.9	41.6	49.7	51.8	58.8	-
Wool kg/ewe	2.36	2.55	2.47	2.33	2.37	-
Ewe mortality (%)	8.3	5.4	5.3	5.7	4.5	4.2
Lamb mortality total (%)	23.0	5.4	1.9	2.6	1.4	1.2

(by permission of Council of Scottish Agricultural Colleges)

Survey of the Island of Mull

In 1973 the Highlands and Islands Development Board published
a report, the *Island of Mull, Survey and Proposals for Development*,
which had been stimulated by the concern, voiced by the Advisory
Panel on the Highlands and Islands, about a decline in farming
through the development of sporting estates and forestry. The
existing and potential resources of the Island of Mull were considered in terms of population and employment, land use (agriculture
and forestry), fisheries, tourism, transport and housing and
services. With the help of (i) a soil map and (ii) a land capability
for agriculture map the Board published a set of proposals encouraging the formation of new tenancies for young farmers and suggesting an increased area of afforestation. Two sites which had been
earlier earmarked for forestry, Antuim and Scoor, were re-allocated
to agricultural tenancies and a large estate, Glenforsa, then

TABLE 7 : Financial summaries for the lamb crop years (1st December
to 30 November) 1976-1980 at Melfort Estate, Argyll,
Scotland. The final column headed No change 1980 is an
estimate of costs and income if no change in management
had occurred after 1976. (By permission Council of Scottish
Agricultural Colleges).

	1976	1977	1978	1979	1980	No change 1980
Ewes mated	701	734	750	757	760	760
Lambs weaned	83	105	126	130	142	75
Income per ewe mated (£)						
Lamb revenue	6.25	14.50	18.93	19.33	24.88	6.99
Ewe revenue	1.78	2.34	6.02	5.22	6.01	3.72
Hogg revenue	0.32	1.29	0.16	1.04	1.22	-
Hill Sheep Subsidy	3.60	3.60	3.60	4.10	5.50	5.50
Less Ram purchases	0.37	0.60	3.23	0.73	0.69	0.69
Ewe purchases	-	-	-	0.36	0.51	-
Total Gross income	13.17	23.32	25.48	28.60	36.41	15.52
Variable Cost per Ewe (£)						
Feed	0.80	1.15	2.00	2.81	3.20	0.82
Wintering of Hoggs	-	0.18	-	-	-	-
Dip, Vet, etc.	0.90	1.36	1.94	3.67	4.27	1.50
Haulage	0.51	0.30	0.40	0.40	0.53	0.50
Casual Labour	0.09	0.14	-	0.50	1.00	1.00
Total Variable Costs	2.30	3.13	4.34	7.38	9.00	3.82
Gross Margin (Cash)	10.87	20.19	21.14	21.23	27.41	11.70
Valuation Change	+0.01	+0.55	+0.52	-0.20	-1.71	± 0
GROSS MARGIN	10.86	20.74	21.66	21.03	25.70	11.70
GROSS MARGIN FOR FARM	8,482	15,223	16,251	15,926	19,535	8,892

Before charging grazing costs.

administered by the Department of Agriculture, was to be divided to
form two more : four new sites were suggested for afforestation.

This case study is different from the Melfort study in many ways.
In this instance, a national body, the Highlands and Islands Development Board (HIDB), set out to suggest a strategy for the development of an area which was owned by a large number of private individuals and two government departments, the Forestry Commission and
the Department of Agriculture. With successive British governments
reluctant to impose their will in matters affecting land use, the
HIDB proposals had to be persuasive. After ten years their most
significant effect seems to have been on attitudes to agricultural
tenancies. The proposals for the Glenforsa Estate and Antuim, both
owned by the Secretary of State for Scotland, were implemented with
the formation of three new tenancies. At Scoor, in the south of
the island, land was divided to increase the sizes of two small
farms rather than create a third. At Torosay another tenanted
holding was created by the landowner and in at least three other
instances farms which might have been repossessed by the landlord
when tenancy agreements ran their full course, were re-assigned.
However this atmosphere of co-operation was not universal; in one
notable case an estate was sold and split between forestry and
sporting interests, an episode that caused the Highlands and Islands
Board to seriously consider changing its persuasive role for one of
compulsion ! As far as forestry is concerned there has been an
extension of planting, principally on the poorer quality soils with
little prospect of improvement for agriculture.

The presentation of data dealing with the potential of soils for
agriculture has not, of itself, been a determinant of any particular
course of action but, in conjunction with other social and economic
criteria, it has been and continues to be a valuable tool in the
hands of the informed planner. The Soil Survey of Scotland has
prepared maps for lowland arable areas and the hill lands at 1:50,000
and 1:250,000 respectively. In future the existing information will
be upgraded while larger scale maps of the hill lands will be produced, both necessitating a better understanding of how soil characteristics affect land use and land users.

10

SOILS IN ARID AREAS - MANAGEMENT AND CHARACTERISTICS

I. Akalan

The Scientific and Technical Research Council of Turkey,
Ataturk Bulvar 221
Kavaklidere-Ankara, Turkey

Developing countries with large rural populations including
Turkey have some important problems in common. These include lack
of energy sources and supplies, fertilizers, irrigation water,
increasing inflation, unemployment, other economic pressures such
as shortage of capital, inadequate supplies of food. Among these,
probably food supply is the most important. Thus the prosperity and
future development of a country must depend on the effective use of
its land and water resources, which not only determine the level of
human sustenance possible, but also to a considerable degree its
economic vitality. Increasing population pressures throughout the
world have made it increasingly important to improve and intensify
agricultural production if a world food shortage with widespread
human malnutrition or starvation is to be avoided.

The Turkish government has given increasing priority to agricultural development. Progress has been rapid in areas well-endowed
with soil and water resources and with improved methods of soil and
water management, productivity has increased consistently. However,
by far the largest part of Turkey's land surface is arid or semi-
arid in which agricultural productivity is generally low and highly
variable from year to year.

Turkey has a rather high food energy consumption (2891 cal.
per capita) compared with most developing countries and even some
developed ones. Maintaining this advantage requires appropriate
birth control measures and increasing agricultural production in
line with the increasing population. The potential for exporting

agricultural products is an exciting possibility that cannot be overlooked.

Turkey is a predominantly agricultural country, having about 65 % of its population living in rural areas. About 45 % of the national income and 85 % of the export trade result from agricultural production. 33 % of the total land area is under cultivation.

The geographical location of the country and its topographical features give rise to a variety of climatic conditions and different kinds of soils. All the major soil groups with the exception of Laterites are found in Turkey, so that a wide variety of high quality crops are grown (Akalan, 1968).

Classification of land-use capability in Turkey has shown that most rural areas are not used effectively. Indeed, some farmers who live near the national forests have cut down the trees to clear the land for cultivation, causing erosion and depletion of the soil.

Suitable crop patterns for the different land classes need to be developed for economic and efficient land-use planning. At the same time, the usefulness of the present land classification system for economic development planning has to be assessed. The criteria presently used may be inadequate and need to be modified in the light of accumulated experience.

With two-thirds of the area under wheat dry-farmed, improvement in the production of this most important crop of the arid and semi-arid areas of Turkey is of the highest priority. The delicate balance between limited water supplies, soil and vegetation means that special soil and water conservation strategies are urgently needed; indeed, dryland regions have proved to be very productive when appropriate technological advances and agricultural developments have been applied.

Central Anatolia is a typical semi-arid region, and Turkey's approaches to its problems may be of use to other countries.

LAND CLASSIFICATION

The land classification for Central Anatolia shows 12.44 million hectares (40.7 % of the total land area) to be suitable for cultivation, being of classes I to IV. This is followed by 9.2 million hectares suitable for grassland (30 %), 6.4 million for forestry (21 %) and 1.1 million wetlands (3.5 %). At present, some 12.6 million hectares are under cultivation - nearly 160 000 hectares are thus unsuited to this purpose. This situation has led to a number of problems, including such severe erosion that some areas will never be cultivated again (Akalan, 1974).

Central Anatolia has three main regions :

- North Central, 4.7 million hectares cultivable (40 % of the land area)
- East Central, 2.4 million hectares cultivable (30 % of the land area)
- South Central, 5.3 million hectares cultivable (50 % of the land area).

SOIL CHARACTERISTICS

The most common residual soils found in Central Anatolia are Chestnut Soils, Brown Soils, Reddish Brown Soils and Sierosems. Azonal soils like Alluvial soils, Lithosols, Regosols also cover considerable areas. The most common characteristic of the Central Anatolian area however, is its aridity, which controls the uses to which the soils may be put. Despite apparently suitable soil classifications, several areas are in fact not suited to cultivation.

The upper horizons of soils developed under imperfect drainage conditions in arid areas have abnormally high concentrations of sodium, calcium, magnesium and potassium chlorides and sulphates, with smaller amounts of bicarbonates and carbonates. The surfaces of these saline or halomorphic soils are often covered with an irregular crust of salts, carried up by capillary action and chrystallizing when the water evaporates.

Soil development in the Central Anatolian Plateau has been through calcification, so that it is not surprising to find that 98 % of the region's soils contain calcium carbonate; indeed, low-lime soils have generally developed from volcanic rock.

The climate and the vegetation that it supports are largely responsible for this situation. The rainfall in these areas is insufficient to leach much of the calcium and the other divalent cation from the soil profile, nor are soluble constituents leached from the surface horizon removed; they are only moved to the lower horizons. Thus a zone of calcium carbonate accumulation occurs at depths approximating the average penetration of rain water. The more soluble $CaSO_4.2H_2O$ will also accumulate in the drier soils, usually below the zone of carbonate concentration, and even soluble sodium and potassium salts may be present at lower depths of the zonal soils.

Calcium carbonate renders the soil slightly alkaline, provides calcium ions in forms available to plants, and plays an important role in the granulation of soil particles. It also causes phosphorus deficiency as it fixes phosphate ions in non-available forms, which means that phosphate fertilizer additions have to be calculated in

terms of the calcium carbonate content of the soil. High calcium
carbonate concentrations facilitate the reclamation of alkaline
soils with sulphur, which flocculates clay particles and increases
the permeability of soils.

The native vegetation has also played a part in the calcification process. It ranges from a variety of grasses to desert shrubs,
the former being particularly effective in selectively returning
bases to the soil surface. The grasses' extensive root systems are
also responsible for the availability of fairly large amounts of
organic matter in the subsurface horizons.

Indeed the influence of organic matter on the physical and
chemical properties of soils is out of all proportion to the relatively small amounts that are present. It generally accounts for
about one half of soil cation exchange capacity, and is probably the
major factor responsible for the stability of soil aggregates.
Furthermore, the humus that develops from the organic matter seems
to be an effective stabilizer of soil colloids, especially in the
presence of calcium, preventing downward movement of these particles.

The amount of organic matter in soils varies widely, although
it is safe to say that its levels are usually low and difficult to
maintain in arid soils because of low precipitation and reduced
vegetative cover. This has a marked effect on the physical, chemical
and biological properties of soils, and over-intensive farming
practices that denude soils of organic matter are becoming a world-
wide problem.

In Central Anatolia, 20 % of the soil has less than one %
organic matter,   63 % has one to two %, and only 4 % has above 3 %
(Soil and Water General Directorate, 1978).

Most of the physical properties of soils, and thus their suitability for a variety of uses, are strongly influenced by the size
of its mineral particles - the soil texture. The relative proportions of particles of different size governs moisture storage,
aeration, drainage, ease of root penetration and a host of other
essential features. Analysis of particle size shows 7 % of
Anatolian soils to be sandy, 33 % loam (the most desirable type) and
some 60 % to be various degrees of clay. This high proportion of
clay soils poses many problems because, although clays tend to store
more water than sandy soils, the pore size is small, air circulation
is low, the tendency to run-off and subsequent erosion is high, and
they are strong absorbants for many exchangeable cations, resulting
in low availability to growing plants. High clay soils are also
more difficult to plough and ploughing machinery requires substantially more gasoline.

Anatolian soils tend to be moderately alkaline - 90 % have pH
in excess of 6.5, reaching as high as 8.5 in cases where high con-

centrations of sodium salts (mainly carbonates and chlorides) are present. Such alkalinity is considerably greater than the optimum at which soil micro-organisms work best and, since the decomposition of organic matter is usually a major source of available nitrogen, it is not surprising that over 80 % of the soils are nitrogen-deficient. Nitrogen is easily lost from the soil, and must be replaced by farm manure, crop residues, legumes or commercial fertilizers; the last is used most extensively (Akalan, 1968).

The second most critical element is phosphorus. While the level of phosphorus is generally high in these arid soils, it is very often not in available forms and, like nitrogen, it has to be artificially applied. Potassium, on the other hand, tends to be available at satisfactory levels, so that no chemical fertilizers are necessary (Akalan, 1968).

Colloidal clays and humus control many of the processes that occur in soils. As the arid soils of Anatolia are humus-deficient, the clay minerals assume greater significance, montmorillonite, illite and kaolinite being the most important. The soils have higher water-soluble salt content than those in more humid regions; sodium, calcium and magnesium chlorides, sulphates and bicarbonates are the most widespread (Akalan, 1973).

Most of these arid soils have a low moisture-holding capacity because of shallow depth (sometimes as little as 30 cm) and the low porosity of clays; in many areas drainage is also poor. Thus the potential for water erosion is high, and climatic and topographic characteristics coupled with farming practices, such as burning stubble in harvested fields, also make wind erosion a severe problem (Akalan, 1973).

SOME APPROACHES TO DRYLAND AGRICULTURE

Central Anatolia has three basic problems: erosion, water supply and water quality. While farming operations must vary according to soil, crop, topography and climate, farming practices, types of cultivation equipment used and crop selection are the keys to dryland agriculture in the region.

In areas where erosion potential is high and rainfall is too limited to permit annual cropping, a 'summer fallow' system has been successfully used to conserve moisture in the root zone and ensure that seed can be planted in moist soil before the autumn rains start. The land is kept free of any crop in alternate years, and weeds are carefully controlled. Not only is water conserved, but available nitrogen and other nutrients are produced by microbiological breakdown of humus during the fallow period, and earlier germination may advance harvesting by as much as two weeks, avoiding hot weather drainage (Zinn, 1972).

Maximum efficiency requires :

- effective weed control;
- maintenance of high infiltration rates;
- slowing the surface movement of water;
- avoiding tillage that exposes soil to rapid drying and
- a moisture barrier or mulch at the appropriate depth.

The drawback to the fallow system is that it reduces the organic content of the soil, so that steps have to be taken to minimize this loss. Eight million hectares are cultivated under this system in Turkey, and recent research has indicated that legume crops might be produced during the fallow period in areas with precipitation greater than 400 mm.

In areas of Central Anatolia where the water quality is poor - generally high in salt content - emphasis is being placed on salt-tolerant forage crops. Suitable for use as fodder, the idea is to shift agricultural production towards sheep and cattle breeding which is more appropriate to the resources available (Akalan, 1971; 1976).

REFERENCES

Akalan, I., 1968, "Soils", University of Ankara, Faculty of Agriculture Publications No; 356, Ankara University Press, Ankara.

Akalan, I., 1971, "Characteristics of salt affected soils of Cumra -Konya area", University of Ankara, Faculty of Agriculture Publication No: 434, Ankara University Press, Ankara.

Akalan, I., 1973, "Soil Physics", University of Ankara, Faculty of Agriculture Publications No: 527/172, Ankara University Press, Ankara.

Akalan, I., 1974, "Soil and Water Conservation", University of Ankara, Faculty of Agriculture Publications No: 532/177, Ankara University Press, Ankara.

Akalan, I., 1976, Salt affected soils of Turkey, their formation, classification and reclamation, in "The Plant production under saline conditions" Proceedings of Cento Symposium, 11-14 May, 1976, Adana Turkey, Office of the United States Economic Coordinator for Cento Affairs, Ankara.

Soil and Water General Directorate, 1978, "Land Resources of Turkey", Soil and Water General Directorate Press, Ankara.

Zinn, T.G., 1972, The Summer Fallow System-Cento Seminar, in: Agricultural aspects of arid/semi-arid zones", Proceedings of Central Treaty Organization Seminar in Ankara, Turkey, Office of the United States Economic Coordinator for Cento Affiars, Ankar

COMMENTARY : SOILS

For most ecologists, soil can be likened to the computer
scientist's black-box. We know that it is essential but have little
knowledge of its complexities; even less about how it functions in
relation to root growth. What hope is there for the non-specialist
decision-maker?

Soils are probably our most important non-renewable natural
resources - as yet, they are indispensable. While members of the
Seminar discussed and argued the technicalities of soil formation
and the relevance of the different horizons within a soil profile,
there was widespread agreement that most people's concept of soil
was erroneous. Perhaps because of the system of classification
presently in vogue we think of soil as being of this or that type or
series. But, in reality, there is a continuum of soil types, the
continuum being arbitrarily broken for convenience into sections with
no more variation than can be readily comprehended. The extent of
variation is too often overlooked not only by decision-makers but
sometimes by soil surveyors. Has sufficient been made of the
presence of natural vegetation as integrated indicators of soil
conditions? Is it not appropriate to consider the potential and
value of multi-variate statistics for defining the soil continuum
and the selection of break-points setting the limits of relatively
homogenous and comprehendable sections of that continuum? It was
suggested that multi-variate analytical techniques might have an
invaluable role to play when moving from soil characterisation to
land classification, two totally different concepts. While the format
is concerned with labelling a type of soil, the latter attempts to
provide an index integrating soil type, topography, climate etc., so
that the classes which are identical can be interpreted in terms of
the actual and potential occurrence of natural assemblages (plants
and their associated animals) and agricultural and forestry crops.
Ideally decision-makers need one system of soil/land classification,
enabling at one and the same time prediction of the performance of
natural vegetation and crops (agricultural and forestry). To an
extent, land classification can be seen as a tool for planning
strategy, with soil characterisation playing a larger part in more
local tactical considerations. Whatever the reason, soils are not
generally thought to be subject to change: but in reality, soils are
continually changing; they are dynamic, with an ever changing flow
of inputs and outputs reflecting constantly changing rates of

173

different processes, such as litter decomposition and nitrogen mineralisation. As is well known, soils planted with conifers tend to become impoverished; agricultural soils subject to intensive cultivation lose some of their organic material. Are these differences real and if they are, should our systems of classification be modified so as to enable changes of this sort to be acknowledged? Simply stated, soil x is no longer x after 'n' years of intensive cultivation.

Considerations of this sort bring into question the different approaches to soil classification. These have mainly been evolved by soil surveyors with a predominant emphasis on textural characteristics. But to be useful in environmental planning, classification of soil should more closely relate to the abilities to supply plant nutrients, to absorb and retain heavy metals, to neutralize the acidity of atmospheric pollutants, to carry structures (buildings) of different dimensions etc. In short, most systems of soil classification don't directly provide the answers to the questions confronting decision-makers. They are mostly insufficiently angled to the capabilities of soil.

Very often we talk about soil improvement when, in reality, we mean that we have increased crop yields. Thus the application of artificial fertilizers may increase yields but it is unlikely to improve soils. On the other hand drainage may improve soil water status and possibly affect soil structure, but is this a real gain if it is achieved at the expense of some plant and animal species? Judgements of better or worse are subjective. In some instances they may be helped by an undisputed monetary gain, but how should allowances be made for other types of arbiter, for example the impairment of natural vegetation? There was repeated reference to secondary, but very important, issues. It was felt that decision-makers had a right to expect soil scientists to give a clear statement of the probable outcome of different planning decisions. They should be able to predict the increased concentrations of nitrogen and phosphorus in freshwater streams following the application of artificial fertilizers to intensively managed land, but as yet they cannot in quantitative terms; they should be able to predict the influences of atmospheric pollutants, but cannot. In short, it seems that soil science has reached the stage where it is possible to discuss in qualitative terms, the likely responses to change but the decision-maker needs to have quantitative measures. In some instances a yes/no answer is appropriate - this or that course of action is or isn't acceptable - but in most instances the questioner needs to know how much of this or that is acceptable. For example, how many days per year is a particular soil workable; a question that inevitably highlights the need to accept that soil is only one of many facets of a plant's environment, the particular question necessitating a knowledge of soils and climate. For their part decision-makers must have sufficient technical competence to appreciate that soil is only one facet of a plant's environment with topography and climate also being of major importance. Further, that "soil improvements" are only part of the armoury of the agronomist, horticulturist, silviculturist etc. Unfortunately some of these other improvements, notably the introduction of more refined vehicular equipment, may prove to be hazardous to some soils. Accelerated land erosion occurs as a result of some modern agricultural cultivation, and, at different stages, notably when preparing sites and at final harvest, in the development of man-made forests. Decision-makers need information about these risks of erosion, and also the risks incurred when erecting buildings on soils with (i) different bearing/carrying capacities, and (ii) different propensities to shrink, both leading to uneven settlement. For the future the present trend towards simple but sophisticated systems of classification needs to be strengthened to enable decision-makers to profit from the wealth of the environmental data already to hand.

A

IDENTIFICATION OF ECOLOGICAL FACTORS CHARACTERISING THE RANGE OF *ECOLOGICAL* HABITATS

ii-Rural
c-Water Resources

ASSESSMENT AND MANAGEMENT OF WATER RESOURCES

M. Newson

Institute of Hydrology
Staylittle
Powys, U.K.

A BACKGROUND

Britain may be classified as a developed nation; yet a third of her land surface, the uplands, is essentially undeveloped. The uplands are also unplanned and visitors from developing nations remark on how little basic resource evaluation and planning has gone on. Britain was, after all, a founding force in topographic mapping and in town planning. The rural nature of the uplands has, however, meant that the urban-dwelling decision makers have preferred to see them unplanned (see Davidson and Wibberley, 1977), except where planning protects them from the urban expansion.

Whilst British geographers have occasionally emerged from their academic regionalizations to evaluate land use or land potential at home (Stamp, 1948; Coleman, 1970), physical resource planning of rural areas has been more a feature of the work of Britons abroad. This is consistent with a resource dependency upon colonial exploitation; only in wartime or during rapid economic changes such as those of the last decade does Britain's vulnerability in this respect force more attention to resources at home.

Thus, there are no established boundaries for the British uplands; even fiscal definitions for the application of European Community funds are hotly debated, and Figure 1 is based on the most recent of these (European Community, 1975). Land above 213 m has been called 'the last great area for development' (Williams and Harding, 1982). Those developments which have affected the uplands so far have occurred as a series of waves, running up the hills and often down again because of subtle changes in the basic harshness of the upland climate and soil environment. The uplands

Figure 1:   Hills and uplands in Britain, after the 'less favoured'
            classification by the European Economic Community (1975).

are marginall in both the physical and economic senses; one need
only trace successive changes in the boundary between clutivated
land and 'moorland' (Parry, 1978; Parry, Bruce and Harkness, 1982).
The approximate sequence of waves of development over the last
century has been mineral extraction, followed by water storage,
afforestation, recreation and conservation. At several stages in
the sequence, especially during two world wars, during the
Napoleonic threat and since British participation in European farm
policies, the use of the uplands to grow food has had a marked
influence on upland land use. Aided by new technology, from four-
wheel-drive tractors to improved grass strains, upland livestock
farming has expanded rapidly, producing unprecedented controversy
because of the simultaneous expansion of demands for recreation
and conservation in the same areas (Shoard, 1980).

    It would be facile to suggest an allocation of land in the
uplands upon the basis of the historical 'first-come, first-served'
sequence. However, it is important, to note that the abundant

rainfall and low evaporation over the British uplands made water a
resource of considerable significance <u>at the very beginning of the</u>
<u>present era of development</u>. During the Victorian era Britain began
to clear up the social and environmental mess of her industrial revolution; the provision of abundant supplies of fresh water was an
integral part of this reforming ethos. Many of the industrial
centres in the north and west of Britain owe their origins to the
availability of water power, and so were relatively close to upland
streams for supplies once the Victorian civil engineers (Binnie,
1981) had mastered the skills of dam building. The central Pennines
are typical of these smaller, early water supply developments.
There were, however, some much more ambitious schemes, for example
the construction of Lake Vyrnwy (1891) to serve Liverpool, and the
Elan Valley dams to impound water for Birmingham (1904). All of
these early reservoirs provided their water by direct supply
through pipelines expensive to construct but cheap to run, only
requiring the energy of gravity and the minimum of purification
because of the <u>low levels of development of land around the upland</u>
<u>catchment areas</u>.

LAND USE IN CATCHMENT AREAS

    Victorian obsession with public health improved both domestic
and industrial water supplies to the point where rare cases of contamination evoked widespread concern, an example being the Croydon
typhoid outbreak of 1937. The main concern of water supply authorities became one of preventing the otherwise pure upland water
stored in reservoirs from biological contamination. The typhoid
outbreak forced a new review of the land use around reservoirs
(Ministry of Health, 1948), and the burden of the review report was
that catchments must be sterilised against pollution from human or
livestock sources. There was thus a <u>presumption in favour of</u>
<u>afforestation</u>, not the least because of the effect of trees as a
physical barrier against public access to the water body. The
merits of trees were also taken to include protection against
erosion of hillsides and resultant sedimentation of the reservoirs;
however, very little scientific guidance was available, and the
review was made in advance of hydrological research under British
conditions.

    Individual water supply managers made their own decisions about
afforestation, often constrained by local conditions, especially
land ownership. In the Elan valley for example, trees were planted
in places round reservoir margins whilst the tenant farmers were
prevented from any intensification of their livestock husbandry.
At Lake Vyrnwy, a village settlement was created with a livelihood
based on both farming and timber, profits from both going to the
water undertaking. Elan and Vyrnwy are examples of "wet estate"
which help make the water industry the second largest land-owner

in Britain (Water Space Ammenity Commission,1977), but there are
many catchment areas upon which land-use decision-making is completely outside the control of the water interest. There is no
planning legislation to cover land-use change in the majority of
cases; only proven pollution could lead to control under more
recent legislation (HMSO, 1974).

Demand for water grew rapidly during the economic recovery after
World War II and a new generation of reservoirs was constructed in
the uplands, larger and with a river-regulating rather than direct-
supply role. Such reservoirs are designed to supply lowland conurbations via rivers, flood waters being contained by the dam for
release during periods of low flow, and supply being obtained
through intakes alongside the river. These have tended to be major
schemes, e.g. Llyn Clywedog, Llyn Celyn in Wales, Cow Green and the
recently-opened Kielder Reservoir in northern England. They have
large catchment areas, too large for it to be economically viable
to secure land-use control through water industry ownership. Indeed
pollution prevention, one of the principle reasons for control in
the past, is now less essential because the water taken from
reservoir-regulated surface rivers needs more extensive treatment
for pathogens and chemical contaminants before supply. The attitude
to recreation on and around reservoirs has relaxed somewhat
(Institution of Water Engineers and Scientists, 1972), and water
undertakings who own their catchments are under economic pressure to
raise profits on farming and forestry as well.

THE BEGINNINGS OF RESEARCH

Whilst it was not until the late 1970's that a major national
question arose over afforestation around reservoir catchments, the
pioneering scientific study of land-use and runoff had already
indicated a potentially serious effect of afforestation in the mid-
1950's. Frank Law, Engineer to the Fylde Water Board in Lancashire,
which took supplies from the Stocks Reservoir in the Forest of
Bowland, instrumented a site at the reservoir to measure rainfall
both above and below forest canopies, evaporation and runoff. His
most important results came from a forest lysimeter, a 450 m^2 block
of mature Sitka spruce trees isolated from the rest of the plantation by a concrete wall. Measurements from the lysimeter convinced
Law that complete afforestation of the Stocks catchment would
reduce the reservoir's yield to supply by nearly half. He estimated
the cost of replacing the 'lost' water at nearly an order of magnitude higher than the benefit accruing from letting the land to
forestry (Law, 1956).

Law's work had more effect on the progress of scientific hydrology than on land-use planning in the uplands. One of the major
criticisms of the calculations at Stocks Reservoir was that they

were based upon the lysimeter, a small block of land relative to the
size of the prescriptions which Law made (i.e. the reservoir catchment). It was catchment studies, therefore, which received a major
fillip. The Hydrological Research Unit, set up in 1962, undertook
the provision of firmer data based on the hydrological effects of
both grass-covered and forest-covered upland, and chose the head-
waters of the Severn and Wye in mid-Wales as the location for comparative catchment studies. Because the Plynlimon catchments involved mature forest, the Coalburn catchment was established in the
northern Pennines to investigate the hydrological effects of the
early stages of afforestation (see Robinson, 1980; Robinson and
Blyth, 1982). Paradoxically, whilst the Institute of Hydrology
(which grew from the Hydrological Research Unit) began at the catchment scale of investigation, it has been through forest lysimeter
techniques identical to those of Frank Law that the results from
the Severn and Wye have been confirmed and extrapolated (Calder,
1976).

THE PRESENT DATA BASE FOR ASSESSING WATER INDUSTRY/LAND USE INTER-
ACTION

Results of the study of the Severn and Wye on Plynlimon, in
central Wales, began to emerge by the mid-1970's (Institute of
Hydrology, 1976). It was apparent after careful statistical checks
on the Plynlimon measurements that, although the grassland Wye converted 80 % of the incident rainfall to runoff, the forested parts
of the Severn reduced this runoff coefficient to 60 %. The indic-
ations were that the direction of Law's conclusions were correct;
furthermore, the process of rainfall interception by the forest
canopy, which he had identified as the cause of the reduced runoff,
was confirmed at Plynlimon and its detailed study has been the
focus of a large expansion of the work (Calder, 1977; 1978). It
was by hydrometeorological study of the interception process that
the results at Stocks, Plynlimon and a growing network of other
sites in Britain were found to conform to a general model which can
be used at either a sophisticated level (e.g. Calder, 1979; Gash,
1979), or simple level (Calder and Newson, 1979) to extrapolate
predictions to all upland areas.

The paper by Calder and Newson produced a good deal of controversy, mainly because of confusion over its aims; the predictions
that loss rates (evaporation from catchment rainfall) would almost
double around most existing reservoir catchments were based on a
rate of change from grass to trees suggested as optimal by the
forest management agency (Forestry Commission, 1977). It was correctly suggested that most of the rainfall would still become runoff, particularly in the high rainfall areas of Scotland (Calder
and Newson, 1980). Foresters claimed that there was consequently
very little disbenefit to the water industry in exchange for the

major benefit to the economy which home-grown timber would represent.
However, there were far fewer criticisms that the results were un-
representative than it was Law's misfortune to bear, and the Forestry
Commission has itself recognised the importance of interception
(Binns, 1980).

Scientific results on the water quality implications of land-use
change around reservoirs were also emerging during the 1970's. It
has been mentioned above that there are fewer water quality constraints for newer regulating reservoirs than is the case for direct
supply types; however, the flush of water released into major rivers
by the major upland reservoirs is still crucial in diluting much of
the pollution which is inevitably introduced in the lowland area.
This is particularly true of nutrients which, if allowed to accumulate could cause eutrophication and rapid deterioration of supplies,
whether distributed via pipeline or by a river system. Both agricultural improvements and successful forestry on the impoverished,
leached soils of the British uplands require large additions of
fertiliser, and research is now being done on the water quality
implications of both activities. The present major interest is the
leaching of phosphates from forest plantations; Gibson (1976) and
Harriman (1978) have confirmed the risk of eutrophication in upland
water bodies, phosphorus being the limiting ion for algal growth
under upland conditions. Youngman and Lack (1981) go on to suggest
that 260 reservoirs in the uplands would be at risk of algal
problems in the event of extensive expansion of afforestation; Kay
and McDonald (1981) stress bacteriological indicators of water
quality, their research indicating that in this respect, improved
agriculture is the greater danger. An experimental study of the
nitrate release from improved pastures within the Plynlimon experimental catchments has shown that peak levels of nitrate release
exceed health limits for water consumption after both the culti-
vation/drainage and fertilizer application phases of improvement.
However, there is a rapid return to a new steady state, with only
slight nitrate enrichment of drainage waters. Clearly, much
depends on the regularity with which foresters or farmers will find
it necessary to fertilise the uplands, and the scale on which they
do it relative to the scale of the water catchment.

The water industry interest in the uplands does not end with
those streams which flow into reservoirs and with the expensively-
created body of water behind the dam. Fisheries are of considerable economic importance, and good fishing streams are also healthy
in a general ecological sense, an indication that conservation is a
worthwhile aim for river management. There is increasing evidence
that afforestation can increase sediment yields in upland streams
(see Newson, 1980; Robinson and Blyth, 1982), which may affect the
spawning of salmonid fish; this is caused by drainage ditching, so
it may also be a feature of upland agriculture. Coniferous forests
also appear to exacerbate the acidity of streams through the impac-

tion of dry or aerosol pollutants on the canopy, the interception
process, and nutrient uptake from the soil. Fish mortality has
increased in acid rainfall areas such as southern Scotland, but
even the Plynlimon catchments, which are remote from polluted rainfall, show reduced temperatures, pH and fish numbers under forest
cover (Crisp, Cubby and Robson, 1980). Mills (1980) has produced
a forester's guide for the management of the physical environment
for fish in forest streams, but data-collection is still in progress
on the stream chemical environment and its effect on fish, and there
is still a great deal of controversy on this issue.

TOWARDS UPLAND LAND-USE PLANNING IN BRITAIN

   Historical precedence has already been dismissed as a facile claim
in justification against the development of other resources proven to
be disadvantageous to the water industry. The quantitative findings
of scientific research would, in a perfect world, be the basis for
objective socio-economic analysis to define balances, for example
between forestry and farming on water catchment areas, with local
analysis of land capability defining where each use of land would
be optimum (e.g. Maxwell, Sibbald and Eadie, 1979). We are at
present, however, far from that position in Britain, although there
are clear signs of a marshalling of facts and figures on behalf of
each of the major upland resources, as well as improvements in
mapping.

   After the publication of the Plynlimon results and their extra-
polation (Calder and Newson, 1979), the water industry understandably
took a defensive attitude to the apparent long-term threat to their
expensively created upland water impoundments. During the preparation of a forest strategy for the United Kingdom (Centre for
Agricultural Strategy, 1980) the central coordinating machinery of
the water industry delimited areas over which the existence of
water catchments should act as a constraint to forest expansion
(Figure 2). There was also an immediate expansion of research by
both foresters and hydrologists into all aspects of the subject
necessary to transform the findings into public policy; most of the
research is continuing (see Figure 3 for its location).

   However, one aspect which has not yet attracted the attention it
deserves is the relationship between silvicultural techniques and
hydrological effects of afforestation. While results at Plynlimon
and elsewhere have shown little variation in the amount lost by
interception between conifer species, techniques such as wider
spacing of trees might bring about both higher timber and higher
water yields (Last, 1981). Even though the financial turnover of
the uplands is but a small proportion of the national budget, the
extensive area of land involved demands a better solution to the
decision-making problem than the present one.

Figure 2:   Water management constraints on the exploitation of land
            suitable for afforestation, after the Centre for
            Agricultural Strategy's evaluation (1980).

Hydrological data are required to :

(i)   quantify the rate of evapotranspiration and interception loss
over the time periods to which the water industry works, rather
than those convenient in the analysis of scientific data (Calder,
Newson and Walsh, 1982). Most reservoirs are designed to yield a
supply over a critical period, which may vary from one summer for
smaller schemes to several years for larger ones; annual or seasonal results, such as those from Plynlimon, need to be broken down
into monthly, weekly or even daily resource management guidelines.
The two processes must also be quantified for significant droughts
and the periods either side of them when water storage is at its
most essential;

(ii) compare loss rates between forestry and other crops. Water
catchment areas are not sharply divided into grass or tree cover;

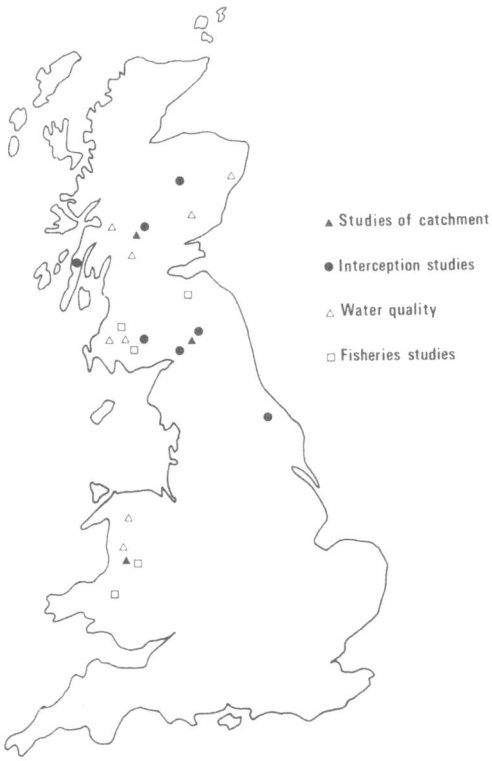

Figure 3: Current major locations of research activity designed
           to refine calibration of the land-use/water management
           interaction in the British Uplands.

some have bracken, heather or other crops of intermediate height.
These plants are now receiving the hydrological attention that
their areal extent (especially in Scotland) deserves;

(iii) evaluate the amount of interception loss in areas such as
Scotland where significant amounts of winter precipitation are in
the form of snow;

(iv) calibrate the significance of water quality changes that
occur as the result of land-use change against the location and
rate of the change. For example, reasons of land tenure and the
uptake of innovation indicate that the main interest in Wales will
be in improved pasture, but in Scotland it will be afforestation.

     These hydrological data will be of little value without the
basic infrastructure for land planning, particularly readily available data and information on current land-use and the potential for

development. Here again there are clear signs of progress: the
Forestry Commission has mapped areas of Britain suitable for
forestry expansion, based upon climate and soil characteristics
(see Locke, 1980); The Ministry of Agriculture, Fisheries and Food
(MAFF) has also revised an earlier land classification to differen-
tiate in some detail between uplands that are improvable and those
that are not (MAFF, 1980). The conservation interest is now required
to specifically identify those sites which are of special interest,
and one benefit of the high levels of youth unemployment has been a
marked increase in the rate at which biological surveys are being
done in Britain. Recreation interests are being served by the
recent structure-planning exercises of local government, which map
preferred areas for recreational development. As recreation generally requires building development, it is controlled by planning,
whereas land-use change is not.

Even these few single-interest statements of upland capability
have led to a significant expansion in the amount of physical
planning information held on maps. There are still relatively few
examples of computer held information on land-use and land-potential,
but the uplands have been the subject of two major exercises by the
Institute of Terrestrial Ecology in Britain (Ball, Radford &
Williams, 1983; Bunce, Barr & Whittaker, 1981). Over the next
five years, the results of what may be termed the second phase of
hydrological research will be available to all interests and will
hopefully permit much more sophisticated prediction than that of
Calder and Newson (1979), incorporating water quality as well as
quantity information. This should allow the water interests to
take a more flexible approach to afforestation and other land-use
changes. Indeed, both the water industry and the urban dweller
will be better served by any moves that put rural resource planning
on a more objective base, thus ending the era of the single-resource
agency approach (Countryside Review Committee, 1976). The data
banks of agencies like the Institute of Terrestrial Ecology can
store, at a national scale, the research results and other information that must form the basis for sound political decision making.

IS THE BRITISH UPLANDS CASE UNIQUE ?

As mentioned earlier, visitors from abroad are often surprised
at the lack of a planning framework for land use change in rural
Britain. Many are unaccustomed to single-agency resource development, especially where their own history of rural development is
relatively short. Since British research work is more in the form
of impact assessment on changes occasioned by market forces than
strategic resource evaluation, it may also be unique.

In the rest of the world there are bound to be specific solutions
to the problem of water management in the face of land use change;

in many countries the major impact is often timber harvesting from
natural forest ecosystems. Bosch and Hewlett (1982) have examined
the results of 94 catchment experiments from the U.S.A., Canada,
East Africa, South Africa, Australia, New Zealand and Japan. Most
of the situations involved deafforestation, and the authors generalize that for every 10 % reduction in cover, streamflow increases by
~40 mm yr^{-1} (conifers and eucalypts), 25 mm yr^{-1} (deciduous hard-
woods) and 10 mm yr^{-1} (brush, scrub).

However, the major concern amongst world resource experts at
present is probably the exploitation of tropical timber resources.
Here the interception process, the main explanation for the results
catalogued by Bosch and Hewlett for temperate and dry climates, is
already under investigation. Interception losses of between 3 % and
38 % have been measured (UNESCO, 1978), and recent use of British
instrumentation in West Java linked the 12 % interception rate
(compared with ~22 % in the British uplands) to the very high rainfall intensities of the tropics (Institute of Hydrology, 1982).
Nevertheless, transpiration rates by such a dense and deep vegetation
system are much higher than in the rest of the world, and timber
exploitation leads directly to increased runoff. More important to
the economies of the tropical world than the change in runoff is the
resulting soil erosion and sedimentation, which drastically reduces
the opportunity to grow commercial timber or improve agricultural
settlements. There are also major climatic implications of an
altered water balance in this "engine room" of world weather.

CONCLUSIONS

This Seminar has the theme of utilising the full potential of
land whilst balancing the environmental constraints. Questions may
well arise as to whether we have enough environmental information
and, if so, how can it be used in decision-making ? The study of
the hydrological effects of land-use change in the British uplands
has revealed some general principles which may be of benefit to
those attempting to answer these questions.

Firstly, it indicates the value of strategic science. Hydrology
needs long data runs and information on extremes. Although the work
of Law (1956) bespoke the direction of the research at Plynlimon it
was conducted for at least a decade, diversifying meanwhile, until
the need arose for a hydrological reaction to forest expansion.
Financial resources for research are now much less plentiful and
research which answers today's problems rather than anticipating
tomorrow's is becoming easier to justify - a dangerous position.

Secondly, the hydrological study of upland land-use change has
indicated that the scientist must anticipate the needs of the
decision-maker, even down to the level of units of measurement;

water engineers in Britain found interpretation of research hydrological data difficult until a dialogue was established and there are
still points of misunderstanding.

Thirdly, the results of environmental appraisal need a formal
relationship with the structures of decision-making. As indicated
above, it is dubious whether a structure exists in the case of the
British uplands but even single resource agencies need a better
service from environmental science than just a pile of scientific
publications or a bill to conduct more work !

It is perhaps not surprising that in the House of Commons on 21st
March 1980, a Government spokesman said ... "As regards afforestation,
its percentage and its effect on catchment areas, I have received
considerable correspondence from certain scientific sources .....
I am advised that there is a lack of clear scientific evidence".
Such is the political filter between science and decision-making !

REFERENCES

Ball, D.F., Radford, G.L. and Williams, W.M. 1983, A land charact-
    eristic data bank for Great Britain, Institute of Terrestrial
    Ecology (I.T.E.), Bangor Occ. Paper 13, I.T.E., Bangor Research
    Station.
Binnie, A.M., 1981, "Early Victorian Water Engineers", Thomas
    Telford, London.
Binns, W.O., 1980, "Trees and Water", Arboricultural Leaflet 6,
    HMSO, London.
Bosch, J.M. and Hewlett, J.D., 1982, A review of catchment experi-
    ments to determine the effect of vegetation changes on water
    yield and evapotranspiration, J. Hydrol., 55: 3.
Bunce, R.G.H., Barr, C.J. and Whittaker, H.A., 1981, Land classes
    in Great Britain. Merlewood Research and Development Paper, 86,
    Institute of Terrestrial Ecology, Merlewood Research Station,
    Grange-over-Sands.
Calder, I.R., 1976, The measurement of water losses from a forested
    area using a natural lysimeter. J. Hydrol., 30:311.
Calder, I.R., 1977, A model of transpiration and interception loss
    from a spruce forest in Plynlimon, Central Wales, J. Hydrol.33:
    247.
Calder, I.R., 1978, Transpiration observations from a spruce forest
    and comparisons from an evaporation model. J.Hydrol.38: 33.
Calder, I.R., 1979, Do trees use more water than grass ? Water
    Services, 83: 11.
Calder, I.R. and Newson, M.D., 1979., Land use and upland water
    resources in Britain - a strategic look. Water Resources
    Bulletin, 15:1628.
Calder, I.R., 1980, The effects of afforestation on water resources
    in Scotland, in: "Land Assessment in Scotland," M.F.Thomas and
    J.F.Coppock, eds., Aberdeen University Press, Aberdeen.

Calder, I.R., Newson, M.D. and Walsh, P.D., 1982, The application
     of catchment, lysimeter and hydrometeorological studies of conif-
     erous afforestation in Britain to land-use planning and water
     management. Proc. Symp. Hydrolog. Research Basins, SonderL.
     Landeshydrologie, Bern, 3:853.
Centre for Agricultural Strategy, 1980, "Strategy for the U.K.
     forest industry", Centre for Agricultural Strategy, Report 6,
     Reading.
Countryside Review Committee, 1976, "The countryside - problems and
     policies", HMSO, London.
Crisp, D.T., Cubby, P.R. and Robson, S. 1980. A survey of fish pop-
     ulations in the streams of the Plynlimon experimental catchments.
     Freshwater Biological Association, Teesdale Unit, Report 3.
     Freshwater Biological Association, Windermere Laboratory,
     Ambleside.
Coleman, A.M., 1970, The conservation of wildscape: a quest for
     facts, Geogr. J., 136:199.
Davidson, J. and Wibberley, G., 1977, "Planning and the rural
     environment". Pergamon Press, Oxford.
European Community, 1975, "Mountain and hill farming and farming
     in certain less-favoured areas", European Community, Brussels.
Forestry Commission, 1977, "The wood production outlook in Britain",
     Forestry Commission, Edinburgh.
Gash, J.H.C., 1979, An analytical model of rainfall interception
     by forests. Quart. J. Roy. Met. Soc., 105: 43.
Gibson, C.E., 1976, An investigation into the effects of forestry
     plantations on the water quality of upland reservoirs in Northern
     Ireland, Water Research, 10:995.
H.M.S.O., 1974, "Control of Pollution Act 1974", H.M.S.O., London.
Institute of Hydrology, 1976, Water balance of the headwater catch-
     ments of the Wye and Severn, Institute of Hydrology Report 33,
     Institute of Hydrology, Wallingford.
Institute of Hydrology, 1982, Research Report 1978-81, Institute
     of Hydrology, Wallingford.
Institution of Water Engineers and Scientists, 1972, "Recreation
     on reservoirs and rivers", Institution of Water Engineers and
     Scientists, London.
Kay, D. and McDonald, A., 1981, A comment on : New problems with
     upland waters by Youngman, R.E., and Lack, T., 1981, Water
     Services, 85:13; Water Sciences, 86:29.
Law, F., 1956, The effect of afforestation upon the water yield of
     catchment areas, Jnl. Br. Waterworks Assn., 38:489.
Last, F.T., 1981, Land use, forest policy and tree biology.
     Biologist, 28:280.
Locke, G.M.L., 1980, Land assessment for forests, in: "Land
     assessment in Scotland", M.F.Thomas and J.T.Coppock, eds.,
     Aberdeen University Press, Aberdeen.
Maxwell, T.J., Sibbald, A.R. and Eadie, J., 1979, Integration of
     forestry and agriculture - a model. Agricultural Systems 4:161.
Mills, D.H., 1980, "The management of forest streams", Forestry
     Commission Leaflet 78, H.M.S.O., London.

Ministry of Agriculture, Fisheries and Food (MAFF), 1980, The classification of land in the hills and uplands of England and Wales, MAFF Booklet No. 2358, MAFF, London.

Ministry of Health, 1948, "Gathering grounds. Public access to gathering grounds, afforestation and agriculture on gathering grounds". H.M.S.O., London.

Newson, M.D., 1980, The erosion of drainage ditches and its effect on bed-load yields in mid-Wales: reconnaissance case studies, Earth Surface Processes, 5:275.

Parry, M.L., 1978, "Climatic change, agriculture and settlements", Dawson, Folkestone.

Parry, M.L., Bruce, A. and Harkness, C.E., 1982, "Moorland Change Project", Department of Geography, University of Birmingham.

Robinson, M., 1980, "The effect of pre-afforestation drainage on the streamflow and water quality of a small upland catchment". Institute of Hydrology, Report 73, Institute of Hydrology, Wallingford.

Robinson, M. and Blyth, K., 1982, The effect of forestry drainage operations on upland sediment yields: a case study, Earth Surface Processes and Landforms, 7:85.

Shoard, M., 1980, "The theft of the countryside", Temple Smith, London.

Stamp, L.D., 1948, "The land of Britain, its use and misuse", Longmans, London.

UNESCO, 1978, "Tropical forest ecosystems", UNESCO, Paris.

Water Space Amenity Commission, 1977, "The recreational use of water supply reservoirs in England and Wales", London.

Williams, H.J. and Harding, D., 1982, Towards a land use strategy for the uplands of Wales. Q.J.Forestry, 76:7.

Youngman, R.E. and Lack, T., 1981, New problems with upland waters, Water Services, 85:13.

12

SOME EXAMPLES OF LAND AND WATER USE PLANNING IN BRITISH COLUMBIA,

CANADA

J. O'Riordan

Planning and Assessment Branch, Ministry of Environment

Victoria, British Columbia, Canada

INTRODUCTION

Although water and land uses are closely inter-related, few administrations have managed to co-ordinate them effectively into a single planning process. As yet, terrestrial and aquatic biophysical classification systems have not been fully integrated, though moves in this direction are now being made in Canada through the ecological systems mapping programme (Welch, 1978). But the main constraint on effective land/water use management is the continued separation of government agencies responsible for respective components of this integrated system. In Canada, the province of British Columbia is no exception to this trait, in that its Ministry of Environment is responsible for water use, allocation and management, while the Ministries of Forests; Lands, Parks and Housing; Agriculture and Food; and Energy, Mines and Petroleum Resources are responsible for various land use activities on Crown Lands throughout the Province.

However, the scope of land and water relationships in resource planning is so diverse that any single Ministry that attempts to encompass all aspects of the system would become too large to be administered effectively. An alternative approach has been tried in British Columbia through the creation of the Environment and Land Use Committee of Cabinet. This Committee, chaired by the Minister of Environment, comprises eight Ministers, responsible respectively for both the resource development (highways, industrial development, energy, mines, settlements) and resource conservation sectors (agriculture, parks, fish, wildlife and water resources). It is supported by a Deputy Ministers'* Technical Committee which advises

---

*In both federal and provincial governments in Canada the deputy minister is the permanent civil service head of a department, not an elected official.

it on all policy issues affecting natural resource planning and
development. Through this Technical Committee land and water management programmes are co-ordinated to a degree in the province
(O'Riordan, 1981).

    This chapter uses case examples to describe some aspects of land
and water use planning now being developed in the province. Obviously
it is not possible to tackle the issues comprehensively in a short
text; the intent is to sample a few of the problems, indicate some
of the solutions being considered, and to introduce the reader to the
broader concepts of strategic land and water use planning.

    The British Columbian Ministry of Environment and some of its
companion resource Ministries such as Forestry have recently developed a strategic planning process. This is typically a hierarchical
system. Planning at the regional scale with long time horizons
(such as structure planning in Britain) provides a context for more
detailed planning for smaller units with shorter time horizons. In
theory the planning process should move down through the hierarchy
from large areas to small areas although this is not so easy in
practice because information at the regional scale is often poorly
organized. The process should also be systematic, considering (i)
demands for supplies of resources both now and in the future, (ii)
the values placed on these by society, (iii) alternative management
measures to attain specified objectives, (iv) selection of preferred
measures and policies according to approved criteria, and finally
(v) monitoring results in preparation for the next round of planning.

    If all resource Ministers develop their draft strategic plans
and sufficient knowledge exists concerning the functional relationships between resource uses (in this case land and water uses), then
explicit trade-offs can be made to produce an overall balance in
resource allocation within acceptable constraints. But again practice
differs from theory, not all resource strategies have been devised
and, more importantly, the interrelationship between resource uses
are still poorly understood. In such circumstances, it is wise to
act conservatively to avoid irreversible changes. Better still, it
is important to create a 'learning environment' through which knowledge of the relationships between natural systems and man's valua-
tion of resource use can be improved by careful monitoring. Such a
process has yet to be fully implemented in British Columbia, and
indeed in most jurisdictions.

    Three case examples are briefly considered. In the first,
limited water availability and conflicts in its use affect agricultural land development. In the second, rapid urbanization in a
small watershed is increasing flood frequency, resulting in losses
in agricultural production. In the third, logging in small watersheds is affecting water quality for domestic use. The solutions
for all three of these problems lie in applying biophysical analyses

in the context of existing socio-economic systems: no planning system
can be effective without such an integration of these values (Hasler,
1975).

THE NICOLA BASIN

   The Nicola basin illustrates the situation where water availability
affects land use, and also where land use affects water quality.
Situated in the dry southern interior of the province, the basin
covers 7280 square kilometers, mainly rolling plateaus dissected by
small river valleys (Figure 1). With elevations ranging from 180 m
in the valleys to over 2300 m in the Coast Ranges that represent the
western boundaries to the basin, precipitation varies from around
25 cm to 50 cm. This moderate precipitation in the higher elevations
together with winter snow accumulation provides a significant freshet
in the spring months (May, June).

   The main water/land use issue is the conflict between maintaining
water in-stream to support migrating and spawning pacific salmon and
trout and the need to divert water from these same streams to irrigate
approximately 12 000 hectares of pasture lands. Irrigated hay crops
are vital to the major cattle ranching operations in the basin, providing forage during the six-month winters. Although ranchers have
developed approximately 57 000 dam^3 of licenced storage to support
their irrigation systems, in dry years (which occur on average twice
a decade) river flows in some watersheds fall below the levels required to permit migrating salmon to reach their spawning beds. This
event occurs between August and October, at a time when irrigation
water is critical for the second crop of hay.

   The values of both water uses are high. If water is not available
for forage, ranchers must import hay from the United States at a
current price of $40 per tonne. The anadromous fishery contains
significant populations of coho, pink and chinook salmon and steelhead trout. The current annual gross value of these fish in the
commercial and recreational fishery on the coast is estimated at
$7.0 million. Based on habitat surveys, salmon production could be
doubled, and steelhead production quadrupled, for a total value of
$12.1 million annually provided that in-stream flows and water quality
are protected, and a fish stocking programme undertaken.

   One purpose of the Ministry of Environment's strategic plan was
to determine the best mix of in-stream and diversion uses for each
of the basin's nine watersheds (Figure 2). It was readily apparent
that it would not be economically efficient to develop additional
storage in each of these watersheds to meet all the potential demands
for fisheries and irrigation. Some trade-offs would have to be made
whereby fishery flows were protected in the most biologically productive of the watersheds, while irrigation use was given more

Figure 1:   The physical features of the Nicola Basin, British
            Columbia, Canada.

emphasis in the less productive watersheds.   In order to develop
such a designation of priority water uses, the following general
objectives were developed:

(i)   Water supplies for domestic and industrial use must be avail-
         able from surface and groundwater sources to meet licenced
         requirements at all times.
(ii)   Water supplies for irrigation use must be available at least
         four years out of five, and 50 per cent of licensed requirements
         must be available in drought years.   This security of supply
         is required to keep the average rancher economically viable.

Figure 2:   The major watersheds of the Nicola Basin, British
            Columbia, Canada.

(iii)   Fishery resource maintenance flows must be available in
        average runoff conditions (return period one in 2.3 years)
        in unregulated or partly regulated streams. This means that
        in wet years, higher flows will be available and production
        levels increase, while in drier years, flows may fall below
        the desired levels and production will decrease accordingly.
        Over time, these production gains and losses will offset
        each other to sustain a viable fishery.
(iv)   Ambient water quality in surface and groundwater sources must
        be at or above levels established for designated uses, such as
        fisheries, irrigation, domestic or recreational use.

Based on these objectives, it has been possible to analyse current and future supply/demand relationships for both in-stream and diversion uses in each sub-basin. They basically fall into three groupings.

(i)   Middle and Lower Nicola,   - all current licenced diversions for
      Coldwater and Spius   irrigation and domestic use meet
                                     objectives.
                                   - fishery maintenance flows meet objec-
                                     tives.
                                   - unlicenced surplus water is available
                                     for storage to support agricultural
                                     expansion.

(ii) Upper Nicola, Guichon,   - all current licenced diversion uses
      Quilchena   meet objectives.
                                   - fishery maintenance flows are met
                                     only in wet years, with significant
                                     loss of production in the Upper
                                     Nicola and Guichon Creeks in dry
                                     years.
                                   - unlicenced surplus water is avail-
                                     able for storage to support addit-
                                     ional diversion uses or in-stream
                                     flows.

(iii) Clapperton, Moore-Stump   - current licenced irrigation diver-
                                     sion needs are not met in average
                                     and dry runoff conditions.
                                   - fishery maintenance flows are not
                                     met at all.
                                   - little or no unlicenced water is
                                     available for storage.

At the strategic (i.e.sub-regional) level of water and land use planning, it is sufficient to determine priority water uses by sub-basin, and identify the most effective management measures required to achieve these resource objectives. Water use priorities were designated for each sub-basin according to the following criteria:

(i)   All existing licenced beneficial uses must be protected;
      thus licenced users are guaranteed their current water require-
      ments.
(ii) Non-licenced (i.e. surplus) water is to be allocated to in-
      stream (fishery) and/or diversion use based on the present and
      future values of these uses.
(iii) Beneficiaries must be willing and able to pay their share of
      capital and operating costs for measures required to increase
      supplies. This criterion often acts as a major constraint for
      even though net benefits are available (criterion (ii) above),
      users (ranchers) or agencies responsible for fisheries (Federal
      Ministry of Fisheries and Oceans; B.C.Ministry of Environment)
      often cannot supply the necessary capital.

(iv)   Public preference - especially where public goods, such as
       sport fisheries, water-based recreation or aesthetics are
       important.

The main management options were focussed upon developing more
storage, making more efficient use of existing licenced water by
changing from ditch irrigation, having large losses due to evaporation
and seepage, to sprinkler systems, or moving diversion points from
tributaries to Nicola Lake and Nicola River downstream from the lake.
In this latter case, costs of irrigation would increase because
pumping would be required to replace gravity flow systems in tributary streams. However, fisheries would benefit from increased in-
stream flows in the tributaries, and in some instances it will be less
expensive for the Ministries responsible for fishery management to
pay the pumping costs than to develop additional storage.

Before a final set of water use priorities could be established,
the question of water quality had to be addressed. There is little
point in designating sub-basins for fisheries and providing the necessary flows, only to find that ambient water quality is below that
required to sustain the anadromous fishery. Here again, there is a
close link between land and water use. One of the main sources of
wastes found in surface water-courses comes from agricultural operations themselves. Sediments, nutrients and pesticides are significant non-point sources of pollution stemming from the ranching
activities in the Nicola basin.

Cattle feeding in or near ditches and water-courses during winter
disturb soil and release large quantities of nutrients to surface
waters (Environmental Protection Agency, 1977). Pesticides and some
fertilizers applied to forage and horticultural crops also find their
way into streams, are assimilated by aquatic biota, and eventually
concentrate in fish that feed on these biota. Sediments can degrade
fish spawning areas by smothering eggs while nutrients promote algal
growth reducing light penetration and thus biological productivity.
Phosphorus from land sources (often a key nutrient for promoting
algae when nitrogen is already abundant from atmospheric sources)
enters water courses in association with eroding sediment (Likens
and Borman, 1974).

Thus there is a direct linkage between water withdrawn for irrigation and the quality of that same water. The more hectares irrigated
and the greater the cattle population, the greater the discharge of
nutrients back to the surface waters. At present, Nicola Lake, which
is fed by three sub-basin streams flowing through cattle country
(Upper Nicola, Quilchena and Moore-Stump - see Figure 2) is eutrophic,
experiencing large scale blooms of blue-green algae in the summer
months. These detract from its aesthetic appeal for water-based recreation and, if continued unchecked, could extract sufficient oxygen
from the surface waters to affect the fish resource. At present

Nicola Lake not only contains the anadromous salmon and steelhead
mentioned earlier, but also rainbow trout and kokanee (land-locked
salmon) which support a large recreational fishery (approximately
30 000 angling days annually). Several of the streams in the Nicola
Basin also have high levels of nutrient and coliforms, potentially
affecting both the fishery and domestic water uses.

Control of non-point waste discharges from land to water courses
is difficult to implement for a number of reasons. First, in many
jurisdictions (and British Columbia is certainly one), there is little
or no legal authority to institute control measures. Second, sources
are difficult to pinpoint, so that direct action cannot readily be
undertaken. Third, in the case of nutrients and even toxic materials
from urban or agricultural runoff, there is limited understanding of
the precise relationships between the pollutants, resultant ambient
water quality, and impacts on water uses (Environmental Protection
Agency, 1980b). For example, there are no ambient water quality
objectives for nutrient concentrations in surface waters for fish
in Canada.

Consequently, controls have to be implemented incrementally in
the Nicola, starting with the most obvious problem areas pinpointed
through a stream water quality monitoring programme. The main
practices for reducing nutrient inflows include :

  - establishing and protecting stream bank vegetation buffer strips;
  - fencing to control livestock access to watercourses;
  - constructing manure containment facilities;
  - control of fertilization techniques.

With little or no regulatory and enforcement powers, environmental protection agencies must convince the rancher himself to
exercise good farming practices. This can best be achieved through
peer pressure from the agricultural community setting their own
standards, and through limited financial incentives. In British
Columbia, the Provincial Federation of Agriculture (a coalition of
farmers) has established a programme of agricultural environmental
control, with guidelines of good practice in controlling all types
of waste discharge to the environment.

As a result of the analysis of water supplies for alternative
uses and water quality objectives, to the extent they exist in
British Columbia, the Ministry of Environment developed the water
use designations and management priorities for each sub-basin, as
outlined in Table 1. An overall balance between fishery and agricultural development was struck, with agricultural expansion limited
in some watersheds (Guichon, Coldwater) to protect fisheries, and
potentially increased in others (Quilchena, Clapperton) where fishery
values are moderate to low (British Columbia Ministry of Environment,
1982).

Table 1: Summary of a water management strategy for Nicola Basin..

WATERSHED	PROPOSED PRIORITY USE DESIGNATION(S)	MANAGEMENT MEASURES	COMMENTS
Upper Nicola	Fisheries and Irrigation	˙Douglas Lake storage ˙Improve irrigation use efficiencies	Tie-in with Nicola River Operational Plan
Mainstem Nicola	Fisheries and Irrigation	˙Decide on Nicola Lake dam ˙Improve irrigation efficiencies ˙Monitor Nicola Lake outlet	Operational Plan to be completed by December 1983
Coldwater	Fisheries	˙Storage reconnaissance ˙Irrigation divisions supported by storage	
Spius	Fisheries	˙Any additional irrigation supported by storage	
Guichon	Irrigation and Fisheries	˙Storage development in Mamit ˙Change diversion points to Nicola River ˙Monitoring of Mamit Lake inflow	Operational Plan to by completed by 1984
Quilchena	Irrigation	˙Storage reconnaissance ˙Additional monitoring	To be completed by 1985
Moore-Stump	Irrigation	˙Storage reconnaissance ˙Additional monitoring ˙Moore Creek diversion to Stump Lake	To be completed by 1986
Clapperton	Irrigation	˙Change diversion points to Nicola River	Evaluated by December 1983

These land/water uses are not always in conflict. In the main-stem Nicola, Upper Nicola and possibly on Guichon Creek, potential storage developments are economically viable only because both agricultural and fishery interests benefit and can share costs. Single purpose storages, unless very small, are simply not economical in today's restrained economic conditions.

THE SERPENTINE-NICOMEKL BASIN

The Serpentine-Nicomekl basin, located just south of Greater Vancouver, illustrates another type of land/water resource management problem - drainage control in an urban setting (Figure 3). The 4800-hectare flood plain of the Serpentine-Nicomekl rivers has some of the most productive farmland in the province. Even with its present flooding and drainage problems, the basin supplies approxim-

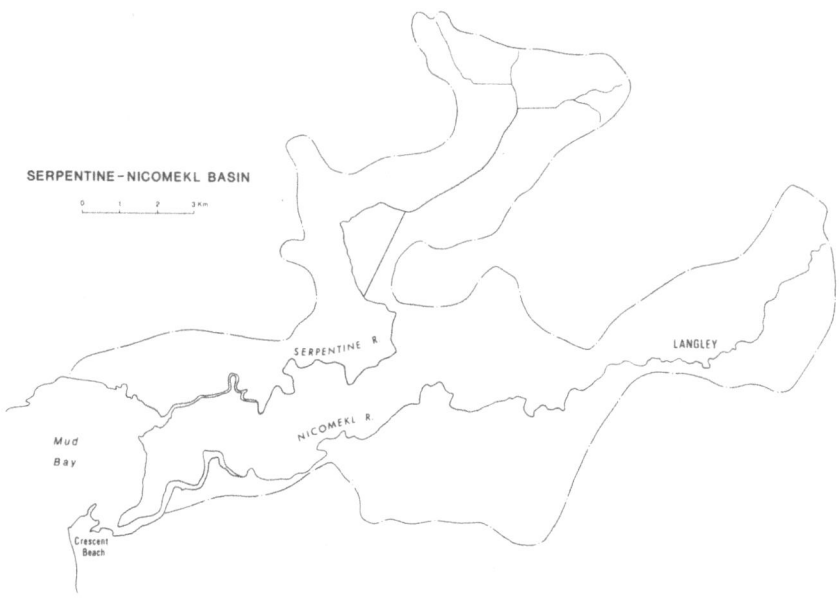

Figure 3:   The Serpentine-Nicomekl basin, British Columbia, Canada.

ately half the horticultural crops consumed by the Greater Vancouver
market (1981 population 1.3 million) and a significant portion of
its dairy produce. However, some 26 400 hectares of low uplands that
surround this flat floodplain are experiencing rapid urbanization as
part of the growth of the metropolitan region. This urban development on non-agricultural land is encouraged by provincial policy,
which protects all highly productive agricultural lands from residential subdivision.

   Urban development increases flooding and drainage problems on
the flood-plain in a number of ways. First, storm drainage systems
- curbs, gutters, catchbasins and pipes - are installed, with subdivisions. These drainage systems discharge directly into natural
watercourses which flow into the Serpentine and Nicomekl rivers.
Thus the quality of water is degraded, peak flows are increased, and
because water is removed more rapidly, there is less opportunity for
ground infiltration and groundwater recharge. Furthermore, urban
developments often encroach onto natural stream courses with dumping
of fill, garden wastes and increased deposition of sediments. This
reduces the capacity of these tributaries to 'store' water, thus
increasing discharge onto the lowlands.

   As a result of this urbanization, the lowlands experience frequent
flooding during winter and spring storms. In this part of the province, annual precipitation averages around 100 cm, but over 70 per

cent falls between October and March. Winter flooding can damage
roots of pasture perennials, while late spring flooding delays
planting of horticultural crops and thus their market availability
in summer, when there is intense competition from vegetables produced
in Washington State, just south of the Canadian border.

Figures 4 and 5 illustrate the implications of urban development
on storm runoff for one small sub-basin on the south side of the
Nicomekl River (see Figure 3). This sub-basin is by no means the most
heavily urbanized in the area, for at least 50 per cent remains covered by natural vegetation. However, it is assumed that the present
level of urban development will approximately double in the future.
The change in runoff conditions caused by this increased development
is illustrated for two recorded storms. One dumped 204 mm in 56 hours
during January 18-20, 1968 and the other provided almost 80 mm in 56
hours during April 4-6, 1970. The U.S. Soil Conservation Service
method for predicting the volume of runoff and the unit hydrograph
method of estimating peak flows were used in the analysis (Wilson,
1982).

If the January, 1968 storm was to reoccur, with increased levels
of urbanisation, it is estimated that the level of saturation of soil
would be higher and the time of concentration (period of time for
precipitation falling at the top of a watershed to move down the main
channel) reduced from 1.44 hours to 0.84 hours. The effect would be
to increase peak flow by about 17 per cent, and the total volume of
runoff by 25 per cent. If the April 1970 storm was to be repeated
with increased urban development, the impacts would be more dramatic.
Peak flows would be increased by 108 per cent and volume of runoff by
about 183 per cent. The main cause of this change is that in the less
urbanized watershed, soil moisture levels prior to the storm would be
considerably drier than under a greater density of subdivisions. Thus
for farmers in the Serpentine-Nicomekl basin the effects of urbanization on storm runoff will be more significant during the growing
season than in winter.

The costs of such flooding can be very large. A storm in July
1972 which deposited 92 mm in 30 hours resulted in the provincial
government paying over $400 000 to farmers from its disaster relief
fund. But the costs of protecting against such flooding using conventional methods of dyking the main rivers, pumping and drainage
controls are much larger. In 1976, it was estimated that a full-
scale drainage and flood control scheme for the basin would cost over
$32 million in capital works, far beyond the financial abilities of
the farmers, even with 75 per cent subsidies from the federal and
provincial levels of government.

The solution to this problem requires much more imagination.
One facet is to design storm water detention measures into subdivisions, for example small catchment basins in car parking lots

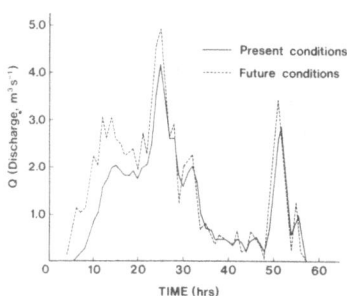

Figure 4:   The probable effect of increased urban development on the
            level of soil saturation and total water volume due to
            upland run-off in the Serpentine-Nicomekl Basin following
            a winter storm.

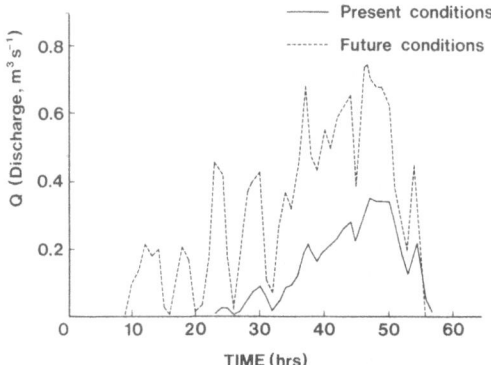

Figure 5:   The probable effect of increased urban development on the
            level of soil saturation and total water volume due to
            upland run-off in the Serpentine-Nicomekl Basin following
            a spring storm.

and small natural depressions, together with better designs of roofs
and gutters that decrease the rate of runoff. Another is to plant
stream banks of tributaries with vegetation so as to increase infiltration and retention time. But such measures cannot alone control
increased discharges: some dyking and pumping schemes will be required to remove water, especially from spring and summer storms.

   The key to this problem is to reduce or eliminate pasture lands
in the floodplain. Present drainage designs aim to reduce standing
water in the winter months to a maximum of 5 days, a criterion
necessary to protect roots from becoming waterlogged. However, winter
flooding is not a problem for annual horticultural crops, because of
the reduction in storm intensity after the growing season commences
(March) the design of flood control works can be less expensive.

Table 2 summarizes daily runoff from storms with a 1 in 10 average return period* in the same small sub-basin that was discussed earlier. If water has to be removed within 5 days in the winter months, an immediate pumping capacity of 2.9 cubic metres per second (cms^{-1}) is required now, rising to 3.4 cms^{-1} with anticipated urban development. These figures drop to 1.3 and 2.2 cms^{-1} respectively, if the design criterion is relaxed to removing water from spring and summer storms within 2 days. This latter criterion is sufficient to encourage economic development of horticultural crops on the Serpentine-Nicomekl.

Table 2: Estimated pumping capacity required to remove storm water within 5 to 2 days during winter, and 2 days in spring/summer, based on present and future upland urban development and drainage criteria. Old Logging Ditch watershed, Serpentine-Nicomekl Basin (Wilson, 1982).

No. of Days of Standing Water	Season	Pump Capacity Present Development (Cubic metres per second)	Future Development
5	Nov. 1 - Mar. 1	2.9	3.4
2	Nov. 1 - Mar. 1	4.2	5.3
2	Mar. 1 - Nov. 1	1.3	2.2

Since pump stations cost $150 000 and $250 000 per cms^{-1} of capacity and are expensive to operate, the economic viability of flood control increases if trade-offs are made in favour of reducing winter forage, and accepting a reasonable level of flood control during the growing season. Additional measures could include the development of natural pondage areas in some parts of the basin so that flooding would occur in major storms, but the water would eventually drain into the rivers. Buildings in such areas would have to be flood proofed by raising them onto pads at least 1.2 m above the surrounding floodplain.

An important component of any solution is to secure an equitable level of cost sharing for both the pumping schemes and the dyking of pondage areas. In the past, few improvements have been undertaken because the farmers on the floodplain were required to pay all of the local government share of capital costs (25 per cent). However,

*A storm of designated total precipitation is expected to occur on average once a decade, or five times over a 50 year period.

municipalities involved in urban development have recently recognized
their responsibilities for the problems and have devised a more
equitable system of cost-sharing. Essentially, they use their property tax base to raise the share of local capital and all operating
costs, and then develop an agreement with farmers to impose a drainage improvement tax to repay some of these costs. In the sub-basin
described, this tax was less than $20 per hectare, well within the
financial resources of the farmers.

THE SLOCAN VALLEY

The third example of co-ordinated land use and water management
involves a frequent concern in British Columbia - the effects of
logging on water quantity and quality. There is a large literature
on forestry hydrology based on numerous case examples in North
America and Europe (Dorcey et al, 1981; E.P.A., 1980a). Newson
(Chapter 11) discusses the implications of afforestation on water
yields in catchment basins in upland Britain; the purpose of this
short section is to touch upon some of the planning approaches that
are evolving in British Columbia to tackle the opposite situation -
the deforestation of watersheds that supply domestic water systems.

The Slocan Valley, situated in the east-central region of the
province (Figure 6), contains a large number of tributary watersheds
where there are conflicts between logging and water quality. Water
quality is affected by the interrelated processes of erosion, transportation and sedimentation. In general, logging will increase peak
flows by reducing the interception and infiltration of precipitation,
increasing the potential erosive power of the water. The effect will
be greater on smaller watersheds with steeper slopes and generally
unstable soils. In addition, logging operations disturb the soil
due to the construction of roads, bridges, landing stations and the
transport of logs to trucks. To offset these problems, logging
methods have been tailored to suit specific conditions of slope and
soil stability (Figure 7). Rubber tyred skidders are used on relatively flat watersheds with stable dry soils, while overground cable
systems are used in the steeper slopes and more unstable soils to
reduce the areas of disturbance.

These logging systems require refinement to avoid water/land use
conflicts. In the Slocan Valley, there are many domestic water
systems that suffer from heavy siltation due to poor logging methods.
Generally, road widths are too wide and their spacing too dense,
resulting in a large amount of soil disturbance. In heavy rains,
water from logged areas tends to run down the roads before reaching
natural channels, greatly increasing the erosion and transport of
materials to these channels. To reduce the costs of logging operations, forest companies often place culverts only in the main and
branch roads. The standard of road construction becomes progressively worse on spur roads and skid trails.

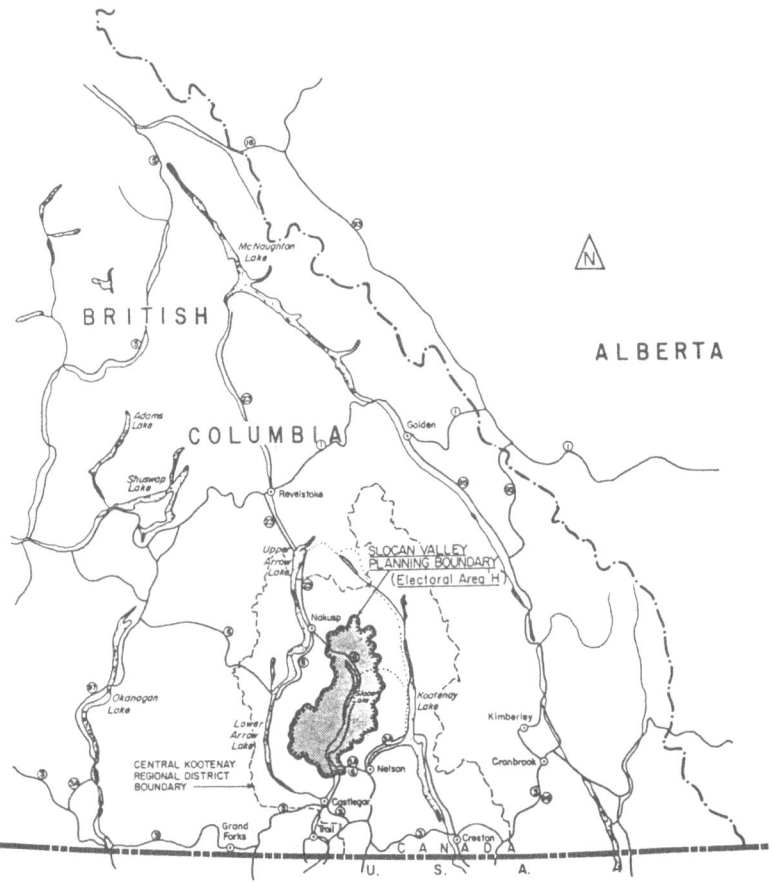

Figure 6:   Location of Slocan Valley (49.46N 117.28W)

As a result, licenced domestic and/or irrigation water users
are faced with two problems - decreased water availability in late
summer due to more rapid runoff in the spring freshet, and a high
degree of sedimentation. Because water managers in the past have
tended to licence use right up to the average monthly low flows,
water shortages are now becoming more common.

To counter these problems, the British Columbian Ministries of
Forests and Environment are developing a watershed sensitivity index
to assess the potential for sedimentation. This index is based on
five variables :
  - sediment yield index
  - mass movement index
  - relief index
  - channel relief ratio index
  - area of watershed.

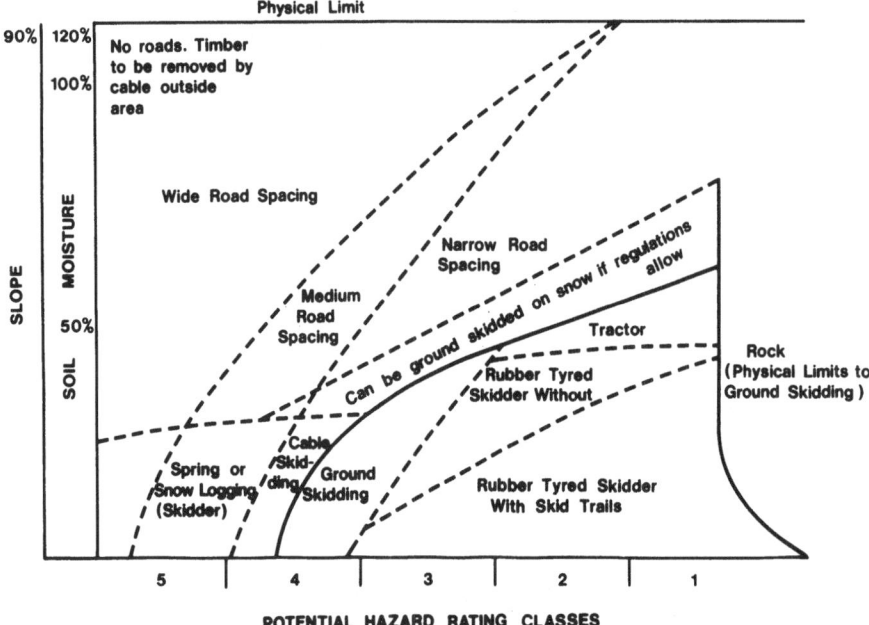

Figure 7:   Logging methods used in British Columbia in different
            conditions of slope and soil moisture and their potential
            hazard rating.

    The sensitivity of water quality to sedimentation is directly
related to the first four of these variables, and indirectly related
to the watershed area (Travers, 1982). Generally, watershed area is
the primary factor in determining the sensitivity index, large watersheds having low indices and small watersheds high indices.

    Clearly, watersheds which have high sensitivity indices and are
heavily used for domestic and irrigation purposes require more controlled logging operations (if approved at all) than do watersheds
with low sensitivity and little or no water use. Through biophysical
analysis of soil types, terrain, bedrock, channel slope and stability,
it is now possible to classify watersheds by this sensitivity index
(Table 3). Based on this information, the next step is to tailor
logging techniques to the watershed. Through strategic planning, an
overall balance in land and water use can be achieved, assuming that
each Ministry is prepared to make trade-offs. In some community
watersheds, logging will not be permitted, while in some highly
sensitive watersheds, no domestic water licences will be granted.
In some instances, it may be cheaper to redesign the community water
systems so they are less sensitive to sedimentation than to impose
higher costs on the logging industry.

Table 3: Classification of licenced watersheds in the Slocan Valley
using a Watershed Sensitivity Index (Fully recorded:
licenced use equals or exceeds average low flows; partially
recorded: licenced use in less than average low flows)
(Travers, 1982).

Watershed Size	Predominant Sensitivity Index	Fully Recorded	Partially Recorded	Total
Large (> $10km^2$)	Low	5	13	18
Medium (5-$10km^2$)	Moderate	6	7	13
Small (< $5km^2$)	High	8	4	12
	Very High	8	4	12
Totals		27	28	55

CONCLUSIONS

All three case examples described concern land and water use
issues on a relatively small geographical scale. However, the policy
decisions that drive many of these conflicts occur at a much larger
scale. For example, urban development on the uplands of the
Serpentine-Nicomekl occurs as part of the metropolitan regional plan
for the Greater Vancouver area. To some extent, urban growth is
increasing most rapidly in these uplands due to other environmental
constraints, such as preservation of farmlands and protection of
wetlands for fish and waterfowl. Forest cuts in the Slocan Valley
are a component of the total annual allowable harvest of wood in the
south-east region of the province. The greater this regional total,
the more pressure it places on individual resources in specific
watersheds. The concern for salmon protection in the Nicola is part
of a much wider goal of increasing salmon stocks in the Fraser River
system to their historical levels (i.e. prior to commercial exploitation and habitat losses).

All these policy decisions are strategic in nature. At present
they are often made without a reasonable understanding of their implications on other resource uses. To improve such decisions, there
must be a hierarchical analysis of resource allocation choices in
the province, starting at the regional level, and breaking down into
the sub-regional and eventually watershed analysis. Information
flows will be two-way - downwards from the policy level to the site-
specific and also upwards, so that the functional inter-relationships
between resource systems are understood (Figure 8).

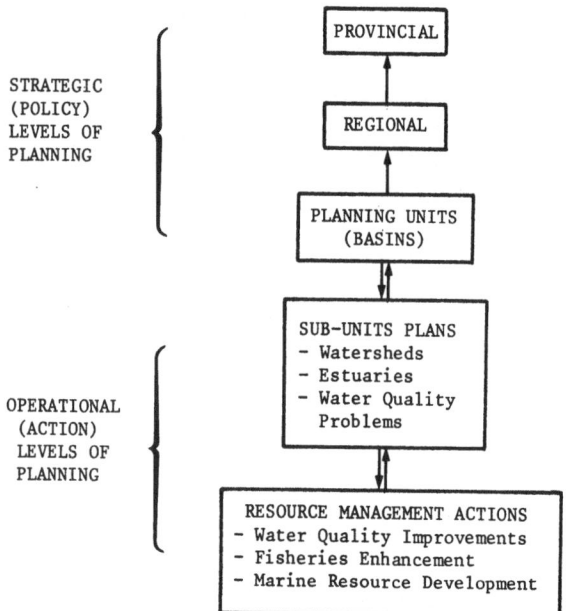

Figure 8: Projected strategic planning hierarchy for the British
           Columbia Ministry of Environment.

As mentioned in the Introduction, if each resource ministry
could develop its own strategic choices - levels of harvest, production of fish, etc., it would be better equipped to bargain with
other interests to develop a resource use balance in the overall
interests of society (Dorcey et al.,1980). To support this approach,
biophysical analysis integrating both terrestrial and aquatic systems
has to be improved and co-ordinated much more closely with the socio-
economic system that represents societal values and institutions.

In the age of the computer, the massive amounts of data that
must be sorted and seived in such an approach are now manageable.
The missing ingredient continues to be man's imagination, innovation
and the will of institutions - too often vertically oriented - to
make it work.

REFERENCES

Anderson, C., 1977, "A Method of Classifying Timber Sites to
    Minimize the Impact of Timber Harvesting", Report prepared for
    the E.L.U.C. Secretariat, B.C. Government, Victoria, B.C.
British Columbia Ministry of Environment, 1982, "Nicola Basin
    Strategic Plan", Planning Branch, Ministry of Environment,
    Victoria, B.C.

Dorcey, A.H.J., McFee, M., and Sydneysmith, S., 1980, "Salmon Protection in the B.C.Coastal Forest Industry", Westwater Research Centre, University of British Columbia, Vancouver, B.C.

Environmental Protection Agency, 1980a, "An Approach to Water Resources Evaluation of Non-Point Silvicultural Sources (A Procedural Handbook)", U.S.First Services, Washington, D.C.

Environmental Protection Agency, 1977, "Non-Point Source-Stream Nutrient Level Relationships: A Nationwide Study", Environmental Protection Agency, Corvallis, Oregon.

Environmental Protection Agency, 1980b, "Restoration of Lakes and Inland Waters", Environmental Protection Agency, Washington, D.C.

Hasler, A.P., 1975, "Coupling of Land and Water Systems", Springer-Verlag, New York.

Likens, G.E., and Borman, F.H., 1974, Linkages Between Terrestial and Aquatic Ecosystems, Bioscience, 24:447.

O'Riordan, J., 1981, The British Columbia Experience in: "Project Appraisal and Policy Review", T. O'Riordan and W.R.D.Sewell eds., John Wiley and Sons, Chichester, England.

Travers, R., 1982, "Slocan Valley Planning Programme", Draft Technical Report, B.C.Ministry of Environment, Victoria, B.C.

Welch, D.M., 1978, "Land/Water Classification", Ecological Land Classification Series No. 5, Lands Directorate, Environment Canada, Ottawa, Ontario.

Wilson, K.W., 1982, "The Conflict Between Urban and Agricultural Water Management in the Lower Fraser Valley, B.C." M.Sc. Thesis, B.C. Ministry of Environment, Victoria, B.C.

13

WATER IN ARID AREAS - BASIC FACTS AND EXAMPLES OF ENVIRONMENTAL

IMPLICATIONS, SUPPLY AND CONSERVATION

A. P. Schick

Institute of Earth Sciences
Hebrew University of Jerusalem
Givat-Ram
Jerusalem 91904, Israel

INTRODUCTION

The arid and semi-arid regions of the world cover somewhat over one third of its total non-glaciated land areas (Walton, 1969). There are several bases for defining the arid zone, and it can be subdivided into semi-arid, arid, and extremely arid areas on widely differing criteria, but its main characteristic of a deficient precipitation input in relation to the often very high potential evapotranspiration output is the ultimate cause of deserts. Thus deserts may be produced by combination of a low but not insignificant annual rainfall with very high temperatures, which results in rates of evapotranspiration so high that all but a minute share of the rainfall returns to the atmosphere almost immediately, making no contribution to usable surface or subsurface water storage. Conversely, regions with low temperatures over much of the year, and therefore relatively low values of evapotranspiration, may have decidedly arid characteristics because of very low precipitation.

In addition to the primary factors of climate, semi-independent factors of geomorphology, soils, and vegetation which are based on fundamental characteristics of topography, geology, hydrology, and history of human occupancy are important when investigating the regional and local character and degree of aridity. The topic of water management in arid areas is therefore a strongly interdisciplinary field which cannot be content with a narrow technical approach.

The Saharo-Sind desert belt is by far the largest arid zone and exceeds all the other deserts in area. Its central and western parts comprise hot desert, but the extensions into central and eastern Asia have below-freezing winters and are cold deserts. The North American desert has many climatic counterparts in the North African-Eurasia dry zone, while the primary features of the even smaller South American dry province are a highly arid coastal strip and an inland dry area, which are paralleled by similar areas in Southern Africa. The Australian desert is unique in its areal predominance in that continent, but it includes no extremely arid core area.

This chapter concentrates on some of the hydrological aspects of the extremely arid zone, having mean annual precipitation of less than 100 mm/year, and in most cases less than 50 mm/year. The insight gained from the studies on which this chapter is based can be transferred more or less directly to less severely arid regions.

The main focus will be on the surface water phase of the hydrological cycle, which is specific to the desert. Groundwater, with its great advantage of storage and protection from evaporation, behaves rather similarly in arid and non-arid regions, but it cannot be used as a viable source without recharge from surface water.

DESERT FLOODS

Unlike humid regions, in which most of the runoff volume is supplied by medium-scale events (Dunne and Leopold, 1978), arid regions tend to be dominated by major floods (Baker, 1977). Such events are rare, tend to be spatially scattered, and their scientific documentation is largely inadequate. Researchers have to rely on surveys made after the event without a prior basis for comparison; some examples of such studies in this line are Beaty (1968), Glancy and Harmsen (1975), Leopold (1946), Schick (1971), and Woolley (1946). Although the great development in remote sensing techniques in recent years has improved the situation somewhat, a good data base for the study of low-frequency, hydrologic and geomorphic events in very arid areas remains a world-wide need.

Settlements and other structures in deserts are widely scattered, so that spatial combination of the area affected by a violent desert flood with an area of settlement is relatively rare. However, such events seem to have been more frequent in recent years than in past decades. One reason for this is undoubtedly the great expansion of economic activity in deserts, mainly various forms of mining. There are currently more structures that might be affected in the desert, they are more widely distributed, and served by more widespread and organized road networks.

Because of the foreseeable further increase in the development of desert resources in future, we need a better level of knowledge and awareness of desert floods and their effects. Besides

their destructive aspects, such processes are an important dimension
in guiding planners in the selection of siting alternatives. If
correctly interpreted, diagnostic landforms such as high flood marks
and terraces on the one hand, and low-flow marks such as inner bars
on the other, can convey valuable information on hydrologic characteristics.

HYDROLOGIC PROCESSES

     In small mountainous watersheds in the truly arid regions,
rainfall and runoff show extreme variations both in time and in
space (Sharon, 1972; Sharon *et al.*, 1976). In Nahal Yael, an 0.5
km^2 watershed in Israel's Southern Negev, mean annual rainfall over
a recent ten-year period was 31.6 mm, with annual extremes of 1.9 mm
and 66.9 mm (Schick, 1977). About 30 % of this ten-year rainfall
became runoff inside the watershed, but ultimately less than 5 %
flowed into the higher-order channel downstream. The remaining 70 %
could not wet the soil to a sufficient depth to escape a more or
less immediate return to the atmosphere by evapotranspiration, and
was thus lost as a usable water source.

     This conclusion was based on observations of the wetting
front in various lithologies, which never exceeded a few decimeters
in depth, well within the reach of capillary upward movement. It is
also supported by calculations based on infiltration tests conducted
with the aid of high-intensity simulated rainfall in the Negev and
Sinai (Salmon and Schick, 1980; Yair and Lavee, 1976) and by documentation of the water-retaining effect of the clayey *Schaumboden*
(Jaekel, 1980).

     The subsequent 25 % loss was by infiltration into channel
alluvium. The first kilometer downstream from the point of alluviation accounted for 15 % of the infiltration loss. This considerable percolation occurred despite the shallowness of the sub-
alluvial bedrock (1-3 m below the channel bed), and suggests that,
for similar catchments 1 km^2 in area located in climatic regions
of 500 mm/10 years of rainfall, several thousands of cubic metres
of water saved from evaporation could be utilized. The significance
of infiltration of such a magnitude is of a major importance in
developing small local sources of water supply. A further finding
from Nahal Yael was that, over the same ten year period, the small
(500 m long) alluvial fan received a total of 36,400 m^3 of water
from upstream, but transmitted downstream only 8,100 m^3.

     The hydrologically significant runoff over that decade was
caused by seven major events. The longest interval between subsequent major events was over 59 months, and the two shortest intervals were 4 and 8 hours. If each of the last-mentioned "double"
events are regarded as a single one, a major event in extremely

arid regions such as Nahal Yael would seem to occur, on the average once in two years. This frequency decreases with increasing drainage area (Schick, 1971).

In the 3100 km^2 watershed of Wadi Watir in Eastern Sinai, a major geomorphic event that activates the entire fluvial system probably occurs only once in 4-5 years. In still more arid areas such as in the core regions of the Sahara, the intervals seem to be even longer (Dubief, 1954).

Most of the runoff events are short-lived and show a strongly peaked hydrograph (Schick, 1971). Peak discharges for small (50-100 ha) upstream watersheds exceeded 10 m^3/sec/km^2 several times during the decade. Peak discharge per unit area decreases with increased drainage area.

The entrainment of suspended sediment by the flow generated by a major precipitation event is rapid. This is remarkable, considering the wind erosion which effectively cleans the desert surface of silt and fine sand. For a major event in Nahal Yael, only the smallest uppermost tributaries have a suspended sediment concentration of less than 10,000 ppm. This concentration increases with event magnitude and with the length of alluvial channel traversed by the flow. At the mouths of intermediate size watersheds (10-100 km^2) in the Eastern Sinai it often exceeds 100,000 ppm. Extreme values around 400,000 ppm were recorded (Schick and Sharon, 1974), and indicate a combination of high-energy flows with water losses by infiltration into the channel bed.

An analysis of suspended sediment in 82 floodwater samples taken from five substations in Nahal Yael during four events showed that considerable amounts of sand are transported in suspension in almost all cases (Lekach and Schick,1982). The partial concentration of the sand has a threshold at a total concentration of 30,000 ppm. Below this threshold, sand amounts to no more than one third of the suspended load, while above it all the sediment added is sand. The fines seem to originate in the slopes, and the sand from the channel. The silt-clay is derived by wash from underneath the monolayer stony cover that protects the slopes, and its availability depends on the length of time that has elapsed since the previous event. The dependence is non-linear and tends to approach a terminal value, probably reached after a preparation time of ten years or more. More work is needed to provide a predictive model for the evaluation of sediment supply from small arid catchments, based on climate, geomorphic parameters and event magnitude, but a promising start has been made (Lekach and Schick,1980).

In the Nahal Yael ten-year sediment budget, bedload accounted for about one-third of the roughly 2,000 tons of material transported out of the watershed (Schick, 1977). Although this figure cannot

be regarded at present as being very accurate, it probably reflects
a reasonable order of magnitude. The source of the bedload is
almost exclusively the scour layer although lateral cutting of older
bars and of terrace slopes also takes place. Bedload particles were
observed rolling down the 5 % channel slope when only partly submerged during a medium-low water stage. Mean distances of transport
of bedload particles (size 20-80 mm b-axis) is about 65 meters per
event. As the main alluvial reach of Nahal Yael is about 1000 m
long, it takes about 15 major events (i.e. 20-30 years) for a single
particle to be evacuated from the watershed. However, a few individual particles were transported up to 8 times the mean distance in
a single event. Such transport may be associated with the "wall of
water" mechanism (Leopold and Miller, 1956) which is probably
sustained by infiltration of the flood front into the alluvial
channel bed (Schick, 1971). Also, events of truly catastrophic
proportions (recurrence interval of 100 years and over) are capable
of completely cleaning nearly all the available sediment out of
small arid watersheds (Schick, 1974).

Silting of reservoirs and other structures is of particular
importance in arid areas. As if to compensate for the shortness of
the burst of activity, desert streams carry enormous quantities of
sediments. Major impoundments, like the Nahal Shiqma recharge
reservoirs in the semiarid-subhumid part of the Southern Coastal
plain of Israel, have lost as much as 40 % of the original storage
during a single event in the 1960's. Two ancient 15 m high dams
located on loess-covered catchments near Mamshit (Kurnub) in the
northern Negev, were restored in the 1920's and silted completely
within a year or two.

Bedload presents special and more difficult problems. Its
transport is dependent on stream power, but recent evidence from
the Nahal Yael experimental dam, as well as from samples, indicates
that bedload tends to come in waves, in part unrelated to water
discharge. Although not really understood and not really specific
to deserts, this phenomenon bears directly on the feasibility of
diverting the most sediment-laden portions of desert floods around
or right through the reservoir.

Landform changes that result from the runoff and sediment
regime described above are very often subtle and difficult to detect
over short human time spans, but on occasion they assume monumental
proportions. Detailed sections repeatedly surveyed between fixed
bench marks showed that the 7800 m^2 fan of Nahal Yael aggraded,
over a six-year period, by a volume of 10.6 m^3, i.e. a mean aggradation of 1.4 mm only (Schick and Sharon, 1974). On the other
hand, the 6200 m^3 alluvial fan of Wadi Mikeimin was formed literally
overnight at the outlet of the 13 km^2 watershed in Eastern Sinai.
This fan was the result of a highly localized intense rainstorm
which created a peak flow of 70 m^3/sec down the Mikeimin, only to

dissipate quickly in the bed of the large Wadi Watir, whose drainage area was unaffected by the event. Wadi Watir was actually impounded by the Mikeimin fan for the duration of 22 months, during which the Watir was geomorphically dormant. During this period only small amounts of underflow collected as a small natural pond behind the Mikeimin fan (alias dam), which extended all the way to the opposite side of the Watir valley. The event of November 1972 caused a peak flow of about 400 m^3/sec at the site of the Watir-Mikeimin confluence and obliterated the 1971 fan (Schick and Lekach, 1981).

Geomorphic evidence shows that remnants of tributary fans that clogged trunk stream valleys in mountainous deserts such as the Sinai are not unique. These obstructions must have served as temporary dams for the subsequent fluvial event in the trunk stream. In fact, damming fans of the Mikeimin type undoubtedly develop as a result of the difference in time lag between rainfall and the runoff in small tributaries compared with the large stream channel, and they are subsequently breached during a single event. This process contributes to the peaking of the hydrograph, as the breach of the dam is in reality a "wall of water". The result in increased erosive energy is obvious.

In February 1975 a major 48-hour rainstorm covered most of the Sinai Peninsula; most of Egypt was also affected (Labib, 1980). A peak flow of 1650 m^3/sec was estimated for a point 30 km south of the outlet of Wadi El Arish - the main drainage system of the Sinai - into the Mediterranean Sea (Gilead, 1975). Here too, huge amounts of sediment caused the almost instantaneous deposition of a delta some 300 m long and 500 m wide. It took several weeks for waves and currents to erode the delta and restore the coastline to its original state. This flood demolished a railroad bridge built in 1914 and strengthened following flood damage in 1925, indicating that we are dealing with an event of a frequency of 50 years or more (Feldman, 1976).

Similar spectacular geomorphic events have been reported from other arid areas. In Tunisia, the widely reported 1969 floods destroyed Roman bridges which had remained intact for about 2,000 years (Stuckmann, 1969). In 1973, a one-in-200-year event in the 21,000 km^2 watershed of Oued Medjerdah in Tunisia resulted in a deposit over an area of 138 km^2 in the lower part of the fluvial system, having a depth of 5-40 cm, and occasionally up to 1.5 m (Claude and Loyer, 1977). Similar events were reported from arid regions in North America (for example, Glancy and Harmsen, 1975).

RUNOFF AND THE GEOMETRY OF ARID CHANNELS

In recent years, attempts have been made to relate discharge characteristics, such as mean annual discharge, to selected para-

meters of the hydraulic geometry. In view of the paucity of available gauging records from deserts, this method is particularly
attractive to workers in arid environments.

The method usually related width, depth, cross-sectional area,
or a combination of these parameters as determined by the low bars
or inner berms of the channel to a selected discharge characteristic.
From semi-arid areas, fair to very good statistical relationships
with discharge characteristics were obtained for ephemeral streams
in Nevada (Moore, 1968), California (Hedman, 1970) and New Mexico
(Scott and Kunkler, 1976). Certain of these studies show remarkable
persistence in the geometry data for repeated measurements at
intervals of several years, despite the entire bed being reworked
by floods (Osterkamp et al., 1983).

Streamflow records are totally unavailable for extremely arid
regions, but good correlations result if drainage area is substi-
tuded for discharge. A consistent and statistically sound regression was obtained between inner-berm cross-sectional area and
drainage area for 22 watersheds in the Southern Negev and Eastern
Sinai (Last 1974). The relationship was found to be especially
reliable for drainage areas between 10 and 100 km^2.

The inner-berm data were found to be quite conservative and
unaffected by the passage of floods. Step-wise and multiple
regressions relating drainage area (representing discharge at some
power between 0.5 and 0.75) to width, depth, slope, and sediment
size on the bars improved the regression coefficients to levels
above 0.9.

The geomorphic mechanism which undoubtedly exists at the base
of these relationships is still unclear and the method seems to
have lost favour in recent years. Even so, the potential of the
method in arid areas for practical purposes is self-evident.

GEOMORPHIC PROCESSES AND DECISION-MAKING IN DESERTS

Inherent in the opening of the desert to modern development is
the problem of perception. Urban planners and highway engineers
tend to be insensitive to the potential, albeit infrequent, dangers
of the dry river beds and closed basin lakes which must carry and
store excess surface water. The alluvial fans that rim all
mountainous areas in the desert become the main object of attention
in this respect. On the one hand, they strongly attract human
occupancy because they represent preferred sites in terms of local
topography, accessibility and, more particularly, water supply. On
the other hand, the alluvial fan is located at the outlet of a
mountain watershed draining into a low-lying rift or other major
valley area, and is the locus of future floods.

The extreme uncertainty of most heavy rainfall events, well
known to desert people everywhere, precludes any efficient forecasting.
Thus, although data limitation still reflects pre-satellite methods,
the chances of developing a warning system for floods in the desert
seem remote. The relatively short flood-to-peak interval (Sheaffer,
1961), characteristic of desert floods, presents an additional
impediment to the efficient use of warning systems. Nearly all
catastrophic desert floods have a flood-to-peak interval of less
than one hour, and often the flood comes as a wall of water. Clearly,
effective ad hoc adjustments or protective measures are out of the
question whenever the rain does not cover the endangered site and
its occupants are unaware of the peril. In the town of Desert Hot
Springs, California, which is located on an alluvial fan, few
citizens were found to be aware of a flood hazard, and when a flood
struck, they took no concerted action to reduce losses in anticip-
ation of the next stream of water, mud, and boulders disgorging on
the alluvial cone (White, 1964).

Long-term detailed observations in Utah showed that the largest
share of water, sediment, and indeed of potential damage results
from small basins of up to 25 km^2 in area (Woolley, 1946). In the
Arava rift valley of the Southern Negev, most settlements are
located on the western edge, close to the foot of the mountains.
They are invariably located on the bajada (an alluvial land form
often encountered in the margins of arid mountains, formed by the
lateral coalescence of alluvial fans) which skirts the Arava along
its entire length. The residential areas of Eilat are distributed
over an area of about 3 km^2, at elevations between 20 and 90 metres
above sea level. Various industrial and transport installations,
public and tourist buildings are located in the more southerly
part (Karmon, 1976), along the western coast of the Gulf of Aqaba
(Figure 1). Most of these buildings are on the coast, at or near
the mouths of canyons that descend to the gulf. Some of the public
and residential buildings were built on flat-topped hills, which
are dismembered parts of an ancient bajada. Some of the kibbutzim
of the Arava, such as Gerofit and Elot, were also sited on similar
topographic features.

From the point of view of flood and erosion hazards, the
buildings on the isolated hills are in a relatively safe position,
but they are not entirely protected from flood damage, because of
the danger that a channel in flood flowing at the foot of these
hills will erode laterally. Indeed, some of the buildings of
Kibbutz Elot were originally planned too close to the bank of the
Roded valley, and the river waters did in fact undermine their
foundations. Channel widening has little chance of being naturally
restored in deserts (Wolman and Gerson, 1978), and diverting
embankments is at best only a partial remedy, as they must be
periodically maintained. They also invite encroachment on to the
"protected" area, and thus increase the danger of catastrophic
damage.

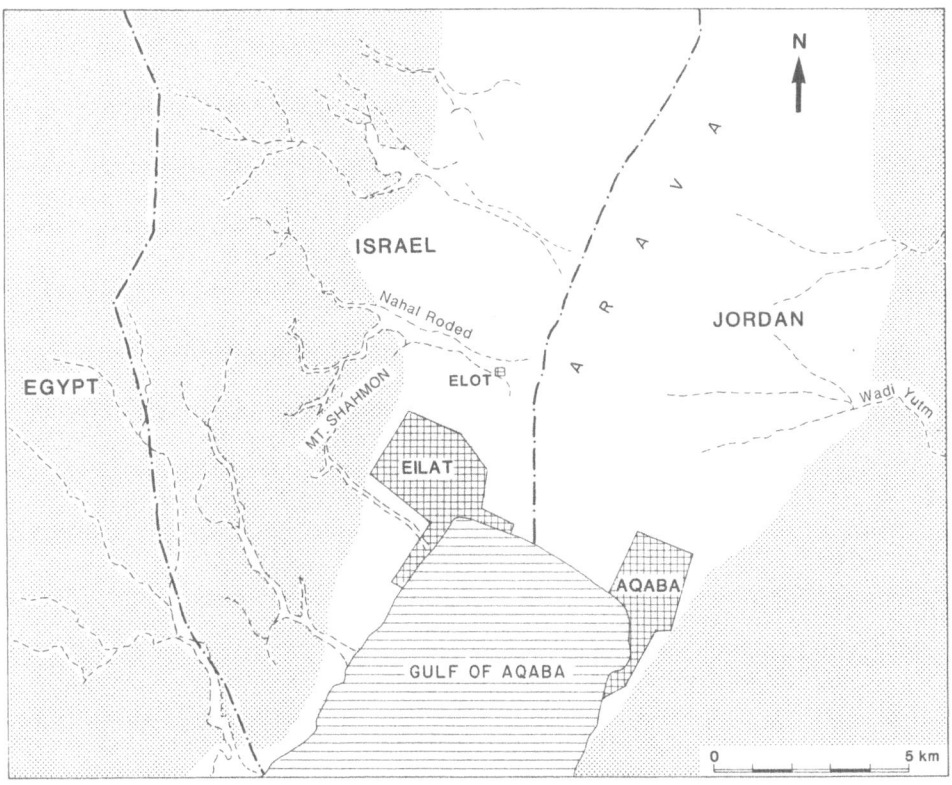

Figure 1 : Map showing the position of Eilat (29.33N 34.57) and
            surrounding high ground above 100 m (shaded area)
            together with the drainage basins east of Mt. Shahmon.

        Buildings on the active bajada area are exposed to a serious
flood hazard. The alluvial fans that make up the bajada constitute,
transport surfaces for water and sediment, in which flow channels
join and divide at high density over time, and changes in course
occur frequently. These changes are due to sediment blocking in
the channels, associated with the bar-forming transport mechanism
and with water losses by percolation.

        It is to be expected that severe floods at Eilat cause serious
damage to houses and roads. In 1963, a sudden storm initiated a
flood which cut a channel a metre deep across the main avenue of
the town, blocking all traffic. Channels were cut also between
houses, exposing their foundations; many buildings were close to
collapsing. Water entered the prefabricated houses through the
doors, and the inhabitants were compelled to make holes in the walls
facing downslope in order to let the water out. The town roads of
Eilat are, as a rule, constructed without culverts, so that at

times of flooding they are certain to be damaged. The airport too,
was flooded and unusable.

The flood hazard to the residential area of Eilat originates
in three small drainage basins that emanate from the eastern and
southern slopes of Mount Shahmon (Figure 1). Each of these is no
more than 2 km^2 in area. Had the storm which caused the catastrophic
1966 Ma'an flood in Southern Jordan (Schick, 1971; Cooke and
Doornkamp, 1974, p. 125) centered over Eilat, three 20 m^3/sec
torrents would have passed through the town; a union of two of these
would have resulted in a 40 m^3/sec peak flow descending right through
the central urban area. Passing between any two adjacent houses ten
metres apart, and assuming the flow gradient to be similar to that
of the fan (approximately 7 %), a torrent of water over half a metre
deep flowing at an average velocity of 7 m/sec would have resulted.

Prior to the rainstorm of October 1981, each and every one of
the local flow events which affected the town of Eilat caused deep
erosive cuts in the upper and middle parts of the built-up area,
especially along the margins of the road pavements. The event of
October 1981, in contrast, resulted in considerable and widespread
deposition. Since the previous similar event had occurred in the
mid-seventies, one possible explanation lies in the change of urban
land use during the intervening period. A detailed study is under-
way at present, so only tentative conclusions can be drawn. However,
it seems that the heavy construction of housing, public amenities,
and roads in Eilat, their layout sometimes planned without due
regard to the drainage needs in such a parched environment, coupled
with a persistent urban expansion upslope in the direction of the
rocky outcrops of the Shahmon catchments, have brought about a
complete shift in the hydrologic response of the terrain. Instead
of an assemblage of small patches of impermeable roofs and asphalt
effectively isolated one from the other by expanses of highly
absorbing deep colluvial sands and gravel, there is now a higher
level of surface runoff connectivity, which requires drainage
facilities several orders of magnitude more efficient than before.
The barren land between the houses and roads is now subjected to
very much higher erosion, with accompanying deposition at every
local opportunity.

Even more rapid rates of flow may be expected when possible
flood damage associated with larger stream channels is considered.
The tragic loss of life in the Ma'an flood was due to the channel
flow and to the flooding of areas adjoining the banks. Observations
at the gauging site in Wadi Yutm indicate that flow during that
event was supercritical, with surface velocities of about 5 m/sec
determined by float measurements. Surges with velocities of approximately 12 m/sec occurred about once a minute during high flow
(Hashemite Kingdom of Jordan Hydrology Division,1966).

Similar velocities were observed during the November 1972 flood in Wadi Watir in the Sinai (University of Jerusalem Geography Department, 1972). Unfortunately, the surging mass of water and rubble that bursts from the mouth of a steep desert watershed may follow any one of many possible courses down the surface of the fan (Beaty, 1963). Course changes also occur in successive flows, sometimes near the apex of the fan and sometimes farther down the slope.

It has been proved in recent years that damage from sheet floods generated solely from a slope area may reach considerable proportions without additional water from any integrated drainage basin; some of the damage to the Timna Copper mines from the 1966 flood was due to slope wash alone. The deposition of sediment in the foundation pits of the Kitan textile factory in Dimona in the late 1950's was also due solely to flooding from hillslopes. The flood-producing potential of slopes and slope assemblages can be assessed by an appropriate geomorphic survey and mapping procedure, which allows systematic and regional identification of the dominant processes, providing the basis for planning.

Recent detailed studies in the Sde Boqer site in the Northern Negev (mean annual rainfall 90 mm) has yielded interesting data on the degree of variability of the runoff and sediment source areas on slopes. The very large local variation shown by measurements of small natural plots (Yair and Lavee, 1981), is corroborated by spatial surveys of vegetation indicating the soil moisture regime (Yair and Danin, 1980) and by animal burrowing activities, which are indicative of both rainfall regime and sediment production. Thus, hydrologic evaluations of catchments in arid areas cannot rely merely on average values drawn indiscriminately from various landscapes, but rather should depend on very careful small-unit process-oriented terrain evaluation.

Beaty (1968) arrived at the following two generalisations from his study of flooding from the White Mountain :

(1)   most of the larger trunk channels resemble steep and narrow
        bedrock flumes with a bed of alluvium, grading into fairly
        steep fans - a set of conditions indicative of hazardous
        debris flows;
(2)   intervention by engineering works is not always beneficial:
        the influence of manmade dikes and ditches on alluvial
        surfaces may concentrate the runoff into larger channels, in
        which it acquires increased volume and velocity and conse-
        quently increased potentially destructive energy.

When viewed against the background information on floods in arid lands, the main conclusion that emerges from careful studies of major desert flood events is that the foremost weapon of defence against loss of life and property is careful consideration of

siting. Ideally, an attempt to determine which activities can
afford to be located in flood-prone areas and still pay the "natural
tax" of flood losses could serve as a yardstick for granting permission to use these areas (Sewell, 1961). In deserts, where alluvial
fans are flood-prone areas, the lack of knowledge of how to evaluate
magnitudes and frequencies on the one hand, and the multiplicity of
choice due to the variety of spacial possibilities on the other,
make such a procedure difficult, and perhaps even unrealistic.

Geomorphic evaluation of the terrain - in itself a relatively
simple and inexpensive procedure - can point to preferences that
have to be considered together with other factors included in the
selection of sites for desert installations such as dwellings and
industrial plants,ports and airfields, roads and railways (Cooke
et al, 1982). The effectiveness of a terrain flood-hazard evaluation
is immeasurably increased if it can be correlated with precise
data, of which there is at present very little. However, even
without basic local knowledge of rain intensity, peak flow, fluvial
parameters and other factors, even an intelligent attempt to try as
little as possible to control the flood, and as much as possible to
avoid it, offers considerable benefits. Above all, this means
resisting the natural tendency of engineers to copy the methods
used in regions where most of their work is done - the non-arid
regions. And even in those regions complete protection is never
possible.

In arid regions, where knowledge of design values is meagre,
erosive forces are immense, and engineering works often semi-improv-
ised, the need for caution based on knowledge applies even more
strongly. As an example, modern highways cross very wide braided
stream channels over long bridges, such as the one on the Minab
River in southern Iran near the northern shore of the Persian Gulf.
This bridge is supported on massive concrete pylons, deeply embedded
in the channel alluvium. However, no one really knows what the
depth of the scour layer in large floods might be. For small and
medium floods it may approximate the depth of water above the
original bed, but the relation probably does not remain linear for
higher flows.

In modern desert development, centralized authority is so
strong that little attention is paid to local attitudes. Sometimes
local authorities demand, and get, a system of dykes and diversions
mainly because government money is available for drainage on a
municipal or regional administrative basis, and an objective
technical evaluation based on intrinsic and comparative considerations is relegated to a secondary role. At the same time, vast
amounts of money are wasted in deserts everywhere by not undertaking elementary precautions as a result of sheer ignorance of
desert features and processes.

DESERT ROADS AND GEOMORPHIC PROCESS

The design of roads in arid areas confronts the engineer with
a range of problems very different from those encountered in temperate regions. The large distances and the low volume of traffic
in deserts usually dictates a low investment per unit length. For
geographic reasons, much of the human activity in deserts occurs
in and around the mountainous areas, so that roads either lead into
the mountains or skirt along the border scarp. In the first case,
the road will invariably use the floor of a major valley for most
of its length. The second case which is more common, presents the
engineer with the problem of crossing numerous ephemeral stream
channels of varying size and different characteristics. The road
is, in effect, a series of crossings of alluvial fans which develop
at the outlet of desert stream channels from a mountain block
(Schick, 1974b). The Arava rift valley road, which follows its
margin and crosses all the alluvial fans deposited by channels that
drain to the Arava from the west, belongs to the second type.
Similar problems confront essentially coastal roads like the Eilat
- Sharm esh-Sheikh road. The railway route to Eilat, which is
planned to traverse landforms similar to those of the road, presents
engineers with much more serious problems. Flood risks to the road,
such as sediment cover, erosion of foundations, or the destruction
of a bridge, can be repaired quite easily. The damage caused to a
railway line by a severe flood, particularly if its location and
bridges had not been carefully planned, could be serious, and
repairs would take weeks or even months.

One of the possibilities of reducing the flood hazard to
transport lines in places where they cross desert washes and fans
is to determine the route locally, on the basis of flood experience
correlated with geomorphic and vegetation features. The location
of the fan crossing must be guided by the willingness to avoid
damage from spills, which are common near the apex, and to minimize
frequent sheet flooding and small diversions near the fan toe.
Since topographically the middle area often has the largest local
relief, economics discourage the use of this belt for roads. This
is especially true for railroads built in association with mining
operations in some deserts.

The engineer has several alternatives at his disposal in his
effort to optimize the protection of such a desert road from the
diverse flood hazards. The first set of alternatives involved the
horizontal location of the roadway, which can be planned to hug
the mountain front, to be distant from the mountains and run along
the foot of the bajada, or to adopt any intermediate position. The
position of the roadway can also be varied locally according to site
consideration.

A roadway built above the general fan level is usually tied to a channel crossing with a culvert. To keep the channel at grade and avoid sedimentation, the culvert floor should be on, or nearly on, the natural profile of the ephemeral stream channel. Where the road surface is at, or below, the channel grade line, the situation is more conducive to the construction of bridgeless crossings, also known as submerged bridges. In crossings of this type, the roadway dips into the channel and its structure is strengthened within the bankfull perimeter by concrete aprons, designed to let the flood-waters pass over the road with minimal damage.

The choices of the horizontal and vertical positions of the roadway are closely interrelated, which is illustrated below by a few examples. Theoretically, a decision could be made to trace the road as a straight line. For a traveller on such a road, the result would be continuous ups and downs due to the recurring passages from high fan areas to low inter-fan depressions. On a smaller scale, the generally elevated fan areas would have short but abrupt dips into active as well as abandoned stream channels and distributaries along the road. Another solution, also only theoretical, is to trace the roadway as a contour. Keeping to a constant elevation, the roadway will then be near the mountain front in its inter-fan segments, and far from it on the fans themselves. Because of the high drainage density characteristic of arid mountain blocks, the spacing between neighbouring outlets of drainage basins at the mountain front would be mostly of the order of 1 km or less. The curves of such a road pattern would therefore be relatively dense and quite impossible for modern traffic.

It is important to realize that flood hazard considerations are not at the top of the road planner's list of priorities. Beside his main objective, which is to provide transportation between certain sites (often settlements located on alluvial fans because of water and land considerations), he has to adhere to predetermined standards of traffic speed and safety. These standards specify horizontal radii of curvature which, for fast modern traffic, must be kept relatively large. Also, the permissible rate of slope change with distance cannot exceed certain limits. These two con-straints make it impossible to implement any one of the two theor-etical examples mentioned above. In practice, a compromise between the two is chosen, subject to additional input items like local cuts and fills, existing site constraints, scenery considerations, and the flood problem.

The alluvial fan, being a well-adjusted transportation surface in equilibrium with the long-term spectrum of flood events, will effectively seek to correct any deviation - positive or negative - from that surface (Beaty, 1974; Bull, 1964; Denny, 1965; Hooke, 1967; Leopold and Miller, 1956; Lustig, 1965). Because of the sporadic nature of desert flood events in both space and time, the equilibrium restoration effect may not be visible during the period immediately following the construction of the road. Statistics showing damage at relatively few places during just a few years can be misinterpreted to be the normal state of affairs to be routinely encountered. However, the road which crosses an alluvial fan constitues a disturbance in the dynamic geomorphic sense. The cross-surface of the road is sub-horizontal, whereas the fan surface on which it rests is sloping. Therefore either the upslope roadway boundary is below grade, or the downslope boundary is above grade; and sometimes both deviations exist.

During the flood event of February 1972 in the Eastern Sinai, an estimated minimum sediment volume of 10,000 - 20,000 m^3 was discharged from any medium-size (50 - 100 km^2) drainage basin, several sediment settling basins, dug upstream of the road, had capacities of only 500 to 2500 m^3. These basins, intended to attenuate the erosive energy of the floodwaters prior to reaching the road, were obviously incapable of performing any useful function; they silted up within minutes.

Engineers, mostly trained in temperate environments, tend to attempt a total control of the drainage problem, which means bridges for the major channels and drainage ditches leading to culverts along the roadway. The engineer is more deterred by the cost of such an enterprise, considerable in relation to the transportation benefit afforded by the road even in this area of universally cheap earthmoving equipment, than by the advice of the geomorphologist. This advice can be summarized as follows :

(1)   The road surface should adhere to the original fan surface as closely as possible. Available evidence indicates that exposure to flood damage increases with vertical deviation of the road structures from the grade line.

(2)   Sediment settling basins located on arid alluvial fans are ineffective. For all but insignificant flows, they are filled with sediment during the first minutes or even seconds of a flood. To make them effective, they must attain a capacity of at least one tenth of the total water volume of a typical flood event. Large holes that size are difficult to dig, have to be re-excavated periodically, and might incur the wrath of nature lovers.

(3)   In all cases examined by the author, bridgeless crossings were preferable to culverts. The crossings are, on the whole, less expensive, and entail a much smaller overall deviation from the grade surface of the fan. Further, it is possible to design them carefully in ways that would locate them (i) on the paths of the most probable flow lines;   (ii) at right angles to these flow lines; and (iii) vertically positioned slightly below the grade surface, so that during flows, they would be covered by a thin

veneer of sediment which helps to protect the road surface from
erosion.

These procedures require the services of a proper geomorphic
survey, which has to precede the detailed planning stage (Brunsden
et al., 1975; Griffiths, 1978).

In contrast to bridgeless crossing, culverts silt up easily,
often require raised embankments, and entail the construction of
lead ditches, which are loci of lateral erosion.

(4) Drainage ditches parallel to the roadway on its up-fan
side do not serve any demonstrably useful purpose except for very
small flows, which can be routinely handled in any case. Necessary
periodic maintenance is a further disadvantage.

HUMAN ACTIVITY AND EROSION IN DESERTS

The environmental balance between erosion and conservation is
highly fragile in arid and semi-arid zones. Slight changes in
climate, land use, or technological impact may tip the balance from
acceptable rates of erosion to detrimental ones. The opposite is
also true: reduction of grazing can contribute significantly to a
diminution of the erosion problem.

The feedback relation involved in the "human activity --
erosion" system is in reality highly complex, and the understanding
of its mechanics is markedly deficient. Several examples cited
below serve to illustrate this complexity.

Comparison of sediment discharge records at the Grand Canyon
gauging station on the Colorado River for two periods (1926-1941
and 1942-1960) show that the suspended sediment loads for the
latter period are only 50 per cent of those for the earlier period
(Hadley, 1974). In the past 20 years, livestock numbers have been
greatly reduced on those arid and semi-arid rangelands, and erosion
control practices have been initiated in many severely eroding areas.
These can reduce sediment yield by as much as 40 per cent. The
earlier erosion period is undoubtedly associated with widespread
aggradation in valley bottoms, documented by numerous studies (for
example Emmett, 1974). This aggradation may signify a reversal of
the widespread trend of arroyo cutting, which began in about 1880.
Observed rates of valley alluviation indicate that valley bottoms
would fill to the level of the old valley floor within a period of
200-700 years.

It is not clear at present whether recent reversals such as
those indicated by Hadley (1974) are a persistent feature, or a

shortlived phase, perhaps aided by small but effective changes in climatic factors, such as rainfall intensity. The period coincident with the advent of heaviest grazing in the American Southwest (1850-80) was characterized by a deficiency of the low-intensity rains which support vegetation, and by more than the average number of heavy rains, which could act upon a weakened vegetal cover and promote erosion (Leopold, 1951).

Considerable evidence points to a coincidence of increasing aridity with degradation, and increasing humidity with aggradation, but there are opposite views. The role of unconcentrated wash erosion, whose major importance was only recently realized and measured by geomorphologists, makes it quite possible for massive aggradation to occur in even very small tributary channels only several tens of metres from the boundary of the drainage basin. Gully erosion, which may have been insignificant during the depos- ition of the main alluvial fill in the valleys, assumes an important role during periods of valley trenching (Leopold et al., 1966). On a large scale, such as in the Sahara, evidence is accumulating to suggest major changes in fluvial regime during the last 10,000 years (Pachur, 1980; many others). On the local scale, the responses to even slight climatic changes are extremely subtle, and depend considerably on environmental factors such as lithology, aspect, and many others (see, for example, Bull and Schick, 1979).

A recent and detailed study documented the acceleration of soil loss under a range of controlling variables in the semi-arid part of Kenya (Dunne et al., 1978). The findings indicate a three to four- fold increase in the erosion rates for large areas since the early 1960's. In a comparable study in a semi-arid part of Tanzania (Rapp et al., 1974), erosion control was found to be mainly a question of better management of the grass cover in the drainage basins and the protection of cultivated fields against splash erosion, with gully erosion being less important. Much of the product of accelerated erosion can be correlated with aggradation forms in the valleys, such as sand sheets.

In semi-arid Tunisia; agricultural areas flourished during Roman times, but they have since deteriorated due to overgrazing, deforestation, over-cultivation, and more recently, mechanical ploughing (Bonvallot and Hamza, 1974). The resulting state of catastrophic erosion is a combination of deflation, sheetwash, erosion by ravines rapidly developing into badlands, and caving of channel banks. In ten years, the proportion of the valley area dissected by ravines increased from 27 % to 35 %, thus the soil from nearly 1 % of the area is lost every year. Parallel to that, a spectacular widening of the main channels downstream takes place: in one case the width increased by 40 % in ten years (Bonvallot and Hamza, 1974).

Vegetation is naturally at a bare minimum in truly arid areas, and geomorphically the role of vegetation as a surface protector is taken up by protective stony pavements formed by vertical concentration and packing of large particles resistant to weathering. Although conventional agricultural disturbance can be practically ruled out in these areas, other forms of human interference with the terrain cause disruption in the state of quasi-equilibrium supported by the widespread armouring. Large scale warfare such as has occurred in the Sinai in the last decades causes systematic breaking of the protective crust by vehicle tracks, which in turn serve as a preferred locus for gullying. Gravel quarrying near desert towns, which is invariably located in active channel beds, leaves large amounts of material of sieved and unused size to lie waiting for entrainment by the next flood, thus increasing the sedimentation effect of the flow further downstream, as well as the potential for lateral erosion. Furthermore, the big ugly holes left behind in fully exploited and abandoned channel areas, are only filled by sediment after relatively long periods and thus become an aesthetic blemish.

Large scale interference with the desert pavement can be due to a variety of factors. In the Central Negev Highlands, hundreds of thousands of stone mounds cover wide areas. These mounds are the result of systematic stone clearing done by ancient farmers in order to increase slope runoff to fields in the valley below (Evenari et al., 1971). This action caused a new cycle of aggradation in the tributary valleys, only in part aided by check dams. Later, the cleared surfaces between the mounds recovered much of their armouring properties, and concomitantly a trenching cycle developed in the valleys, which was aided by the deterioration of the check dams due to human neglect.

Large areas in the Negev, notably on Pleistocene fan surfaces but also on other lithologies, are covered with trails formed by grazing animals, which were more abundant during times of greater vegetation (Bull and Schick, 1979). Thus the wetter conditions such as occurred in even the driest deserts during the Pleistocene and early Holocene (for example, Gerson and Yair, 1974), produced contrasting geomorphic effects: the trails are remarkably stable in terms of erosion, as they are composed of gently sloping tightly packed benches separated by scarplets of steep angular stones, much like a system of conservation terraces. The implications for modern parallels of land use and management are obvious.

Quite apart from regional changes in environmental conditions or in land use which affect the geomorphic balance, localized interference can have effects of very large magnitude. With the closure of the High Dam at Aswan, 132 million tons of sediment annually are prevented from travelling 'naturally' through the lower Nile Valley to the delta, and the mean annual deposition 1.0 mm thick on the

Nile floodplain has stopped (Shalash, 1977). Although no serious changes in erosion of the riverbed and banks have so far been recorded (Rzoska, 1976), the results obtained in numerous studies on degradation below dams in the United States and elsewhere (Leopold et al., 1964, p. 454-456; Dolan et al., 1974) are virtually certain to become apparent sooner or later in the geomorphology of the lower Nile Valley.

WATER SUPPLY AND CONSERVATION IN DESERTS

There are basically two approaches to the problem of augmenting water supply in arid lands: increasing the supply of usable water, and reducing the demand for water. Supply and demand, as well as delivery, have to be considered as an integral system (National Academy of Sciences, 1974).

Likewise surface water and groundwater must be considered together because they are actually different manifestations of one and the same thing (Leopold, 1967). Groundwater sources in the desert, unless replenished by subsurface flow from mountains outside the desert, are often found to be inadequately recharged and sometimes provide water collected under more humid conditions that existed in the geological past. Others use preferred percolation loci often located far from the trapping site.

Several of the many methods available for enhancing water supplies and improving water conservation are based on intimate onsite knowledge of climatic, hydrologic, and terrain conditions. Rainwater harvesting and runoff agriculture, for example, conserve water by minimizing the distance of overland flow from the point of impact of the raindrops to the point of useful disappearance below the surface, to be saved from the omnipresent evaporation. In the microcatchment systems developed for the northern and central Negev (Shanan and Tadmor, 1979), the "*negarin*" areas are so delimited as to provide a mean length of surface flow of no more than a few metres, terminating at the roots of a single fruit tree. A variant of the same principle of conservation is trickle irrigation - a system of plastic pipes placed on the soil among the plants. Water carried in the pipes drips onto the soil beside each plant at a rate carefully matched to the plant's needs. Longer pathways of surface flow require more elaborate conveyance systems, or special terrain assemblages, with concomitant problems of erosion, seepage, and maintenance being more difficult to overcome, though not insurmountable (Evenari et al., 1971; Billy, 1981; Wildenhahn, 1981; many others). Some of the simplest water-harvesting systems in Arizona collect 20 - 40 % of the precipitation for later beneficial uses, while a more elaborate system can collect more than 90 % (Flug, 1981).

Many interesting and efficient terrain adaptations have been
tried with the intent of increasing water-harvesting efficiency,
rainfall collection off galvanized iron farmhouse roofs in arid
Australia being a well-known example. The ancient inhabitants of
the Negev highlands cleared catchments of stones, causing an increase
of average annual water yields by 24 - 49 % (Shanan and Schick,
1980). Treating the terrain with sodium salts increases the conduc-
tivity of water, decreases the infiltration rate, and enhances runoff. Treating with other materials like wax compounds, or covering
with stone-covered polyethylene sheets are other possibilities.
For sandy areas typical of parts of the Sahel, a buried plastic
membrane collector has been developed; water yields of 30 - 60 %
are suggested by simulated rainfall trials. If validated in the
field, this well distributed supply system would significantly
improve the existing imbalance in water availability between a few
favoured, and consequently overgrazed points and the large intervening ungrazed areas (Shanan et al., 1981).

Water conservation by reduction of evaporation from water
surfaces using liquid chemicals or floating barriers, the reduction
of seepage losses by various methods, and reducing evaporation from
soil surfaces with various mulches are all technically feasible and
sometimes economically justifiable. However, these and other conservation measures, as well as all methods of supply enhancement, rely
on an intimate knowledge of the basic inventory of climate, water,
and terrain characteristics and processes. The ancient Negev farmer
could not have succeeded in his agricultural exploits without an
undoubtedly painstaking learning process provided by his natural
environment, and sometimes managing to improve on it (Evenari et al.,
1971). Today we have a much better scientific base, but this is
often more than counterbalanced by the overwhelming pressure of a
lot of water being required quickly, no matter the monetary and
environmental cost. As a relatively simple example of recent and
highly applicable research effective rainfall on wind-facing slopes
can be as much as three times greater than on lee slopes (Sharon,
1980). Such a difference can mean either success or total failure
for many rainfall harvesting catchments. Considering the marginal
conditions of arid zones, the statement "if a technology looks as
though it is immediately transferable, look again" (Dregne, 1979)
seems particularly applicable to deserts.

REFERENCES

Baker, V.R., 1977, Stream-channel response to floods, with examples
    from Central Texas, Bulletin of the Geological Society of
    America, 88: 1057.
Beaty, C.B., 1963, Origin of alluvial fans, White Mountains,
    California and Nevada, Annals of the Association of American
    Geographers, 53: 516.

Beaty, C.B., 1968. "Sequential study of desert flooding in the
     White Mountains of California and Nevada," US Army Natick
     Laboratories, Mass.

Beaty, C.B., 1974, Debris flows, alluvial fans, and a revitalized
     catastrophism, Zeitschrift fur Geomorphologie, Supplementband,
     21: 39.

Billy, B., 1981, Water harvesting for dryland and floodwater farming
     on the Navajo Indian Reservation in: "Proc. U.S.-Mexico
     Resource Workshop on Rainfall Collection for Agriculture in
     Arid and Semiarid Regions, Tucson, Ariz.," 1980, G.R.Dutt,
     C.F.Hutchinson, and M.A.Garduno, eds., Commonwealth Agric.
     Bureaux, Slough, UK.

Bonvallot, J. and Hamza, A., 1977, "Causes et modalites de l'erosion
     dans le bassin versant inferieur de l'Oued El-Hadjel (Tunisie
     centrale)," International Association of Hydrological Sciences,
     Paris.

Brunsden, D., Doornkamp, J.C., Fookes, P.G., Jones, D.K.C., and Kelly,
     J.M.M., 1975, Large scale geomorphological mapping and high-
     way engineering design, Quarterly Journal of Engineering
     Geology, 8: 227.

Bull, W.B., 1964, "Geomorphology of segmented alluvial fans in
     western Fresno County, California," US Geological Survey,
     Washington D.C.

Bull, W.B., and Schick, A.P., 1979, Impact of climatic change on
     an arid watershed: Nahal Yael, Southern Israel, Quaternary
     Research, 11: 153.

Claude, L., and Loyer, J.-Y., 1977, "Les alluvions deposees par
     l'Oued Medjerdah lors de la crue exceptionelle de mars 1973
     en Tunesie: aspects quantitatif et qualitatif du transport et
     du depot," International Association of Hydrological Sciences,
     Paris.

Cooke, R.U., Brunsden, D., Doornkamp, J.C., and Jones, D.K.C., 1982,
     "Urban geomorphology in drylands", Oxford University Press,
     Oxford.

Cooke, R.U., and Doornkamp, J.C., 1974, "Geomorphology in
     Environmental Management", Clarendon, Oxford.

Denny, C.S., 1965, Alluvial fans in the Death Valley region,
     California and Nevada, U.S. Geological Survey, Paper,
     Washington D.C.

Dolan, R., Howard, A., and Gallenson, A. 1974, Man's impact on the
     Colorado River in the Grand Canyon, American Scientist, 62:
     392.

Dregne, H.E., 1979, Technological limitations to arid zone develop-
     ment in: "Advances in Desert and Arid Land Technology and
     Development", A. Bashay and W.G.McGinnies, eds., Harwood,
     Chur.

Dubief, I., 1954, "Essai sur l'hydrologie superficielle au Sahara",
     Birmandreis, Alger.

Dunne, T., Dietrich, W.E., and Brunengo, M.J., 1978, Recent and
     past erosion rates in semi-arid Kenkay, Zeitschrift fur Geo-
     morphologie, Supplementband, 29: 130.

Dunne, T., and Leopold, L.B., 1978, "Water in Environmental
     Planning," Freeman, San Francisco.

Emmett, W.W., 1974, Channel aggradation in Western United States
     as indicated by observations at Vigil Network sites, Zeit-
     schrift fur Geomorphologie, Supplementband 21: 52.

Evenari, M., Shanan, L., and Tadmor, N., 1971, "The Negev, the
     challenge of a desert." Harvard, Cambridge, Mass.

Feldman, Y., 1976, "The 1975 flood of Wadi El-Arish," Unpublished
     report, Department of Geography, The Hebrew University of
     Jerusalem, Jerusalem.

Flug, M., 1981, Production of annual crops on microcatchments in
     "Proc. U.S.-Mexico Resource Workshop on Rainfall Collection
     for Agriculture in Arid and Semiarid Regions, Tucson, Ariz.,
     1980", G.R.Dutt, C.F.Hutchinson, and M.A.Garduno, eds.,
     Commonwealth Agric. Bureaux, Slough, UK.

Gerson, R., and Yair, A., 1974, Geomorphic evolution of some small
     desert watersheds and certain paleoclimatic implications
     (Santa Katherina area, Southern Sinai). Zeitschrift fur
     Geomorphologie,Supplementband, 19: 66.

Gilead, D., 1975, "Preliminary hydrologic appraisal of the Wadi
     El-Arish Flood", Israel Hydrological Service, Jerusalem.

Glancy,P.A., and Harmsen, L., 1975, "A hydrologic assessment of
     the September 14, 1974, flood in Eldorado Canyon, Nevada",
     US Geological Survey, Washington D.C.

Griffiths, J.S., 1978, "Flood assessment in ungauged semi-arid
     catchments as a branch of applied geomorphology", Geography
     Department, King's College, London.

Hadley, R.F., 1974, "Sediment yield and landuse in Southwest
     United States", International Association of Hydrological
     Sciences, Paris.

Hashemile Kingdom of Jordan Hydrology Division, 1966, "Floods in
     Southern Jordan on 11th March, 1966", Unpublished report,
     Hashemile Kingdom of Jordan, Hydrology Division, Amman.

Hedman, E.R., 1970, "Mean annual runoff as related to channel
     geometry of selected streams in California", US Geological
     Survey, Washington D.C.

Hooke, R. LeB, 1967, Processes on arid-region alluvial fans,
     Journal of Geology, 75, 438.

Jaekel, D., 1980, Current weathering and fluvio-geomorphological
     processes in the area of Jabal as Sawda' in: "The Geology
     of Lybia, Vol. III", M.J.Salem and M.T.Busrewil, eds.,
     Academic Press, London.

Karmon, Y., 1976, Eilat, problems of a port on a desert coast in
     "Geography in Israel", D.H.K. Amiran, and Y. Ben-Arieh, eds.,
     Israel National Committee, International Geographical Union,
     Jerusalem.

Labib, T.M., 1980, Soil erosion and total denudation due to flash
     floods in the Egyptian eastern desert, Journal of Arid
     Environments, 4: 191.

Last, Y. 1974, Inner-berm discharge and drainage area in the eastern
    Sinai, unpublished M.Sc. thesis, Department of Geography, The
    Hebrew University of Jerusalem, Jerusalem.

Lekach, J., and Schick, A.P., 1980, The relationship between rain-
    fall and erosion in Nahal Yael, in: "Arid Zone Geosystems -
    A Research Report", A.P.Schick, ed., Department of Physical
    Geography, Institute of Earth Sciences, Hebrew University,
    Jerusalem.

Lekach, J., and Schick, A.P., 1982, Suspended sediment in desert
    floods in small catchments, Israel Journal of Earth Sciences,
    31: 144.

Leopold, L.B., 1946. Two intense local floods in New Mexico, Trans-
    actions of the American Geographical Union, 27: 535.

Leopold, L.B., 1951, Rainfall frequency, an aspect of climatic
    variation, Transactions of the American Geophysical Union,
    32: 347.

Leopold, L.B., 1967, Man and climate, in: "Arid and Semi Arid
    Lands - a Preview", International Centre for Arid and Semi-
    Arid Land Studies, Lubbock, Texas.

Leopold, L.B., Emmett, W.W., and Myrick, R.M., 1966, "Channel and
    hillslope processes in a semi-arid area, New Mexico",
    US Geological Survey, Washington D.C.

Leopold, L.B., and Miller, J.P., 1956, "Ephemeral streams: hydraulic
    factors and their relation to the drainage net", US Geol-
    ogical Survey, Washington D.C.

Leopold, L.B., Wolman, M.G., and Miller, J.P., 1964, "Fluvial
    Processes in Geomorphology", San Francisco, Freeman.

Lustig, L.K., 1965, "Clastic sedimentation in Deep Springs Valley,
    California", US Geological Survey, Washington D.C.

Moore, D.O., 1968, Estimating mean runoff in ungaged semi-arid
    areas, Bulletin of the International Association of Scientific
    Hydrology, 13: 29.

National Academy of Sciences, 1974, "More Water for Arid Lands -
    Promising Technologies and Research Opportunities", National
    Academy of Sciences, Washington.

Osterkamp, W.R., Lane, L.J., and Foster, G.R., 1983, "An analytical
    treatment of channel-morphology relations", US Geological
    Survey, Washington D.C.

Pachur, H.-J, 1980, Climatic history in the late quaternary in
    southern Lybia and the western Lybian desert, in: The Geology
    of Lybia", Vol. III, M. J. Salem and M. T. Bursewil, eds.,
    Academic Press, London.

Rapp, A., Murray-Rust, D. H., Christianson, C., and Berry, L., 1974,
    Soil erosion and sedimentation in four catchments near Dodoma,
    Tanzania, Geografiska Annaler, 54A: 255.

Rzoska, J., 1976, A controversy reviewed: the arguments over the
    Aswan High Dam on the River Nile continued, Nature, 261: 444.

Salmon, O., and Schick, A.P., 1980, Infiltration tests, in: "Arid Zone Geosystems - A Research Report", Department of Physical Geography, Institute of Earth Sciences, Hebrew University, Jerusalem.

Schick, A.P., 1971, A desert flood: physical characteristics, effects on Man, geomorphic significance, human adaptation - a case study of the Southern Arava watershed, Jerusalem Studies in Geography, 2: 91.

Schick, A.P., 1974, Formation and obliteration of desert stream terraces -- a conceptual analysis, Zeitschrift fur Geomorphologie, Supplementband, 21: 88.

Schick, A.P., 1974b, Alluvial fans and desert roads - a problem in applied geomorphology, Abhandlungen der Akademie der Wissenschaften in Gottingen, Mathematische-Physikalische Klasse, III, Folge, 29: 418.

Schick, A.P., 1977, A tentative sediment budget for an extremely arid watershed in the southern Negev, in: Geomorphology in Arid Regions, Proceedings, 8th Annual Symposium in Geomorphology, D.D. Doehring, ed., Binghamton, N.Y.

Schick, A.P., and Lekach, J., 1981, High bedload transport rates in relation to stream power, Wadi Mikeimin, Sinai, Catena,8: 43.

Schick, A.P., and Sharon, D., 1974, "Geomorphology and climatology of arid watershed", Department of Geography, The Hebrew University of Jerusalem, Jerusalem.

Scott, A.G., and Kunkler, J.L., 1976, "Flood discharges of streams in New Mexico as related to channel geometry", US Geological Survey, Albuquerque, N.M.

Sewell, W.R.D., 1961, Human response to floods, in: "Water, Earth and Man," J.R.Chorley, ed., Methuen, London.

Shalash, M.S.E., 1977, "Erosion and solid matter transport in inland waters with reference to the Nile basin", International Association of Hydrological Sciences, Paris.

Shanan, L., Morin, Y., and Cohen, M., 1981, A buried membrane collector for harvesting rainfall in sandy areas, in: "Proc. U.S.-Mexico Resource Workshop on Rainfall Collection for Agriculture: Arid and Semi-arid Regions, Tucson, Ariz., 1980", G.R.Dutt, C.F.Hutchinson, and M.A.Garduno, eds., Commonwealth Agric. Bureaux, Slough, UK.

Shanan, L., and Schick, A.P., 1980, A hydrological model for the Negev Desert Highlands: effects of infiltration, run-off and ancient agriculture, Hydrological Sciences Bulletin, 25:269.

Shanan, L., and Tadmor, N.H., 1979, "Micro-catchment Systems for Arid Zone Development - a Handbook for Design and Construction, Hebrew University, Jerusalem and Centre of International Agricultural Cooperation, Ministry of Agriculture, Rehovot, Israel.

Sharon, D. et al., 1976, "A model for the distribution of the effective rainfall incident on slopes and its application at the Sdeh Boqer experimental watershed", Department of Geography, The Hebrew University of Jerusalem, Jerusalem.

Sharon, D., 1972, The spottiness of rainfall in a desert area, Journal of Hydrology, 17: 161.

Sharon, D., 1980, The distribution of hydrologically effective rainfall incident on sloping ground, Journal of Hydrology, 25: 165.

Sheaffer, J.R., 1961, Flood-to-peak interval, in: "Papers on Flood Problems," G.F.White, ed., Department of Geography, University of Chicago, Chicago.

Stuckmann, G., 1969, "Les inondations de septembre-octobre 1969 en Tunisie: Partie II: Effets morpholgiques". UNESCO, Paris.

University of Jerusalem Geography Department, 1972, "Floods of 23rd-25th November, 1972 in relation to the Eilat-Sharm-esh-Sheikh road, (Floods in Eastern Sinai Mimeographed report No. 3), Hebrew University of Jerusalem, Department of Geography, Jerusalem.

Walton, K., 1969, "The Arid Zones", Hutchinson, London.

White, G.F., 1964, Choice of adjustment to floods, Department of Geography, University of Chicago, Chicago.

Wildenhahn, E., 1981, Hydrological and agricultural conditions in south-west Saudi Arabia, Applied Geography and Development, 17: 57.

Wolman, M.G., and Gerson, R., 1978, Relative scales of time and effectiveness of climate in watershed geomorphology, Earth Surface Processes, 3: 189.

Woolley, R.R., 1946, "Cloudburst, floods in Utah 1850-1938", US Geological Survey, Washington D.C.

Yair, A., and Lavee, H., 1976, Run-off-generative process and run-off yield from arid talus-mantled slopes, Earth Surface Processes, 1: 235.

Yair, A., and Danin, A., 1980, Spatial variation in vegetation as related to the soil moisture regime over an arid limestone hillside, northern Negev, Israel, Oecologia (Berl.), 47:83.

Yair, A., and Lavee, H., 1981, "An investigation of source areas of sediment and sediment transport by overland flow along arid hillslopes", International Association of Hydrological Sciences, Paris.

14

BIOLOGICAL ASPECTS OF FRESHWATER RESOURCES

B.R.S. Morrison

Freshwater Fisheries Laboratory
Department of Agriculture and Fisheries for Scotland
Pitlochry, Perthshire, PH16 5LB, Scotland

INTRODUCTION

Man's ability to adapt to his environment has enabled him to survive in extreme conditions on almost every part of the earth's surface. More recently he has developed machines in which he can explore the depths of the oceans, and by transporting himself to the surface of the moon he has been able to investigate regions not known to contain any living organism. Other plant and animal species cannot adapt so readily and must live within a limited range of environmental conditions. For any given taxonomic group, however, such as insects, birds, mammals and flowering plants, individual species have evolved which can make optimum use of the conditions in a given environmental niche. If one or more of these conditions becomes too extreme, the organism dies. Man's ability to invade these niches and destroy them while adapting them to his own needs has led to the extinction throughout the world of large numbers of animal and plant species. Unlike many of the physical features of man's environment, once an organism has become extinct, it cannot be replaced.

This chapter reviews a few of the changes at present being brought about in the freshwater environment by man's activities and the effects these are having on the flora and fauna. As a basis for discussion, animals found in many of the freshwater lochs (lakes) in central Scotland will be examined. Although these are all species found in western Europe, they may be regarded as representatives of animals in the same taxonomic groups elsewhere, since they have many similar environmental requirements. A typical environment for all the organisms included in Figure 1 is the Lake of Menteith near Stirling. The lake is about 17 m above sea-level,

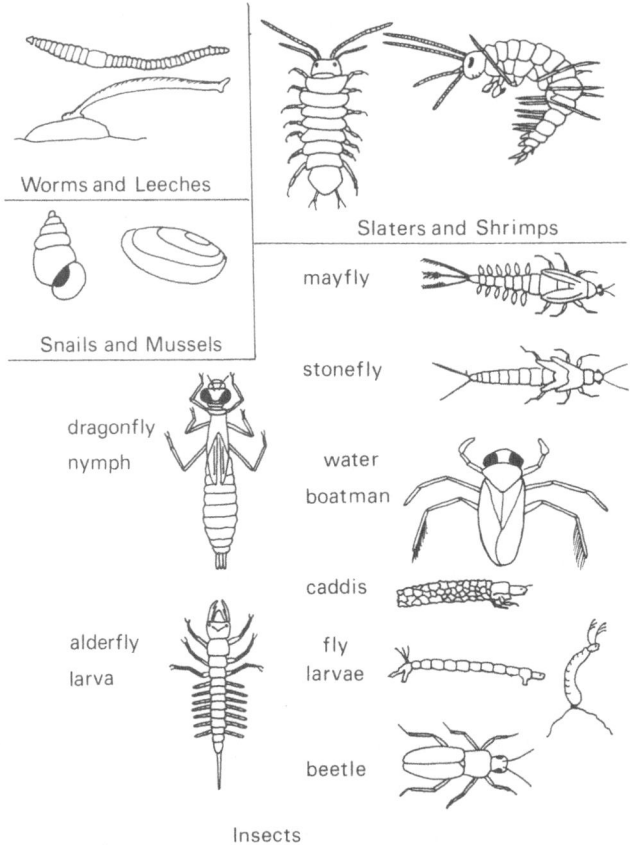

Worms and Leeches

Snails and Mussels

Slaters and Shrimps

mayfly

stonefly

dragonfly
nymph

water
boatman

caddis

alderfly
larva

fly
larvae

beetle

Insects

Figure 1:   Common invertebrate groups in the Lake of Menteith in
            central Scotland.

and lies on superficial deposits partly composed of boulder clay in
an area of Old Red Sandstone. It is about 260 ha in area, with an
average depth of 6 m and with about one third of the bottom under
less than 3 m of water. There is an extensive stony littoral
(< 2 m) zone on the east, west and north shores with beds of floating and submerged macrophytes, and a broad sandy littoral zone along
the south shore, which is bordered by coniferous woodland. A high
proportion of the remainder of the catchment is grassland used for
grazing. The pH of the Lake is about 7.2 and it has a calcium
carbonate alkalinity of about 24 mg/l. (see Figure 2, and diagram
of similar lake type Figure 5B).

   The lake has many features that enable it to support a wide range
of plants and animals; these include the broad littoral zone and
the availability of nutrients like phosphates and nitrates.
Aquatic macrophytes provide shelter for young fish and a substrate

Figure 2: Loch Ard and the Lake of Menteith, contrasting geology
and physical characteristics. Inset shows the lines of
the geological faults separating the low-lying, agriculturally more productive area of central Scotland from
the hillier Highlands and Southern Uplands.
▲ Loch Ard, ▼ Menteith, △ L. Leven.

for many of the invertebrates. A stony bed whether in a lake or
stream is more productive of species than a sandy bed because
spaces formed by irregularities in the shapes of the stones form
micro-habitats which are ideal for shrimps (*Gammarus* spp.) and many
stoneflies, while snails, mayflies, stoneflies, midge larvae
(Chironomidae) and beetles will browse on the algae and moss attached
to the stones themselves. In a sandy environment, on the other hand,
pea-shells (*Pisidium* spp.), midge larvae and worms form the bulk of
the fauna. The burrowing nymphs of the mayfly (*Ephemera danica*
Muller) are also sometimes found in a sandy substrate. The shallowness of the lake enables light to penetrate to the bottom over much
of the area so that plant life can flourish some distance from the
shore. Plant growth is rapid because the water warms quickly during
the summer months. Pike (*Esox lucius* L.), Perch (*Perca fluviatilis*
L.) and Roach (*Rutilis rutilis* (L.)) use the plants for spawning,
while trout (*Salmo trutta* L.) spawn in gravelly areas of the inflow-
ing streams. The lake is also a habitat for the minnow (*Phoxinus
phoxinus* (L.)), stickleback (*Gasterosteus aculeatus* L.) and eel
(*Anguilla anguilla* (L.).

The Lake of Menteith might be described as a *mesotrophic* or moderately productive lake. Although it has a wide range of invertebrates and the water chemistry is such that it can support a plant community rich in species, the quantity of living material per unit area is less than in a *eutrophic* or highly productive lake such as Loch Leven in the Fife Region of central Scotland. As a crude indicator of the production potential of a lake the alkalinity, expressed as mg/l calcium carbonate is sometimes used since it has been found that the more alkaline lakes generally support a greater biomass. The alkalinity of Loch Leven varies between 30 and 70 mg/l) (Holden and Caines, 1974), and is generally higher than the range (16 - 30 mg/l) for the Lake of Menteith. By contrast, many of the lakes in the upland areas of Scotland, in northern Europe and America are *oligotrophic*, that is they are less productive in terms of aquatic life, are acid (pH often < 6.5) and have low levels of many of the chemical elements normally found in fresh water, particularly phosphorus, nitrogen, calcium and magnesium. Most lakes of this type are in areas of slow-weathering rock such as granite or schist. An example in central Scotland is Loch Ard (Figure 2), about 7 km west of Menteith, which is situated in a steep-sided, U-shaped, glaciated valley just north of the Highland Boundary Fault (see Figure 5c). It has a very narrow littoral zone for much of its length, less than 5 m wide in many places, and 35 % of the bottom is at a depth of more than 15 m. The maximum depth is just over 30 m. This means that plant life is limited to small areas of reeds (*Phragmites communis* Trinius) and sedges (*Carex* spp.) along the shore, with patches of submerged weeds, such as *Littorella uniflora* L. and *Lobelia dortmanna* L. in the shallow water. There are no extensive beds of lilies or floating pondweeds such as are common in Menteith. The alkalinity of Loch Ard water is around 5 mg/l, i.e. about one fifth of the mean level for Menteith. Calcium, essential for building the molluscan shell and the crustacean exoskeleton, is scarce and this is reflected in the softness of the shells of the two species of snail present and the very local distribution of the shrimp *Gammarus*. The only mayflies found are acidophilic species. Not all oligotrophic lakes are as steep-sided or as deep as Loch Ard, but they are all characterised by a scarcity of available calcium and low levels of chemical nutrients.

A comparison of the biological, chemical and physical characteristics of Loch Ard and the Lake of Menteith shows differences due primarily to geography and geology. When man alters his environment he may bring about similar changes or, as we shall see, more extreme changes which may result in the elimination of one or more animal or plant species.

ACID RAIN

The causes and effects of acid rain have recently attracted the attention of scientists and the general public in Europe and North

America. If the acidity of rain were due solely to carbon dioxide
and water in the atmosphere the pH of rain would be about 5.6.
Other natural factors, e.g. volcanic activity, influence rain acidity,
temporarily at least. Rainfall analysis indicates that present day
rain in Scotland is often less than pH 5 (Figure 3), and when due
account is taken of the proportion of different ions derived from
sea salt it is found that there is an excess of sulphate and to a
lesser extent of oxides of nitrogen (Harriman and Morrison, 1982).
These originate in the fossil fuels (mainly oil and coal) used in
manufacturing processes. In areas of slow-weathering rock such as
granite, gneiss, schist and slate, where the soil covering is peaty
and very shallow, there is insufficient calcium and magnesium in
the soil to combine with the excess sulphate in the rain, which
instead becomes linked with the more readily available hydrogen (to
produce sulphuric acid) and aluminium. Streams and lakes in such
areas are already deficient in calcium and bicarbonate ions and
cannot neutralise the incoming water, so that in the Loch Ard area
and in parts of south-west Scotland many waters have a mean pH of
4.5 or less. Nitrates derived from nitric acid in the rain are
readily utilised by plants and in the U.K. at least contribute less
to the acidification processes in soils and water. Freshwater
molluscs and crustacea are absent from such areas and only one
species of mayfly nymph has been found in the most acid streams.
The fauna which remain are generally air-breathing (beetles and
water bugs) or have tough chitinous exoskeletons (alderfly larvae
and dragonfly nymphs). A few species of caddis can survive in waters
of this type.

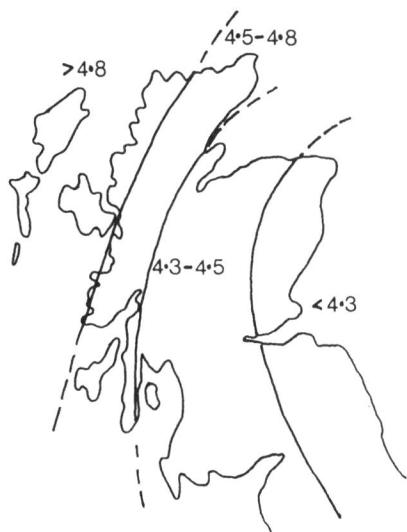

Figure 3: Distribution of mean pH of rain in Scotland.

The most noticeable effect of increasing acidity is a decline in fish catches. This has been observed in south-west Scotland, in Norway (Leivestad et al., 1976), in Sweden (Bengtsson et al., 1980), in Canada (Beamish et al., 1975) and elsewhere. Fish are unable to maintain the correct salt balance within their bodies in acid conditions and they may also suffer from the toxic effects of aluminium ions which are released from the soil in greater quantities in acid conditions. Concentrations of 0.2 - 0.3 mg/l Al at pH approximately 5.0 are known to be toxic to salmonid fish (Cronan and Schofield, 1979; Dickson, 1978), which react to the presence of the aluminium by secreting excess mucus around the gill filaments, thus hindering respiration. Finally, fish eggs may not hatch successfully since some hatching enzymes function best in slightly alkaline conditions (Haya and Waiwood, 1981), and water of very low pH entering an egg in the process of hatching soon renders the enzyme inactive.

As a short term measure to combat the increasing acidity of lakes several countries, for example Norway, Sweden and Canada, are experimenting with lime applications either directly to the water or to the catchment as a whole. Such operations are very expensive, and may be effective for only a few months in shallow waters with a rapid turnover. The long term solution is to reduce the quantities of sulphur dioxide released into the atmosphere, and this must be an important consideration when planning industrial complexes, in particular power stations.

FORESTS

Recent research (Harriman and Morrison, 1982) has indicated that coniferous forests planted in areas sensitive to the effects of acid rain (thin peaty soils on slow-weathering rock) may increase the acidity of the water entering streams and lakes. The trees are more effective than grass-covered moorland as collectors of atmospheric pollutants, which are washed off by the rain and eventually find their way into rivers and lakes. Differences in stream acidity in forested and non-forested catchments are illustrated in Figure 4. When forests are grown on richer soils, any increase in acidity is much less marked, presumably because there is sufficient sodium, potassium, calcium and magnesium present to exchange with the incoming hydrogen ion in the rain. In such areas, a full range of invertebrate species and healthy fish populations may be found. However, the uptake of water by trees reduces the amount entering streams and rivers, which can be particularly serious in dry weather. Adverse effects may also follow if the land is ploughed before being planted; the rate of surface water runoff increases and is accompanied by soil erosion, which in turn may lead to the silting up of spawning streams. Some of these problems can be avoided by contour-ploughing. Finally, when trees are planted close to the edge of streams light is greatly reduced as the canopy closes, thus reducing the productivity level of the stream.

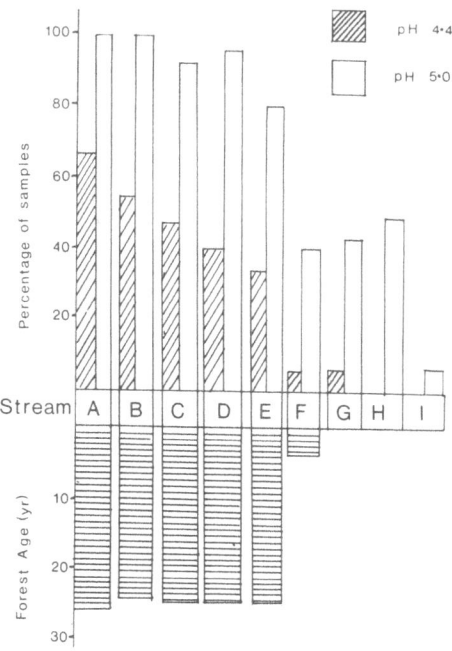

Figure 4: Comparison of the acidity of streams draining forested and non-forested catchments.

FLOODING

The creation of new reservoirs by flooding river valleys immediately produces a change from a flowing water to a still water
environment (Figure 5). (A discussion of this topic is given in an
Institute of Biology Symposium report; Lowe-McConnell, 1966).
Under these circumstances, species adapted to live only in running
water will be eliminated and replaced by still water animals and
plants. Many species of invertebrates can survive in both environments, but in still water they are likely to be confined to the
littoral zone where wave action and high levels of oxygen correspond most closely to river conditions. Perhaps the greatest effect
is seen in the restrictions placed on migratory fish by the dam
required to contain the water in the new reservoir. A fish ladder
or lift is then necessary to enable migrants to circumvent the
barrier.

In new reservoirs the change in depth of water is significant
for both animals and plants, particularly if the flooded valley is
similar to the steep-sided one already described in Loch Ard
(Figure 5c). A river is often not more than 2 - 3 m deep, except
perhaps in its lower reaches, and light is able to penetrate to the

river bed except in very turbid conditions. Plant life can therefore grow over most of the bed provided the substrate is suitable,
and air-breathing beetles, water bugs and snails have only a short
distance to travel to the water surface. The temperature and oxygen
content at all depths in a fast-flowing river are relatively uniform.
(Figure 5A).

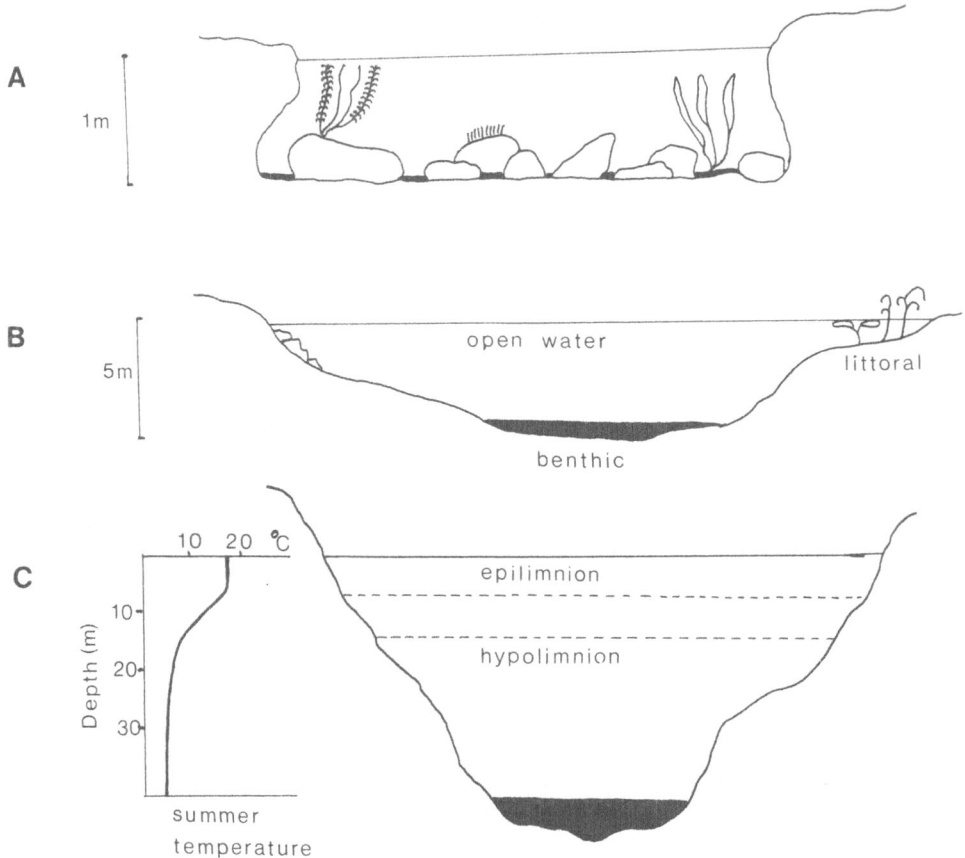

Figure 5: Differences in some of the physical characteristics of
rivers and lakes.

A. Cross-section of a river in its upper reaches.
B. Cross-section of a shallow lake with the names of
the main zones.
C. Cross-section of a deep, steep-sided lake formed by
damming a mountain river valley and showing strat-
ification of the open water zone. Temperature
graph shows the thermocline (change in temperature)
which develops during the summer months.

In contrast to this, reservoirs (and lakes) are often so deep that sufficient light for plant growth can penetrate only the upper layers of the open water zone and this limits the distribution of the microscopic phytoplankton, the main form of plant life in this zone. Zooplankton, also in the main microscopic, feed on these plants or on each other and are consequently most abundant in these upper layers, although some detritus feeders are able to survive at greater depths. In lakes of 15 m depth or more there may be a pronounced difference in temperature between the warmer surface waters (epilimnion) and cooler waters at greater depth (hypolimnion) during the summer months and this, together with corresponding differences in oxygen concentrations, further limits the distribution of many freshwater organisms. The lake bottom, or benthic zone, is an area of organic muds with low oxygen concentrations inhabited by midge larvae and worms whose blood contains haemoglobin, the most efficient oxygen carrier of all the blood pigments. At depths greater than 10 m plant life may be absent unless the water is very clear. (Figure 5c).

Although the littoral zone of a lake or reservoir is normally regarded as the most productive, when regular abstraction of water takes place, for example for generating hydro-electricity, the drawdown of water leaves the littoral zone dry. Too frequent drying eventually destroys aquatic life in the area, and this in turn deprives many species of fish of their main source of food. In such situations, it may be necessary to consider stocking the reservoir with plankton-feeding species.

When land is first flooded, fish generally show an increase in growth rate due to the sudden availability of new food sources. In a year or two, however, the growth rate slows down and, depending on the physical characteristics of the new reservoir, may stabilise at a new level or may continue to decline.

FERTILISERS AND EUTROPHICATION

A further long term effect of man's actions follows the application of fertiliser to a catchment or the release of sewage or detergents (which contains phosphates) into a water course. Agricultural fertilisers are designed to improve plant growth on land, and when they are washed into rivers and lakes by heavy rain, or leached from the soil, they have the same effect on aquatic plants. Initially, macrophytes such as lilies (*Nuphar, Nymphaea*) and pondweeds (*Potamogeton* spp.) may show an increase in growth rate, but if dense algal blooms develop, these reduce light penetration and the larger plants eventually die back. Accompanying this process may be an increase in pH of the water, which may prove fatal to more sensitive fish species such as trout (*Salmo trutta*).

In agriculture, fertilisation of arable or grassland may be an annual event and leaching may therefore continue for many years. Research on the effects of a single application of fertiliser in a forest showed that certain elements may be detected in streams at higher than normal levels for up to 3 years (Table 1).

Table 1:   Nutrient leaching into streams after fertilisation of a coniferous forest (after Harriman, 1978).

Element	% loss from forest soils	Time taken to return to pre-treatment levels in the streams
Phosphorus	15	3½ years
Nitrogen	4	3 years
Potassium	20	2 years

The examples of human activities just described can produce damaging effects lasting for many years and, in some cases, irreversibly change the environment and the range of plants and animals living in it. The following two examples illustrate shorter term effects which may disturb the ecology of a water body for a few weeks or perhaps a year, but there is then apparently a return to a situation similar to the one which existed before treatment took place.

INSECTICIDE APPLICATION

The application of insecticides and herbicides is common practice nowadays, the primary aim being to control only the target species without producing too drastic a reduction in the number of beneficial or harmless animals and plants.

In 1978 and 1979 the Forestry Commission was given permission to spray 5,000 ha of coniferous forest in the north of Scotland with the insecticide fenitrothion to control an outbreak of Pine Beauty moth (*Panolis flammea* Schiff.). A condition attached to this was that the spraying should be monitored to determine the effects on wildlife and human health. Many organisations, government, local authority and volunteer bodies worked together on this co-ordinated exercise.

The Freshwater Fisheries Laboratory at Pitlochry was responsible for monitoring the effects of fenitrothion on aquatic life. Samples of water from inside and outside the treated area were collected regularly throughout the initial observation period of 1 month; drift nets were set in streams to detect changes in the number of insects drifting downstream, and fish were held in cages in the stream, samples being taken at regular intervals for fenitrothion analysis. The results showed that as fenitrothion levels in the water decreased, levels in the fish also diminished, as did the numbers of insects caught in the drift nets. Quantitative sampling of the benthic fauna indicated that there was no obvious change in numbers before and after treatment, and electro-fishing results one week after the application showed that the trout population was still well distributed within the treated area. (Holden and Bevan, 1979, 1981; Morrison and Wells, 1981).

The use of this insecticide apparently had little effect on aquatic fauna, which may have been because of the low level of fenitrothion present in the water (max. 48 µg/l). Terrestrial insects other than Pine Beauty moth were affected however, and it was hoped that because the spraying was limited in extent, recolonisation of non-target species from neighbouring untreated forests would take place. The effect on birds and mammals dependent on insect life has still to be determined.

PISCICIDES

Introducing toxic substances into a river or lake in order to kill fish for food has been practised for centuries in Asia, Africa and South America. More recently these same poisons, in particular rotenone, a derivative of *Derris* root, have been used in Canada, the United States and Europe to eradicate unwanted fish species prior to stocking lakes with species more acceptable to the angler.

In many lake ecosystems, fish form the intermediate stage between the smaller animals and plants found in the littoral and open water zones and predatory birds and mammals. For example, removal of fish will have serious consequences for the osprey (*Pandion haliaetus* L.) and the otter (*Lutra lutra* L.) for which fish is the main item of diet. It has been found that rotenone kills most of the zooplankters in the open water at the time of application, but since many of these have resting stages in the bottom mud or are more resistant to rotenone at some stage in their life cycle, recovery of the population usually takes place within 1 - 2 years (Anderson, 1970). The effect on benthic fauna is much less serious (Morrison and Struthers, 1975; Morrison, 1977). The more recent discovery of the piscicidal properties of antimycin A has led to formulations being developed for use in fresh water. Studies have been made of its effects on fish and invertebrates (Lennon and Berger, 1970; Morrison, 1979).

The gap in the food chain created by poisoning is normally filled by the introduction of a different species of fish. Ideally, if disturbance to the ecology of the lake is to be minimised, the new species should have similar feeding and behavioural characteristics.

In Scotland, pike and perch are in general of less interest to the angling public than trout, and present-day fisheries based on trout in many cases started with the eradication of pike and perch. Trout in a lake have food requirements similar to those of the perch, and although the role of the fish-eating pike may be filled partly by large trout, which are also an acceptable food for native fish-eating birds and mammals, the main predator in the new situation is likely to be man. A fishery created for introduced species totally different from those eradicated is likely to produce greater changes in the environment, and an extreme case would be the introduction of the grass carp (*Ctenopharyngodon idella* Val.) into a water previously inhabited by insectivorous or carnivorous species. In this case the carp's preference for vegetable food would lead to considerable environmental changes.

The topics discussed above show how man's activities can upset the natural equilibrium. This is a dynamic equilibrium based on annual variation in numbers of different species, changes in weather conditions, migratory movements etc., which has become established in many freshwater environments and can be altered in such a radical way by man that it may never again return to its earlier state. A new equilibrium develops, but often even from the human point of view we may consider the new balance less desirable than the old as in the case of eutrophication due to leaching of fertilizer, or the destruction of fish populations by acid rain. These are important considerations in any planning programme for industry, agriculture or recreation. However, by regular monitoring of animals and plants, and by using our knowledge of their requirements we can often detect changes before it is too late to save the situation.

REFERENCES

Anderson, R.S., 1970, Effects of rotenone on zooplankton communities and a study of their recovery patterns in two mountain lakes in Alberta, J. Fish. Res. Bd. Can., 27: 1335.
Beamish, R.J., Lockhart, W.L., Van Loon, J.C. and Harvey, H.H., 1975, Long-term acidification of a lake and resulting effects on fishes, Ambio, 4:78.
Bengstsson, B., Dickson, W., and Nyberg, P., 1980, Liming acid lakes in Sweden. Ambio 2:34.
Cronan, C.S., and Schofield, C.I., 1979, Aluminium leaching response to acid precipitation, Effects on high elevation watersheds in the north west. Science 204:304.
Dickson, W., 1978, Some effects of the acidification of Swedish lakes, Verh. int. Verein Limnol. 20:851.
Harriman, R., 1978, Nutrient leachings from fertilised forest watersheds in Scotland. J. Appl. Ecol. 15:933.
Harriman, R., and Morrison, B.R.S., 1982, Ecology of streams draining forested and non-forested catchments in an area of central Scotland subject to acid precipitation, Hydrobiologia 88:251.

Haya, K., and Waiwood, B.A., 1981, Acid pH and chorionase activity
    of Atlantic salmon (*Salmo salar*) eggs, Bull. Environ. Contam.
    Toxicol., 27:7.
Holden, A., and Bevan, D., 1979, "Control of the Pine Beauty Moth
    by Fenitrothion in Scotland 1978", Forestry Commission, Edinburgh.
Holden, A., and Bevan, D., 1981, "Aerial application of insecticide
    against Pine Beauty Moth". Forestry Commission, Edinburgh.
Holden, A., and Caines, L.A., 1974(1972/73), Nutrient Chemistry of
    Loch Leven, Kinross. Proc. Roy. Soc. Edin. (B) 74:101.
Leivestad, H., Hendrey, G., Muniz, J.P., and Snekvik, E., 1976,
    Effects of acid precipitation on freshwater organisms, in:
    "Impact of acid precipitation on forest and freshwater ecosystems
    in Norway", F.H.Braekke, ed. SNSF Project Report No. 6/76, NISK,
    1432 As-NLH.
Lowe-McConnell, R.H., ed. 1966, "Man-made lakes", Institute of
    Biology Symposium No. 15, Academic Press, London.
Lennon, R.E., and Berger, B.L., 1970, "A resume of field applications
    of antimycin A to control fish", Investigations into Fish Control
    40, 1 - 19. U.S.Bureau of Sport Fisheries and Wildlife, Washington
    D.C.
Morrison, B.R.S., 1977, The effects of rotenone on the invertebrate
    fauna of three hill streams in Scotland, Fish. Mgmt.8:128.
Morrison, B.R.S., 1979, An investigation into the effects of the
    piscicide antimycin A on the fish and invertebrates of a
    Scottish stream, Fish. Mgmt. 10:111.
Morrison, B.R.S., and Struthers, G., 1975, The effects of rotenone
    on the invertebrate fauna of three Scottish freshwater lochs,
    Fish. Mgmt. 6:81.
Morrison, B.R.S., and Wells, D.E., 1981, The fate of fenitrothion
    in a stream environment and its effect on the fauna, following
    aerial spraying of a Scottish forest, Sci. Total Environ., 19:233.

COMMENTARY : WATER RESOURCES

The discussion showed clearly that, while conflicting demands
for competing uses of water resources crucially affect planning
options, there are generally no satisfactory institutional mechanisms
to enable appropriate decisions to be made for water resource allocation. 'Streamflow' defines the magnitude of the usable water resource, while its quality, at any location, depends on the uses and
stresses to which it has been subjected. 'Political' boundaries
are rarely in tune with the natural logic of flow systems, which
are delineated by drainage basins. Indeed rivers often form the
boundaries between management and planning authorities, and as a
result the stage is set for inevitable allocation conflicts that
arise from a multiplicity of different independent agents seeking to
serve a wide variety of interdependent interests that affect each
other - agriculture, forestry, industry, urban development, energy
conservation and recreation, to mention some of the more significant
uses. The key to logical planning and development is widely agreed
to be integrated and comprehensive basin management, covering major
drainage basins at the macro-level, and individual river basins at
regional and local levels. For example, in Scotland planning might
be concentrated on the Forth, Clyde, Tweed, Nith, Dee, Tay catchments
whereas in Canada the logic would focus on the major basins that
drain into the Pacific Ocean (containing the Columbia and Skeena
Rivers), the Arctic Ocean (essentially the Mackenzie-Laird system),
Hudson's Bay, and the Atlantic Ocean (such as the vast Great Lakes/
St. Lawrence River system). In reality, planning and management in
Canada are complicated by arbitrary international and provincial
boundaries, generally attuned to latitude and longitude rather than
natural features.

With institutional problems posed by arbitrary boundaries,
planners seldom have access to the full trans-boundary information
needed to combine, in a rational manner, the spectrum of social,
economic and scientific knowledge needed to develop the "best"
planning options, nor would the decision-makers have the full authority to implement them on both sides of that boundary.

Attempts to come to terms with arbitrary boundaries and the
consequent need to share responsibilities by the enactment of national
legislation have only been partly implemented in Canada, although some
provinces have made significant advances in establishing mechanisms
to rationalize competing interests.

253

The types of problem experienced when allocating water resources vary widely from country to country and according to climate. They are particularly severe in arid regions, even where water losses have been minimized. In Israel 90 - 95% of the available water is used; it is available to industry, agriculture and local authority users through a quota system. In temperate climates, forestry can have a profound affect at a distance; the afforestation of upstream areas can alter the quality of water available for downstream consumers. In addition to interception losses, biomass resulting from afforestation requires increasing quantities of water, the loss of which can alter the hydrological balance of an entire catchment affecting, in turn, the amount and quality of water available for other purposes. In some areas, where soils overlie slowly weathering granitic and gneiss bedrocks, the acidity of the low alkalinity water increases some years after planting conifers. This change has led to the application of lime so as to decrease the degree of acidity - a short term expedient sometimes used to sustain fish farming.

Like afforestation, the allocation of land for flood protection measures can trigger a series of problems in addition to those concerned with competing demands. These may involve legal liabilities if subsequent damage occurs to surrounding land due to alterations in hydrological balance. Whatever the solution to these problems may be, it is essential to realise that knowledge of terrestrial ecosystems has to be combined with that of aquatic ecosystems, the two being brought together to provide a full ecological basis for making decisions about resource allocations and land uses.

In the tropics the range of issues that need to be taken into account would always include the distribution of disease, for example, prevalence of onchocerciasis (river blindness) near fast flowing streams, the breeding ground of the black fly vector. By damming the river for agriculture or the generation of electrical power, the flow of the river, and its suitability for black fly would be decreased, but the mosquitoes and snails, the vectors and alternate hosts of malaria and bilharzia, flourish in stagnant water.

While there appear to be quality, quantity and use problems in most sectors of a river system, from those associated with forests, through farming and its associated problems of fertilizer run-off and damage done to non-target species by pesticides, to urban sewage and industrial waste disposal, the conflicts appear to be most severe at the urban/rural interface. Nevertheless, to be effective the planning process should address the entire river system if we are to cope with the different perceptions of different users. For example, if the objective is biomass production, the water over the dam is "wasted", if it is the preservation of agricultural land, then all development may be forced into the hinterland. While undeniably useful, cost benefit analysis, as a tool to resolve problems, must be approached with care because comparisons among alternative uses are not always judged on the same basis: inputs tend to be looked at quantitatively and outputs qualitatively, as for example when considering the need to install sewage treatment or the desirability of providing recreational facilities. But, more fundamentally, many environmental attributes are not, and never will be amenable to this form of analysis. Other methods, including decision analysis, should be considered.

A

IDENTIFICATION OF ECOLOGICAL FACTORS CHARACTERISING THE RANGE OF *ECOLOGICAL* HABITATS

ii-Rural
d-Conservation/Landscape

RATIONALES FOR CONSERVATION

R. Goodier

Nature Conservancy Council
12 Hope Terrace
Edinburgh, Scotland

INTRODUCTION

The Oxford English Dictionary gives two definitions of the word "rationale". The first defines it as - "a reasoned explanation of principles, an explanation or statement of reasons, a set of reasoned rules or directions", while according to the second it is - "a fundamental reason, the logical or rational basis of anything". Because no single rationale will suffice to provide an informative explanation of the processes in all fields of conservation action, I have chosen to consider conservation rationales, according to the first definition, as manifested in four main activities in Scotland which exemplify important classes of the total conservation enterprise. In the concluding section of the chapter I turn to the second definition and attempt to consider how the rationales which have evolved in particular fields of conservation action relate to more fundamental assumptions, the 'logical or rational basis' of the conservation enterprise generally.

Passmore (1974) points out that "to conserve is to save and the word conservation is sometimes used to include every form of saving, the saving of species from extinction, wilderness from land developers or fossil fuels and metals for future use". He attempts to limit the term to the saving of natural resources for future consumption and prefers to use the term preservation where saving is primarily a saving from, rather than a saving for. However, he later concedes that the utilitarian approach to preservation is a special case of the argument for conservation. The original broad concept of conservation as "wise use" of natural resources in a very general sense was first influentially expounded at the 1908

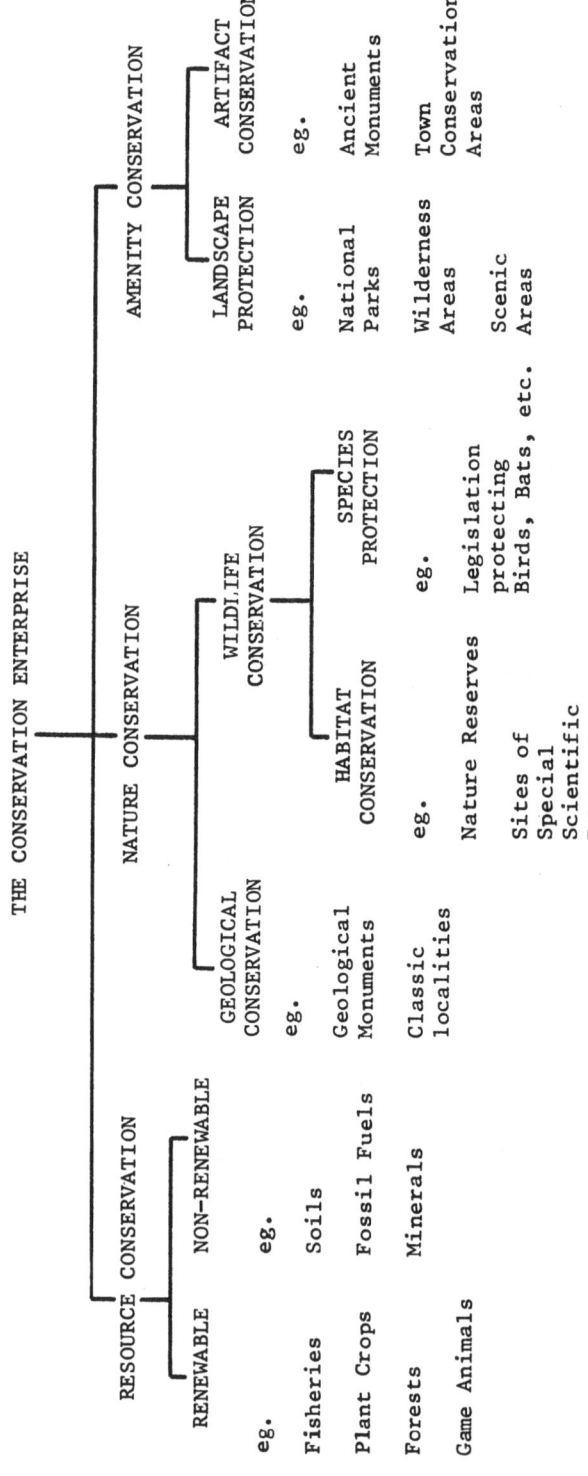

Figure 1: Structure of the conservation enterprise.

White House Conference under President Theodore Roosevelt. In a
later interpretation by Aldo Leopold (1949) conservation, as
applied to land, was held to be "our effort to understand and
preserve the capacity of land for self renewal". According to
Robert Arvill (1967), "modern conservation is a dynamic evolving
concept of co-partnership between man and nature. It requires a
strict management of each resource - land, air, water, and wildlife
- to ensure optimum value and continuity of supply. Conservation
of wildlife cannot be separated from resource management whether
for food supply, forestry, sport, or any of the other basic needs
of man".

In the most general terms, conservation is a way of using the
resources (in the widest sense) available to man, assuming that
humanity will continue to need and value the resources that we do
(- a biological/sociological assumption) and that we have a duty to
try to avoid making things worse for ourselves, for other men and
posterity by the way in which we use resources (- an ethical
assumption).

While Leopold's and Arvill's definitions of conservation quite
rightly emphasised what might be called the essentially holistic
character which appears to be demanded of the conservation
enterprise, because of ecosystem inter-relatedness, the fact remains
that conservation has, in practice, become partitioned into a
number of separate spheres of action.

THE STRUCTURE OF THE CONSERVATION ENTERPRISE (Figure 1)

The partitioning of the conservation enterprise, while in some
sense regrettable, derives understandably from its extreme complexity
and the diversity of its many components. The whole enterprise
appears to fall into three main sectors, the reality of which is
amply shown by the frequency with which these sectors are handled
by separate institutions, placing very different emphases on
conservation aims and on ideas underlying the conservation rationale.
The three main sectors relate to resource conservation, nature
conservation and amenity conservation. While these divisions are
real they are not necessarily absolutely distinct; indeed conservation projects within any of these three divisions often have
implications for the other two. Thus, from some perspectives nature
and amenity conservation can be seen as special cases of natural
resource conservation; alternatively nature conservation is sometimes
viewed as a special case of amenity conservation. While there is a
certain unity of conservation 'style' within the 3 major sub-divisions,
they are each capable of further refinement so that in discussions of
conservation matters, natural resources are often divided into those
which are potentially renewable, such as biological products which

Figure 2:   Main conservation land categories in Scotland.   Prime
            agricultural land as designated A+, A and B+ by the
            Scottish Development Department (1981).
            Nature conservation sites as graded 1 and 2 (over 200ha)
            by the Nature Conservation Review (Ratcliffe, 1977).
            National Scenic Areas as designated by the Countryside
            Commission for Scotland (1978).
            Preferred Coastal Conservation zone as designated
            by the Scottish Development Department (1974).

may be either natural, like marine fisheries, or cultivated, like
agricultural crops or tree plantations, and those which are not
renewable, such as fossil fuels. Alternatively, they may be
divided into those which are basic life-support systems, such as
soil, air, and water, for which substitutes do not exist, and
others where there is at least a hope that some utilizable
alternative can be found. An allied but slightly different perspective to the later is taken by the World Conservation Strategy
(International Union for the Conservation of Nature, 1980), where
attention is directed to 'living resource conservation for
sustainable development' within which it identifies three specific
objectives, firstly the maintenance of essential ecological
processes and life support systems, secondly the preservation of
the genetic diversity of living organisms and thirdly, the
continuing sustainable utilization of species and ecosystems.

To bring some of these subjects into sharper focus, particularly
those concerning conservation in land-use planning, I propose to look
at them as exemplified in four topics covered by the National
Planning Guidelines for Scotland (Scottish Development Department,
1981) and to consider in turn, the rationale which underlies the
conservation enterprise in these fields. I have adopted this
approach because it seems of value to try to analyse the rationales
underlying what is actually being done in selected conservation
fields rather than those related to theoretical proposals. This will
lead later to an examination of the interesting relationship of the
rationales with their premises on the one hand, and results of their
implementation on the other.

CONSERVATION ENTERPRISE WITHIN THE NATIONAL PLANNING GUIDELINES FOR
SCOTLAND

The National Planning Guidelines are concerned with the use of
land, with the different prescriptions being clearly conservation-
oriented even where the term conservation is not employed. The key
map accompanying the 1981 edition of the Guidelines (Figure 2),
leaving aside the relatively small areas of land reserved for large
industrial sites and petrochemical development, is dominated by four
categories:

1. Prime agricultural land.
2. Nature conservation sites.
3. National scenic areas.
4. Preferred conservation zones on the coast.

In the following sections I consider these four topics in turn
and, in so doing, try to give a "reasoned explanation of the
principles" governing conservation action in each.

The Conservation of Prime Agricultural Land

The National Planning Guidelines indicate a presumption against
the development of prime agricultural land in Scotland for purposes
other than for agriculture. Government, (Central Office of Information, 1970) has acknowledged that "one of the main objects of land
use policy is that agriculturally valuable land should not be taken
for other purposes when less productive land can reasonably be used
instead". The protection of high-class agricultural land against
development that would destroy or diminish its use for agricultural
purposes, with the corollary of its conservation for these purposes,
is one particular aspect of the wider field of soil conservation
which developed early on within the conservation tradition. This
took place in the United States, where the ravages of exploitive land
use became all too apparent during the 19th century and led to the
study of soil conservation and the establishment of a soil conservation service within the Department of the Interior in 1933. In
Britain, a more favourable climate and a relatively stable
agricultural system provided less impetus for action, and the
agricultural recession of the inter-war years delayed the recognition
of the need for conservation. The latter finally came about when it
was seen that the highest quality agricultural land was being lost
to urban and industrial development. This realisation was one of the
factors leading to the 1947 Planning Acts. It was given further
emphasis in the more recent 1971 and 1972 Planning Acts, which
introduced Structure Plans in which the proposed land allocations are
subject to prior consultation with the Agricultural Ministries. As
noted by Davidson (1980), this mechanism seems to have worked
reasonably well during the 1970s in conserving high quality agricultural land in England and Wales but less so in Scotland, where
a disproportionately high loss from agriculture, of the better
quality land, has continued. The assumptions behind the conservation
of agricultural land can clearly be related to the general conservation premises outlined in the introduction. Thus it is presumed,
not unreasonably, that there will be the same or increasing need for
food grown in the UK and that it is preferable to produce this on
the best agricultural land. It is furthermore assumed that the
benefits of doing this will, in general, outweigh the use of this
land for other purposes. The Land Use Summary Sheet on Agriculture
accompanying the National Planning Guidelines stresses that "agricultural land is a vital resource, giving the ability to produce
food in perpetuity and its value is not therefore to be measured
simply by comparisons between the monetory return of a short period
between food production and other uses". This defence of the
rationale hints that the premises on which it is based are not
unchallenged. This will be taken up in a later section of this
chapter, where the relation of conservation rationales to their
premises is examined.

The Conservation of Nature and Landscape (Amenity)

Before considering the separate cases for nature and landscape conservation which are exemplified in the National Planning Guidelines, it is valuable to consider how they have come to be considered as separate issues in Britain (in contrast to many other countries), enshrined in separate legislation and, so far as Government action is concerned, executed by separate agencies. This dichotomy is interesting because it represents two different ways of looking at the environment and a segregation which is only now being seriously questioned in the UK. There was a period, about 1945, when this dichotomy might have been avoided and, in order to understand how the present situation has come about, it is useful to examine the argument of that time, which have been usefully summarised by Cherry (1975) and Sheail (1976).

During a Parliamentary debate on 30 November, 1943, the then Minister of Town and Country Planning confirmed that the Government had accepted responsibility for preserving the natural beauty of the countryside. Included within the concept of natural beauty were the animals and plants of the countryside. At this period, the two strands of nature and landscape conservation were becoming closely entwined in the course of the promotion of the idea of National Parks for Britain. The Scott Committee, established to advise on the impacts of building and industrial development on the countryside, reported in 1942, and among its findings considered that the creation of a series of National Parks was long overdue. It also gave support to the establishment of National Nature Reserves, having taken into account the work of a conference on 'nature preserves'. This led to Dower (1945) being asked to make a study on the whole issue of National Park establishment and the wildlife preservation needs associated with them. Dower had strong views on the importance of establishing National Parks and on how they should be managed. He was, for example, very much in favour of them being governed by a national body. He was also of the opinion that, because of the much greater public interest at that time in National Parks than in nature conservation, progress in nature conservation would depend very much on its being associated with the development of the National Park system. This premise was widely accepted, as can be seen from the fact that the Wildlife Conservation Special Committee for England and Wales, under the Chairmanship of Julian Huxley, and the Scottish Wildlife Conservation Committee, under James Ritchie, were set up under the auspices of the Hobhouse and Ramsay National Park Committees, respectively. However, Dower's hopes and predictions were thwarted for two main reasons. Firstly, the acquisition by Local Authorities of far-reaching development control powers under the Planning Acts, which were passed while the National Park Committee were sitting, led to an emasculation of the powers which had originally been envisaged for National Park Authorities and, secondly, the increasing recognition of the science of ecology, the stature of its chief proponents at that time and their deep concern with conservation issues strengthened their position while that of the National Park lobby was weakened. These changes led to the proposals, emerging from the Huxley Committee, being more far-reaching than originally envisaged, and the link with science being emphasised. Thus instead of nature conservation being linked with town and country planning through allying it with the National Park Commission, lobbying ensured that the proposed 'biological service', the Nature Conservancy, was allied with the Department of Education and Science. Once this dichotomy was established it tended to become self-perpetuating and from that time, the conservation of nature in the UK came to be treated as a quite separate issue from the conservation of amenity and provision for recreation. With this separation came the development of a different "rationale" of conservation within the two fields of operation, even though, not infrequently, the same areas of land might be the subject of mutual interest. In what follows, we will examine the rationale of the two enterprises and the strengths and weaknesses in the conservation effort which have resulted from this dichotomy.

Nature Conservation

The map accompanying the National Planning Guidelines for Scotland illustrates only one of the two main aspects of nature conservation – that relating to the protection of specific land areas which lend themselves to cartographic representation. While this is the aspect of nature conservation that is most relevant to the subject of this seminar, it is vital to remember that nature conservation cannot be achieved by site protection alone. For nature conservation to be successful in the longer term, the protection of special sites has to be augmented by care for nature conservation interests being embodied in land uses in the wider countryside.

To understand the rationale behind nature conservation as it is practised in Britain, it is again helpful to go back to the discussions in the 1940s which sought to promote Government action. These discussions were not the first of their kind, but nonetheless they are of particular interest because of the systematic way the arguments were mustered to persuade effective political action and because of the subsequent influence which these arguments have had on the character of nature conservation as we find it implemented in Britain today. The key document in this context is Command 7122, the Report of the Wildlife Conservation Special Committee under the Chairmanship of Julian Huxley (1947) entitled "Conservation of Nature in England and Wales".

In the introduction to their report members of the Committee acknowledged finding themselves in a position of some difficulty

because "the problem of nature conservation had already been
approached, publicly and independently along two different lines of
thought which not unnaturally led to somewhat different conclusions.
The one which may loosely be described as the aesthetic approach,
based the main emphasis on preserving, at least in selected areas,
the characteristic beauty of the landscape and upon providing ample
access and facilities for open-air recreation and for the enjoyment
of that beauty within those areas..... The other, the scientific
approach, while in no way underestimating the importance of aesthetic
values and of their appreciation by the public, was primarily
directed to the advancement of knowledge as such, as well as to the
application of that knowledge to the affairs of the nation". The
Committee denied that there was any essential conflict between these
two approaches; its members recognised "that there are many sections
of the public with particular interests, whether recreational,
economic or scientific which are, within their own spheres, entirely
legitimate and must be taken into full consideration". However,
they also acknowledged that satisfying these diverse interests while
at the same time safeguarding the natural conditions upon which they
all ultimately rest, "present a problem not without its difficulties".
The Committee's deliberations convinced them that the "problem not
without its difficulties" was in essence a scientific problem. Thus
they concluded that "whatever the approach to this problem, whether
by the scientific or amenity planner, it must inevitably lead to one
simple question: what has to be done to enable man to control nature
so as to maintain or establish a series of varied and most delicately
balanced conditions?" The Committee considered that the solution
to this question would only be achieved through research and experiment done by a Government "biological service" adopting a "broad
scientific approach".

     Having opted for a science-based and 'control' approach to nature
conservation, the Huxley Committee considered land needs for
conservation and arrived at seven categories: National Parks,
National Nature Reserves, Conservation Areas, Geological Monuments,
Sites of Special Scientific Importance, Local Nature Reserves, Local
Education Reserves, of which the first, National Parks, was not a
nature conservation category per se but the others were. From these
proposals we can follow the emergence of the categories we see today
in the National Planning Guidelines.

     National Parks were not established in Scotland (see below).
National Nature Reserves were to be "most carefully selected to give
a balanced representation of the scientifically most important set
of conditions obtaining in this country" and were to cover
"biological, physiographical and geological requirements". The
land requirement of this category was to be relatively modest,
covering only 70,000 acres in England and Wales (later augmented
by 109,000 acres proposed in Scotland by the Ramsey and Ritchie
Committees (Ramsay and Ritchie, 1949). Conservation Areas were

defined as "attractive country, the existing character of which it
is desired to preserve as far as may be possible". It was expected
that the case for designating conservation areas would rest as much
on amenity, as on scientific desirability, the latter being justified
because "there are many important types of plant and animal community
which are not demonstrable only on a scale suitable for strict
reservation". Great importance was attached to conservation areas
"as a means of supplementing, with the least interference and
expense, a small range of conditions which have been included in
the small list of National Reserves". Thirty-five areas were
recommended in England and Wales (to which the Ramsay Committee later
added 22 in Scotland).

The Huxley Committee recommended that the most important areas
for geological science should be designated Geological Monuments
primarily on scientific and educational grounds, and be protected by
methods analogous to those applied to Ancient Monuments. Other
sites of importance for geology should be listed as Sites of
Special Scientific Importance. These latter were among the most
innovative and significant of the Committee's recommendations and
it is worth paying close attention to them. The Huxley Committee
considered that, in addition to the geological sites, there were
"many hundreds of small sites of considerable biological and other
scientific importance, the great majority of which could easily be
safeguarded from destruction if their value and interest were but
known to their owners and appropriate authorities". The Committee
went on to say that "it is not suggested that the existence of any
such site should hold up plans for development, but that there should
be machinery by which its existence could be made known at the early
stage of planning, so that such action as may be possible can be
taken for its protection". Local Nature Reserves were intended to
supplement the national effort and provide a stimulus for Local
Authorities' initiative. Local Education Reserves were to be set
up to provide "opportunities for routine collection and supervised
experimental interference as are necessary in the training of
students".

The Scottish Wildlife Conservation Committee under Professor
Ritchie (Ramsay and Ritchie, 1949) had, in some respects, a more
difficult job to do than the Huxley Committee and lacked the ground-
work that had been provided for the latter by the work of the Nature
Reserves Investigation Committee. For these reasons the members of
the Scottish Committee did not produce their final report until two
years after the Huxley Committee; they were clearly influenced by
the work of the latter so that their findings were largely
corroborative of it.

As we have already seen, members of the Huxley Committee were
particularly successful in persuading the government of the day to

align nature conservation with science rather than with town and
country planning, and most of their recommendations were adopted in
those parts of the National Parks and Access to the Countryside Act
1949 which initiated government action in nature conservation. A
"biological service" was set up under the Privy Council in the form
of the Nature Conservancy; conservation of land for nature was enabled
through the establishment of National Nature Reserves (NNRs), Local
Nature Reserves (LNRs) and the notification of special areas called
Sites of Special Scientific Interest (SSSIs). However, there were
important omissions from the Huxley Committee recommendations: in
the 1949 Act, there was no provision for geological monuments,
education reserves or more importantly, conservation areas. It was
left to the Nature Conservancy to try to accomplish these aims
through the two mechanisms of National Nature Reserve establishment
and SSSI notification. The 1949 Act therefore crystallised the
proposals of the Huxley and Ritchie Committees for nature conservation through land management into a two pronged rationale -
strict management of a limited number of nationally important
"representative areas" - the National Nature Reserves, and
"light touch" influence through advice to Planning Authorities on
developments on SSSIs.

    The drawbacks of such limited approach soon became apparent.
In the absence of "conservation areas" and with only meagre powers
to control land use in the National Parks, and with no parks
established in Scotland, it soon became clear that the proposed
nature reserve areas listed by the Huxley and Ritchie Committees
were nowhere near enough to provide a firm basis for nature conservation. This situation led to the initiation of the major
exercise of the Nature Conservation Review in the late 1960s to
provide what has been called a "Doomsday Book" of the areas of
greatest importance for nature conservation in Britain and which
resulted in about 750 key sites being recognised. The Nature
Conservation Review (Ratcliffe, 1977) brought with it a more highly
developed rationale for nature reserve selection, which drew upon
the accumulated experience of 20 years' work in the Nature
Conservancy and the concurrent refinement of ecological ideas.
However, even with the NCR nature conservationists in Britain
recognised that the site protection aspect of nature conservation
could not be effectively achieved through NNRs and that these had
to be backed up with a much larger number of SSSIs.

    The Huxley Committee thought that the interests of SSSIs could
be adequately preserved through advice provided by the Nature
Conservancy to Local Planning Authorities and the owners and
occupiers and this was implied by the consequent legislation in
the 1949 Act. One can see from the way in which the term
"development" was used by the Huxley Committee and its consequences
in the 1949 Act, that the main concern at that time was to afford

protection against urban and industrial encroachment. While the
Huxley Committee recognised potential conflicts between forestry and
agriculture on the one hand and nature conservation on the other, it
indicated, prophetically that, in the case of agriculture, the
danger "springs from efforts to bring under temporary cultivation an
increasing area of marginal land, with the result that sites of the
greatest scientific and cultural value are irrevocably destroyed for
the most meagre short term returns in crops", they nevertheless
concluded that "competition for land between scientific and agricultural interests, though it may occur, is hardly an important
factor".

The last 10 years have seen this assumption fundamentally
challenged and a radical re-appraisal both of the rationale for
SSSIs and, in the 1981 Wildlife and Countryside Act, means for their
protection. The present rationale for SSSI selection has been
recently described by Moore (1982): the current principle underlying
biological SSSI designation (and under the 1981 Act NNRs also become
SSSIs), is that they should include within them, the "basic minimum
of habitat necessary to conserve the flora and fauna of Britain in
something like its present diversity and distribution". It is
emphasised that "this minimum should be supported as far as possible
by conservation in countryside outside the notified areas". Moore
outlined the problems encountered when choosing criteria for the
selection of SSSIs under the new rationale. Decisions have to be
made in relation to the size of the geographical units in which the
search for the best example of each type of site is made, and the
level of subdivision of the main types of site within which the best
examples are sought, the number of examples of a particular rare or
threatened habitat which should be selected, and on the characteristics which should be used to select the best examples of a
particular type of site - should it be its area, its number of species
or the number of rarities it contains?

The present system for site selection uses criteria that allow
tentative solutions to these and many other questions. It must be
emphasised, however, that the system provides a rationale, not a
mechanical formula for the resolution of these questions - while
it reduces the arbitrariness of the system, at the same time it opens
it to discussion and criticism.

The rationale for site protection has also undergone a
substantial evolution since it was first envisaged by the Huxley
Committee and incorporated in the 1949 Act. This can be attributed,
in the main, to the recognition over the last ten years that the
quality of wildlife habitats in Britain, including that of SSSIs,
has been declining seriously, largely due to the direct and indirect
effects of the main rural industries, agriculture and forestry,
rather than to urban or industrial expansion. The provisions in

the 1949 Act for notification of SSSIs to planning authorities were
largely irrelevant to the activities of these rural industries, which
were not subject to planning control: informal consultations with the
agricultural and forestry industries failed to significantly decrease
environmental damage. Increasing evidence of unacceptable declines
in the quality of the wildlife resource (Goode, 1981) led to provisions
within the 1981 Wildlife and Countryside Act for much more active
conservation of SSSIs through consultation and management agreements
with owners and occupiers. It must be remembered, however, that the
Act is largely concerned with (i) nature conservation, through site
protection, against specific local developments and (ii) species
protection through prevention of unlawful killing or disturbance.
There remains a large and important aspect of nature conservation,
such as prevention of wildlife loss through widespread application of
chemicals in agriculture and forestry, industrial pollution, etc.,
that cannot be achieved through site safeguards.

Landscape conservation

    The third major area of conservation action identified in the
National Planning Guidelines for Scotland is the conservation of
landscape, an aspect of 'amenity' conservation that features in the
Guidelines in relation to the protection of National Scenic Areas.

    In Britain the conservation of landscape quality as an amenity
has been very closely related to the use of landscape as a recreational resource. This is particularly true in relation to national
park developments in England and Wales. Other than through national
park establishment, the main contribution to the conservation of the
quality of landscape in these countries has been through the normal
exercise of planning legislation in regulating urban and industrial
developments in rural areas, and by the designation of green belts
and areas of outstanding natural beauty (AONBs). Need for Government
action in this area was recognised in 1942, when the then Minister
of Works and Planning considered that "no national planning of the
use of the land would satisfy the country if it did not provide for
the preservation of extensive areas of great natural beauty, and of
the coastline", and in 1943 the Minister of Town and Country Planning
expressed his intention that "the amenities of the countryside shall
be preserved, and I have in mind that we should set apart certain
areas as national parks". Dower (1945) defined the two dominant
purposes of a national park, in the sense appropriate to Britain,
as the strict preservation of the characteristic landscape beauty and
the provision of access and facilities for the general public, but
to these two purposes he added suitable protection of wildlife and
places of architectural and historic interest as well as the
effective maintenance of established farming.

It is easy to see how the conservation of landscape quality
and the provision of recreational facilities become associated, to
the extent that the public on the whole favours taking its countryside recreation in areas of high landscape quality. However, in
exploring the rationale which has led to the designation of national
scenic areas in Scotland and their appearance in the National
Planning Guidelines, it can be seen that the relationship is not
simple. On the one hand there is the complexity of recreational
needs, while on the other there is the great variation in the
capability of landscape resources to fulfil these needs.

Action in relation to the conservation of landscape quality was
slow to develop in Scotland, partly because the need was seen as
less urgent than south of the border. Thus it was not until 1968
that a Countryside Commission for Scotland was established with
powers introduced by the Countryside (Scotland) Act 1967, nineteen
years after its southern counterpart began its life as the National
Park Commission. However, this has enabled lessons from the south
to be noted, and we are fortunate in being able to trace the
development of t'ie rationale for the national effort in landscape
resource conservation in Scotland through several key documents
produced by the Countryside Commission for Scotland. Of these,
the most notable is the one entitled 'A Parks System for Scotland'
(Countryside Commission for Scotland, 1974).

In this Report, the Countryside Commission for Scotland reviewed
the then current arrangements for landscape conservation and outdoor
recreation provision and concluded that a radical revision of the
framework was needed to establish a better rationale. They
recognised that an outdoor recreational system needed to be
comprehensive, embracing a spectrum from urban to remote wild land,
and should cater for the enjoyment of the countryside by people
practising 'minority land-hungry activities, as well as by those
practising gregarious activities'. They also recognised that
'better resource conservation is an essential objective of a nationwide recreation plan' and must be applied to imposing restraints
in recreation terms, every bit as stringent as for other conservation
values'. This led the Commission to envisage the establishment of
a countryside management system having two parts. The first part
proposed the creation of different types of park which would both
attract and accommodate the various types of recreation activity.
These would include urban parks of the type already established and
having an already well understood and appreciated function.
Country Parks - a 'park or pleasure ground in the countryside which
by reasons of its position in relation to major concentrations of
population affords convenient opportunity to the public for the
enjoyment of the countryside and for open air recreation'. It was
anticipated that these would generally be between 10 and 400 ha in
extent and would generally be subject to an integrated management
programme, usually under local authority ownership. The next

category, Regional Parks, would generally be larger and would
include much land which, while constituting the landscape enjoyed
by visitors, would not actually be available for recreational use.
These Parks would be intended to provide a comprehensive system of
public access to the countryside, ranging from areas of no or low
recreational use to sites used intensively. The fourth category,
Special Parks, were seen as satisfying a national demand for
recreational opportunities, rather than just regional or local
demands. In general they would be 'situated in countryside of
considerable natural beauty and amenity such as would in any event
single them out as national assets requiring particular care and
protection'. The Countryside Commission for Scotland (CCS)
identified three areas as likely candidates for Special Park status -
Cairngorms, Glen Nevis-Glencoe and Loch Lomond-Trossachs. It is
interesting to note that these correspond with three out of the
five areas proposed for National Park status by the Ramsay
Committee in 1945, and in which development proposals were subsequently subject to referral to the Secretary of State.

   The second part of the countryside management system proposed
by the CCS related to the conservation of areas of outstanding
scenic interest which are part of the national heritage. In the
parks system document, the CCS emphasized that the proposals for
the provision of recreation, with its incorporated landscape
conservation element, did not mean that there was no concern to
protect landscape outside parks in Scotland. To ensure this wider
protection of outstanding landscape of national importance, the
report proposed the identification and designation of areas of
special planning control under Section 9 of the Countryside
(Scotland) Act 1967. The survey for these areas was undertaken by
the Countryside Commission (1978) and their report, entitled
'Scotland's Scenic Heritage' identified 40 National Scenic Areas
covering about 1,000,000 ha, as shown on the National Planning
Guideline map, and described the rationale for their selection.

   After careful consideration, the Countryside Commission for
Scotland rejected so-called objective assessment methods and opted
for a selection of areas based on the subjective judgement of
assessors. The Commission decided not to analyse 'scenery in terms
of its geology, geomorphology, pedology, climate, natural history
or cultural history ..... not because we think these things are
unimportant in the influence on the scene, but because we believe
that enjoyment of fine scenery is based on a perception of the whole
which does not depend on more formal kind of analysis'. The CCS
sought to identify scenery which best combines those features which
are most frequently regarded as beautiful. On the whole this means
that rich diverse landscapes which combine prominent landforms,
coastline, sea and fresh water lochs, rivers, woodlands, with some
a mixture of cultivated land, are generally the most prized. Not
all these features occur, however, in all the areas we have

identified. Diversity of ground cover may be absent in some
but compensated for by especially spectacular landforms or
seascape'.

The report on Scotland's scenic heritage identified national
scenic areas but made no prescription for their conservation. The
Parks System for Scotland document suggested that securing these
areas against undesirable development would be adequately achieved
by designating them as areas of special planning control under
Section 9 of the Countryside (Scotland) Act 1967. In the event,
the Secretary of State implemented the proposals for National
Scenic Areas by establishing a procedure under which the planning
authority had to (i) consult with the CCS over certain specified
developments within National Scenic Areas, and (ii) notify the
Secretary of State of any intention to grant planning permission
against the advice of the Countryside Commission for Scotland. An
'Article 4' direction order was issued under the Town and Country
Planning Act (General Development) 1975 which brought certain of the
developments which were normally permitted within planning controls.
The specified developments included major residential, industrial
and road works, caravan sites and vehicle tracks over 300 m altitude,
except where the latter was agreed by the planning authority as
part of an afforestation proposal. Perhaps the main point of note
in relation to the conservation of National Scenic Areas is the
absence of proposals to regulate agricultural and forestry
operations. This may appear surprising in view of the experience
in the English and Welsh National Parks, and in Sites of Special
Scientific Interest throughout Britain. However, this is not by
any means an oversight but is part of the rationale developed by
the CCS in relation to National Scenic Areas. Thus, Dr Jean
Balfour, the Chairman of the CCS commented in a Countryside
Commission for Scotland news release (1981) "I know that, partly
because of experience in English and Welsh National Parks,
foresters in particular are apprehensive that the identification
of special scenery means opposition to afforestation, and that it
is an endeavour to fix the appearance of the countryside. The
CCS does not share this view. Much, though in Scotland not all,
of the countryside depends for its appearance by its use by man.
.....a prosperous countryside and a reasonable freedom to change and
develop, is 'good' for scenery. The safeguarding or conservation
of scenery is about maintaining the overall character, but not the
precise pattern ..... In the hills, with their mountain peaks and
ridges and large expanses of moorland, forests can increase,
providing more diversity while still retaining the overall
character. Safeguarding scenery is not about fossilisation".

Thus development of the rationale in relation to amenity
conservation in the Scottish countryside has, as in the case of
nature conservation, lead to a twofold approach: one aspect is a
development policy, with conservation elements incorporated, for

recreational provision in the countryside through the Parks System
for Scotland, the second aspect is the safeguarding of scenery
through the protection of designated National Scenic Areas. As
with the other topics that have been considered, many questions can
be raised in relation to this rationale and the premises on which it
is based. Some of these will be examined in the final section of
this chapter.

Conservation of the Coast

     In considering the conservation of the coast, a rather different
perspective is being adopted from the previous three cases. Here,
instead of starting from a subject and exploring its application in
geographical terms, we are starting with a geographical entity - the
coastline - and looking at the conservation issues that affect it.
The consideration of coastal conservation therefore provides an
opportunity to see what integration can be achieved between the
different conservation enterprises. The Hobhouse Committee
addressed itself to the problems of coastal conservation, considering
a memorandum submitted by Steers which suggested that the whole
coastline of England and Wales should be planned as one unit. The
Committee debated whether a separate coastal authority was required,
comparable to the proposed National Park Commission, but decided in
favour of a Coastal Planning Advisory Committee. The Huxley
Committee report (1947) considered the problem of nature conservation
on the coast, but confined its recommendations to supporting the
Hobhouse Committee's proposal for a Coastal Planning Advisory
Committee within the Ministry of Town and Country Planning. It
also suggested that (i) scientific advice should be made available
to the National Parks Authorities and to Local Advisory Committees
for those National Parks or Conservation Areas which included
stretches of the coast, and (ii) coastal development should not be
allowed to obscure important geological sections. However, in
spite of the production of many reports on the problems of coastal
planning an integrated approach to coastal conservation has not
been achieved in England and Wales except through the promotion of
'Area of Outstanding Natural Beauty' AONB designation and, more
recently of heritage coasts.

     In Scotland, the development of the North Sea oil resources
in the early 1970s provided a strong stimulus to reconsider
coastal planning policies, and the need for conservation. This
led the Scottish Development Department, in collaboration with
the Countryside Commission for Scotland and the Nature Conservancy
Council, to commission a coastal survey (Scottish Development
Department, 1976). On the basis of the factual information
collected for this, together with an analysis of development needs,
in 1974 the Scottish Development Department published Coastal
Planning Guidelines in relation to oil and gas development. This

document contained guidelines which the Department considered went as far as was practical at that time towards setting out a national strategy for coastal development related to oil and gas exploitation. It was indicated in the report that the Secretary of State would use the Guidelines in reaching a decision on any planning applications which came before him, and that he expected the planning authorities similarly to take them into account in preparing their development plans and considering submitted applications. In the guidelines the Scottish coast was divided into preferred development zones and preferred conservation zones. The 16 preferred development zones were selected in order to provide a range of suitable sites for major oil developments, and developers were advised to select sites within these zones, rather than in the conservation zones.

By far the greatest part of the mainland coast fell into 22 preferred conservation zones in which there was to be a general presumption against developments because of the damaging effects this would have on the environmental, scenic and ecological importance of the area. The planning authorities were encouraged to prepare detailed conservation policies together with recreational plans for such areas. Developers were warned that major development proposals within the conservation zones would require special authorisation involving scrutiny by the Secretary of State and probably a public inquiry. Any intrusion into such areas would have to be justified by compelling arguments, including a demonstration that no suitable site existed outwith the preferred conservation zone. As mentioned above, identification of the preferred conservation zones was achieved by a synthesis of information relating both to ecological and scenic amenity considerations. The distribution of coastal SSSIs was considered together with the new information on coastal sites which was accumulating as a result of the Nature Conservation Review.

This did not mean however, that areas chosen for preferred development zones were necessarily without scenic or ecological importance. Both the Sullom Voe and Flotta Oil Terminals were within areas of considerable ecological importance, even though not directly affecting nationally important scenic areas or SSSIs; in both instances extensive precautions were taken to minimise environmental impact. In fact some of the earliest informal environmental impact studies in Scotland were made in relation to these sites. The Cromarty Firth preferred development zone also coincided with an area of considerable ecological importance – a NCR Grade 1 site of outstanding importance as a wildfowl feeding area. Inevitably there was a conflict of interest between the oil terminal development proposals and those of nature conservation which eventually resulted in a more fine-grained arrangement for industrial and nature conservation zoning within the preferred development zone.

At the same time as the coastal planning guidelines were being prepared and implemented other initiatives related to coastal conservation; were taking place. An extensive survey of areas of ecological importance at risk through oil developments was being made by the Institute of Terrestrial Ecology, potential oil-rig construction sites were being surveyed, and coastal scenery was being studied by the Countryside Commission for Scotland in relation to the definition of the national scenic areas already mentioned. Additionally, comprehensive contingency planning for the treatment of coastal oil pollution involved a blend of information concerned with coastal fisheries, nature conservation, scenic and recreational resources in order to determine the appropriateness of different oilspill treatments in different types of coastal area.

Underlying the rationale for coastal conservation as expressed in the coastal planning guidelines, were importance premises concerning the values and sensitivities of coastal areas. In some respects it might appear that coastal conservation is hardly a priority issue in Scotland where only about 12% of the 2,400 mile mainland coastline is developed for urban or industrial purposes, as compared with 25% in England and Wales. However, the Scottish coast is of outstanding ecological and scenic quality and even localised developments can have quite extensive effects, not only scenically but ecologically through pollution effects on coastal waters in hitherto unpolluted areas. Furthermore, coastal areas may contain biological resources of even greater significance than has hitherto been generally assumed. Lovelock (1979) suggested that coastal areas such as saltmarshes, mudflats and seaweed communities may be sites of atmospheric exchange processes of fundamental importance to the biosphere, so that conservation of these types of area may have long term significance.

The way the rationale for coastal conservation in Britain has developed, compared for example with the United States approach through the Coastal Zone Management Act, has to be seen within the context of the British planning system, which has already given much attention to the regulation of coastal development through development plan zoning.

THE PREMISES RE-EXAMINED

The Premises Questioned

In this final section, questions raised in relation to conservation rationales will be explored more closely, and this critique then used as a basis for considering rationales in the sense of - the "fundamental reasons or logical basis for anything".

To examine every way in which each of the foregoing rationales
have been questioned, and indeed all have been, is beyond the scope
of this chapter. For this reason examples will be chosen to illustrate the main questions that have been raised, and in particular
those which give important insights into fundamental conservation
issues.

Earlier in this chapter the defensive note, struck in justification of conserving the best agricultural land, suggested that the
safeguard of such land could be challenged. Indeed, it was
challenged by the Scottish County Councils in evidence to the
Select Committee on Land Resource Use in Scotland (Select
Committee on Scottish Affairs, 1972), because on current short term
social accounting practice, such as cost-benefit analysis,
conservation often appears less advantageous than some forms of
urban development. Challenge has also emerged because recent
figures indicate that exhortation to conserve high class
agricultural land is proving inadequate (Davidson, 1980). It
should be recognised that the premise for agricultural land
conservation, in common with those for other conservation action,
assumes a particular land use structure which is itself a
consequence of certain social attitudes, concerning the urban/rural
distinction, that are largely taken for granted. A changed
attitude to food production, for example as suggested by Richard
Fordam (1971), and exemplified in the propositions of many radical
conservationists (Goldsmith et al., 1972) who propose a much more
diversified urban-rural mix while maintaining the premise that high
class agricultural land should be conserved, would nevertheless
substantially transform the present rationale governing agricultural
land conservation action.

Special Areas Versus the Wider Countryside

We have already seen, in the account of the evolution of land
conservation for wildlife in Britain, how the original premise,
that land use practices for farming and forestry would have little
adverse effect on the wildlife conservation, has required
substantial correction as evidence to the contrary accumulated.
This crucial change in premise has led to the development of
alternative strategies based on somewhat different rationales.
The first, and currently dominant, strategy recognises the need
to take account of the environmentally damaging effects of
technological developments in forestry and agriculture, and seeks
to achieve this by seeking increased safeguards for wildlife on a
relatively small land area. At the same time, it requires
strenuous searches for ways of ameliorating the effects of modern
agriculture and forestry in the rest of the countryside. A
second, more radical strategy, advanced by those who claim that
the former approach is failing, proposes that conservation

objectives should become dominant over a much larger land area, with the remainder of the countryside being as it were, "written off" to technology. Thus Green (1981) maintains that in addition to specific nature and amenity conservation sites, much larger areas of land must be managed primarily for amenity conservation purposes, and that this requires the designation of "big conservation zones where there is control of agricultural activities and the application of appropriate amenity and land management techniques". Green considers that "it is probably unrealistic to expect much control of agriculture in the remainder of the countryside, so that in the short term some polarisation of land use must be accepted". In some respects, Green's proposals are simply an extension of the current rationale governing the safeguards for SSSIs. They represent a "refuge" approach to conservation, though Green qualifies his adherence to it by suggesting that, in the "longer term, food surpluses, changes in dietary habits and other factors presently unforeseen, may well force a return to agricultural practices that are less inimicable to wildlife and landscape". If the area of conservation zones and reserves set aside has been adequate "species from them might then re-populate the new countryside".

Current practice in wildlife conservation in Britain tends to make pragmatic use of both these strategies. There is an attempt, through SSSI notification, to get more positive conservation management over large enough areas of land to prevent the continued decline in the diversity of the native animals and plants. Additionally an effort is being made to ensure that greater account is taken of conservation needs in activities in the wider countryside by placing, for example, a responsibility for advising on the protection of the natural beauty and amenity of the countryside upon Government sponsored agricultural advisers (Section 41 of the Wildlife and Countryside Act, 1981). It is unlikely that these strategies will displace each other; both are needed to meet the varying circumstances even in so small a country as Britain. Thus, though trends up to the present time have not been cause for optimism, it is highly unlikely that the public at large would be content to see all the countryside, outside special conservation areas and reserves, abandoned to unrestrained industrial agriculture or forestry.

Many conservationists would reject such a despairing solution, maintaining that people would have to live their lives and work in these dispiriting environments, and that to adopt this approach would be to make the same mistakes in the countryside as were made earlier in the cities (namely a failure to create tolerable human environments).

The problem of conserving by site designation revolves around the fate of the land which remains undesignated. This is

as true for amenity sites (eg. national scenic areas) as for nature
conservation sites, and it is one of the reasons why the National
Park concept was not adopted in Scotland. In his recent review of
wildlife conservation in Britain, Mabey (1980) has emphasised this
dilemma in relation to SSSIs, commenting that "site protection -
especially when it is based on partial or specialised criteria -
is neither an adequate nor very appropriate approach to the
conservation of the wider countryside and for meeting the vast
range of local and personal interest attached to it. For that we
need a quite different kind of approach, not imposed from above
but worked out by a continuous process of debate and adjustment
between all involved parties from the level of local community up
to the international".

Conservation Strategies: Integration in the Conservation Enterprise

While Mabey's idea of a "continuous process of debate" across
the whole complex field of conservation action is daunting to the
conservation bureaucrat accustomed to the present structure of
separate institutional frameworks and responsibilities, such is
indeed implied in most conservation strategies which have emerged
in recent years.

The last decade has seen several attempts to formulate
comprehensive global conservation strategies, of which perhaps the
most notable were the radical "Blueprint for Survival" (Goldsmith
et al., 1972) and the somewhat more orthodox "World Conservation
Strategy" (International Union for the Conservation of Nature,
1980). The "Blueprint", taking its cue from the forbidding
picture of world development presented by Forester (1970) and
Meadows (1972) and others, started from the premise that "if
current trends are allowed to persist, the breakdown of society
and the irreversible disruption of life support systems on this
planet, possibly by the end of the century, certainly within the
lifetimes of our children are inevitable". As a result of an
analysis of the causes of this situation, the authors of the
"Blueprint" saw the need for a move away from a growth oriented
industrial ethos towards a more stable society "which can be
sustained indefinitely while giving optimum satisfaction to its
members". This "post-industrial" society would provide
"satisfactions that would more than compensate for those which,
with the passing of the industrial state, it will become
increasingly necessary to forego". On its publication, the
blueprint received acclaim, criticism and ridicule. Because it
attempted so much, some of its diagnoses and prescriptions were
erroneous or oversimple. However, as against the notorious
Nature editorial (Maddox, 1972) which considered it reprehensible
that the Blueprint supporters among scientists should "lend
their names to attempts like these to fan public anxiety about

problems which have either been exaggerated or are non-existent",
the contrary view of Southwood et al (1972) that there is "now no
escape from the necessity for a fundamental rethinking of all our
working assumptions about human development in relation to the world
we live in" received much support, even from those who contested the
detailed prescriptions.

The more recent "World Conservation Strategy" was published by
the International Union for the Conservation of Nature in 1980 with
the help of the United Nations Environmental Programme and the World
Wildlife Fund, and with endorsement by the Food and Agriculture
Organisation and UNESCO. It is primarily directed to promoting
"living resource conservation for sustainable development" through
the maintenance of essential ecological processes and life support
systems, the preservation of genetic diversity, and the sustainable
utilisation of species and ecosystems. In its analysis of the present
situation, it attributes the·failure to achieve conservation to
complex causes, notable among which are - "the belief that living
resource conservation is a limited sector, rather than a process that
cuts across and must be considered by all sectors with the consequent
failure to integrate conservation with development"; inadequate
environmental planning, poor organisation, lack of awareness and
"failure to deliver conservation based development where it is most
needed, notably the rural areas of developing countries". It is
perhaps this last cause that has led to the unwarranted assumption
by many that the World Conservation Strategy is primarily directed
to the developing countries. Undoubtedly conservation based
development is needed in them, but it is also vitally important that
the "developed" countries put their house in order and reduce their
exorbitant, asset stripping demands on the world's resources. As
Dasmann (1976) has pointed out, "the rain forests of Indonesia are
not being cut down because the Indonesians have an incredible
appetite for wood. The wood and other forest products are being
sold to the US, to Japan, to countries in Europe and to other
developed nations".

How should the proposed global conservation strategy and a
reassessment of the different premises affect the present structure
of conservation action in Scotland, as set out in the National
Planning Guidelines? It would appear that the evolution of the
separate conservation enterprises, while perfectly understandable
in terms of dividing the field of action into manageable components
reflecting sectors of interest, has made it difficult to achieve an
integrated approach. Experience has shown that the present
structure is a recipe for only limited (though nevertheless real)
success in some areas that could, in the longer term be outweighed
by failures in others. The rationale underlying the conservation
enterprises within the National Planning Guidelines goes some way,
but not far enough, to give sufficient hope of success in the
conservation enterprise as a whole. This is because conservation

is related primarily; but not wholly, to protection; it hardly
features in a positive way in the major landuse activities. For
example, close study of the land use summary sheets for Agriculture
and Forestry, and indeed of the practices of these two major land
uses in Scotland, show that serious attempts have not been made to
achieve proper integration through furthering forms of agriculture
or forestry that are conservation oriented. While the Guidelines
do serve a useful purpose in highlighting the need for greater
integration, this has not yet been achieved. This need for better
integration was fully recognised by the Select Committee on Scottish
Affairs study on Land Resource Use in Scotland (1972), which proposed
the formation of (i) a Land Use Council to act as a central forum
for discussion of rural land use affairs and (ii) a professional
land use unit to provide technical support. These recommendations
were not implemented, and in any case fall short of the type of
integration that the World Conservation Strategy envisages, which
would involve the major rural industries incorporating conservation
measures in their development programmes. But as yet the available
development mechanisms are inadequate as shown by the recently
inaugurated Western Isles Integrated Development Programme which
allows no funds for the development of conservation oriented land
uses, even in so appropriate an area. In theory, given that a high
proportion of rural activities are dependent on grant aid, Section
66 of the Countryside (Scotland) Act 1967 which exhorts government
departments and other public bodies to "have regard to the
desirability of conserving the natural beauty and amenity of the
countryside" could have been used to effect greater integration of
conservation and development if the organisations had been so
inclined, and if the Treasury had been sympathetic. However, where
there has been any identifiable response to this clause the
commitment has been inadequate and largely cosmetic. Section 41 of
the recent Wildlife and Countryside Act (1981) provides for advice
that could encourage agriculture to be more conservation oriented,
and Section 28 of the Act provides a mechanism which may be used to
promote positive management to this end on Sites of Special Scientific
Interest, but not in the wider countryside.

Changing the Ethos

     The dilemma the conservation movement finds itself in - wide-
spread sympathy but little commitment of resources for effective
action - has not arisen by accident; the fully integrated ecology-
based attitude towards the use of resources, challenges the practice
of resource management current in our society. Prevailing practice
is largely determined by the application of modern technology by
industry, including agriculture and forestry, under the control of
governments or multi-national corporations, banks and insurance
funds, and many kinds of public or private concerns that are often
quite remote from the field of action. Nicholson (1981) considers

that "few people in these worlds (governments, multi-national corporations) know or care anything about the natural environment and
they are advised by economists, accountants, engineers and other
professionals who, if anything, are even more ignorant". Dasmann
(1976) has emphasised the problem of accommodating conservation
within the present dominant ethos of society, suggesting that
"there is reason to question whether conservation can ever by
anything except a trivial sideline in political/economic systems
which are geared to continued economic expansion and to a growing
consumption of material and energy resources and have as a central
concern the enrichment of those who are in favourable circumstances
within it". These are somewhat radical, and possibly exaggerated,
criticisms of the status quo, valid up to a point but providing no
more than hints of the way in which we can move from the present
situation where the conservation-outlook fights an increasingly
effective defensive but still losing battle against the strategies
of exploitation. The present diverse structure of conservation-action
with its accompanying rationales is a reflection of very real
differences in deep seated attitudes and aims that seek fulfilment
within society and which have determined the character of the
institutional structures designed to fulfil them. These divergencies,
as noted earlier, are confirmed by the very institutional structures
to which they have given rise, and perpetuated through forms of
education which strongly influence the conservation movement itself.
Even within nature conservation, changing educational fashions in
biology have also affected the movement to the extent that leading
academic ecologists now play less part in the conservation movement
in Britain than they did twenty years ago. Another instance of the
influence of educational structures is the attitude of "scientists"
towards "sociologists". Relatively few conservationists of
"scientific" ecological background will admit that sociological
structures are an essential part of the "ecology" of the environment.
This led, for example, to heated debates within the UNESCO Man and
Biosphere Programme where, particularly in developing countries, the
accusation has been levelled that the "ecological" solutions to
environmental problems proposed by scientists from the developed
countries are sometimes irrelevant to the local human situation.
This of course parallels the imposition of alien technologies in
agriculture and industry on developing countries. The conservation
solutions should be appropriate to the societies in which they are
to be implemented - this is as true in the Scottish Highlands as in
the Sahel or the Himalayas, but it is a hard lesson that we are only
just beginning to learn; it presents problems that are easier to
acknowledge than solve. The World Conservation Strategy at least
provides a framework within which these solutions can be explored.

Providing the Ethic?

    Discontent with the prevailing developmental ethos, with its attachment to <u>growth</u> (of GNP) as a yardstick of human achievement, has provoked some conservationists to suggest that an ethic of <u>resource use</u> should be promoted as a basis for conservation rationales. We have noted earlier that ethical considerations are deeply embedded (to the extent of sometimes being hidden from view) in the rationales of conservation. Many discussions on this topic take as a starting point Aldo Leopold's (1949) statement that

> "There is as yet no ethic dealing with man's relations to land and the animals and plants which grow on it. Land .... is still property. The land obligation is still strictly economic, entailing privileges but not obligations."

> "The extension of ethics to this third element" (the first and second elements dealing with relationships between individuals, and between individuals and society respectively) "in the human environment is, if I read the evidence correctly, an evolutionary possibility and an ecological necessity".

Recent claims that a conservation ethic is required, (Ratcliffe, 1981; Page and Warren, 1982), however, contain the implication that something new is needed. Such an ethic undoubtedly calls for novel elements but hardly for an entirely new moral stance. Leopold makes clear in his discussion that he is talking about an extension of ethics, prolonging the evolution of ethical concern in inter-personal relationships, through the relations between the individual and society, to man's relations with his environment. Passmore's (1974) thesis was that there was no need to reject the heritage and tradition of western ethical thought because the latter did contain a basis which could be extended to encompass environmental concern, concluding that "to that extent conventional morality, without any supplementation whatsoever, suffices to justify our ecological concern, our demand for action against the polluter, the depleter of natural resources, the destroyer of species and wildernesses".

There is an implication in these recent statements that we do not have a "conservation ethic" now, but the frequent use of the words "should" and "ought" in the exhortations of conservationists would seem to indicate the contrary. Every time these words are used they imply a moral evaluation - conservation is operating on the basis of an implicit "scheme of moral science". To clarify this difficult area, without which all conservation rationales are incomplete, would be to explore the implicit ethical bases which have underlain conservation evaluations, exhortations and actions in recent years, a phenomenological analysis, as it were, of behaviour in nature conservation. As Hare (1952) observes "if we were to ask the person what are his moral principles, the way in which we could be sure of a true answer would be by studying what

he did". We should not be surprised if logical inconsistencies and
paradoxes were revealed in this, as in other fields, of human
endeavour.

Because the evolution of some important areas of conservation
thinking have been associated with the development of ecological
science, there has sometimes been an assumption that conservation
rationales can be deduced from ecological facts. This is not the
case - it would overlook the inseparable link of an ethical element
with conservation rationales. This does not mean however that the
information provided by science has not an essential role to play:
it is absolutely essential to enable the connection to be made
between conservation-rationales and fruitful conservation-action,
whether in relation to physical resources, nature or landscape
beauty.

CONCLUSIONS

I have attempted to critically review the "explanation or
statements of reasons" relating to the practices within four fields
of conservation in Scotland and then, through re-examining the
premises underlying them, to consider their "fundamental bases".

Although common themes can be identified within the several
fields of conservation discussed, justifying their inclusion within
the broader conservation enterprise, it is clear that they have often
had quite separate origins and have, up to now, been pursued as
relatively independent enterprises by separate institutions. This
has brought its benefits in terms of depth of understanding and
expertise in specific fields of interest, but in doing so it has
often made it more difficult to effect the necessary inter-linking
if conservation is to be effective in the longer term. The
individual rationales which have been developed to guide and explain
the several fields of conservation-action are based on premises
embedded in traditions which influence action within the particular
fields and which themselves reflect, usually implicitly rather than
explicitly, ethical principles often related to such concepts as
'value' and 'duty'. Both the rationales and the premises on which
they are based are subject to change with time. Changing circumstances can bring about a change of premise: conservation problems
relating to monoculture forestry may be insignificant when this
land use occupies only a small area, but become crucial when monoculture forestry becomes the predominant land-use. Similarly when
particular strategies are found to be defective, the rationale is
modified to accommodate their substitutes - or perhaps we should
say "should be modified", because there is usually a time-lag
between a recognition that all is not well with the rationales which
govern action (or of the premises on which these are based) and the
substitution of new rationales. Thus a rationale for wildlife

conservation which places too much stress on the protection of
natural or semi-natural habitats in favoured special sites may leave
the commonplace, the wider countryside, including many man-made
habitats rich in wildlife, at risk.

Conservation initiatives are constantly encountering, in their
own and others' actions, the well-known ecological principle that
"it is impossible to do one thing only", yet for practical reasons
some 'boundaries' have to be drawn and distinctions made within this
complex enterprise. Many different approaches are being attempted
in overcoming this dilemma of integration. The National Planning
Guidelines for Scotland encourages integration; Section 66 of the
Countryside (Scotland) Act 1967 instructs Government Departments and
public bodies to have regard to the "desirability of conserving the
natural beauty and amenity of the countryside", while Section 37
of The Countryside Act 1968 instructs conservation agencies, inter
alia, to "have due regard to the needs of agriculture and forestry
and to the economic and social interests of rural areas". The
Wildlife and Countryside Act 1981 goes beyond exhortation and directs
Agricultural Departments to give advice to farmers on (i) the
conservation and enhancement of the natural beauty and amenity of
the countryside as well as (ii) the diversification into other
enterprises of benefit to the rural economy, while Section 48
instructs Water Authorities (statutorily only in England and Wales,
but extended in principle to equivalent bodies in Scotland) that
they are to exercise their function so as to further the conservation
and enhancement of natural beauty and the conservation of flora,
fauna and geological or physiographical features of special interest.
Encouragement to achieve better integration within the conservation
enterprise is also provided by the World Conservation Strategy, and
is reflected in the determined attempt by both governmental and
voluntary conservation agencies to work with developers to achieve
approaches to development which are more compatible with conservation
aims.

The emergence of rationales for conservation is a relatively
new human cultural development. While there has been a rapid spread
in their acceptance in recent years, they are still relatively
shallow rooted within most cultural traditions. They are, as it
were, vigorous young growths whose roots have not yet penetrated
the deeper soil which will give them the necessary strength to
compete successfully with long established plants. The ethical
attitudes on which they are based tend to be those which place
difficult demands on those who adopt them, demanding sacrifice of
short-term material gain for the long-term benefit of man and the
biosphere. However it is becoming increasingly clear that maintenance of 'tolerable' conditions for human life on earth will
increasingly demand the effective embodiment, within approaches to
the use of the world's resources, of conservation rationales.

REFERENCES

Arvill, R., 1967, "Man and Environment", Penguin Books, Harmondsworth,
    Middlesex.
Central Office of Information, 1970, "Land Conservation in Britain",
    Central Office of Information, London.
Cherry, G. E., 1976, "Environmental Planning 1939-1969: Volume II.
    National Parks and Recreation in the Countryside", HMSO,
    London.
Countryside Commission for Scotland, 1974, "A Parks System for
    Scotland", Countryside Commission for Scotland, Perth.
Countryside Commission for Scotland, 1978, "Scotland's Scenic
    Heritage", Countryside Commission for Scotland, Perth.
Countryside Commission for Scotland, 1981, "National Scenic Areas:
    Planning Consultation Arrangements", News Release.
Dasmann, R., 1976, "The Threatened World of Nature", Horace M.
    Albright Conservation Lecture, University of California, Berkeley,
Davidson, D. A., 1980, "Soils and Land Use Planning", Longman, London
    and New York.
Dower, J., 1945, "National Parks in England and Wales", Command 6628,
    HMSO, London.
Fordham, R., 1971, Turning our farms into gardens, New Society, 8.
Forrester, J., 1970, "World Dynamics", Wright Allen Press, Cambridge.
Goldsmith, E., Allen, R., Allaby, M., Dowell, J., and Lawrence, S.,
    1972, "A Blueprint for Survival, The Ecologist, 21, 1-43.
Goode, D., 1981, "Countryside Conservation", Allen and Unwin, London.
Hare, R. M., 1952, "The Language of Morals", Oxford University Press,
    London.
Huxley, J. S., 1947, "Conservation of nature in England and Wales:
    report of the Wildlife Conservation Special Committee", Command
    7122, HMSO, London.
International Union for the Conservation of Nature, 1980, "A World
    Conservation Strategy", Morges, Switzerland.
Leopold, A., 1949, "A Sand County Almanac", Reprinted 1966, Oxford
    University Press, London.
Lovelock, J. E., 1979, "Gaia, a new look at life on Earth", Oxford
    University Press, Oxford.
Mabey, R., 1980, "The Common Ground. A place for nature in Britain"s
    future?", Hutchinson, London.
Maddox, J., 1972, The case against hysteria, Nature 235, 63-65.
Meadows, D. L., 1972, "The Limits of Growth", Earth Island, London.
    1972.
Moore, N., 1982, What parts of Britain's Countryside must be
    conserved?, New Scientist 93, 147-149.
Nicholson, M., 1981, Rallying to the Call of the Wild, The Guardian,
    (4 March 1981).
Page, H. and Warren, A., 1982, More about the purpose of nature
    conservation, Ecos 3, 1, 27-29.
Passmore, J., 1974, "Man's Responsibility for Nature: Ecological

Problems and the Western Tradition", Duckworth, London.

Ramsay, J. D. and Ritchie, J., 1949, "Nature Reserves in Scotland:
    Final Report by the Scottish National Parks Committee and the
    Scottish Wildlife Conservation Committee", HMSO, Edinburgh.

Ratcliffe, D. A., 1981, The Purpose of Nature Conservation, Ecos 2,
    3, 8-13.

Ratcliffe, D. A., 1977, (Edit.), "A Nature Conservation Review",
    Cambridge University Press, Cambridge.

Scottish Development Department, 1974, "North Sea Oil and Gas:
    coastal planning guidelines", Scottish Development Department,
    Edinburgh.

Scottish Development Department, 1976, "The Coast of Scotland",
    Scottish Development Department, Edinburgh.

Scottish Development Department, 1981, "National Planning Guidelines:
    Priorities for Development Planning", Scottish Development
    Department, Edinburgh.

Select Committee on Scottish Affairs, 1972, "Land Resource Use in
    Scotland", Command 5428, Scottish Development Department,
    Edinburgh.

Sheail, J., 1976, "Nature in Trust", Blackie, Glasgow.

Southwood, T. R. E. and others, 1972, "A Blueprint for Survival,
    The Times (letter, 25 January 1972).

16

CONSERVATION PLANNING IN A CHECKERBOARD WORLD : THE PROBLEM OF SIZE

OF NATURAL AREAS

T. E. Lovejoy

World Wildlife Fund
1601 Connecticut Avenue NW
Washington DC 20009, U.S.A.

In addition to outright habitat destruction, selective removal of certain species (e.g. whaling), introduction of exotic species and wholesale application of toxic substances and other chemicals to the environment, one of the major ways in which people are affecting the biology of the planet is through fragmentation of natural landscapes (Da Costa Lobo, Chapter 3; Burgess and Sharpe, 1981). This has a pervasive effect throughout the terrestrial portion of the globe and probably occurs, but to a much more limited extent, in the benthic portions of freshwater and marine habitats.

Nowhere has the process progressed further than in the Atlantic forest regions of Eastern Brazil, of which less than two per cent remain. The progression of habitat fragmentation and disappearance that has taken place in Sao Paulo State (Figure 1) is representative of the Atlantic forest region as a whole. The forest remnants are inhabited by a number of endemic species: 20 species and subspecies of primates, three quarters of which are endangered (including the largest New World primate, the muriqui or wooly spider monkey, *Brachyteleles arachnoïdes*), 171 species of birds, and obviously even greater numbers of plants and invertebrates. These forests, together with those of Madagascar, could easily become the locale of some of the large numbers of extinctions which are projected for the coming decades (e.g. Lovejoy, 1980a; Myers, 1980; Raven, 1980).

What makes all of this more worrying is the growing awareness that the dynamics of isolated fragments of wild lands clearly differ - and in ill-understood ways - from the dynamics of the same pieces of forest when still surrounded by similar habitat. When the surrounding environment is grossly altered, making a forest island in a man-dominated landscape (e.g. of wheatfields), the remnant is

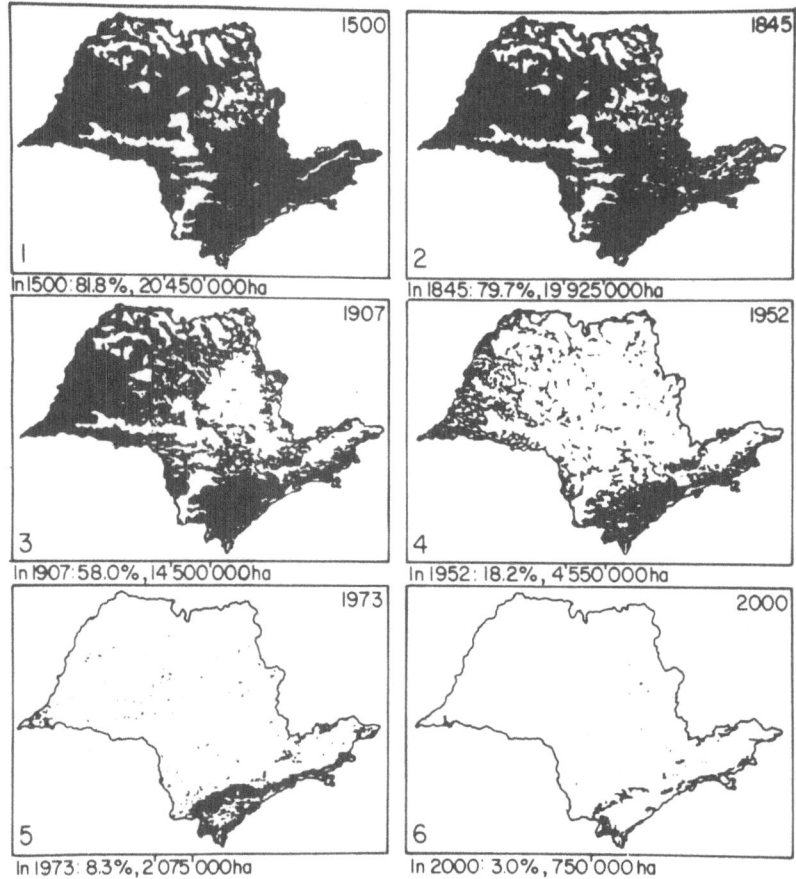

Figure 1:   Maps of Sao Paulo State, Brazil showing the depletion,
            since 1500, of forest cover and the probable situation
            in 2000. The forest cover is shown in black, its per-
            centage and hectarage is also provided.

unable to maintain all the plant and animal species it contained at
the moment of isolation and many of these are lost from the fragment.
The situation is analogous to the decay of a radioactive mineral,
only in the case of ecosystem decay the loss is of species. A
simpler state, more stable in species number, is reached ultimately
(Lovejoy and Oren, 1981).

    Given the pervasive spread of habitat fragmentation, ecosystem
decay is probably a dominant theme in the current biology of the
planet. It has important implications for the management of all wild
areas for whatever purpose, whether forestry, subsistence hunting,
firewood production or conservation. In the case of conservation,
it raises the question as to which species may be lost from nature

reserves, and whether many national parks really will, in the end, fulfil the purpose for which they were set aside (Lovejoy, 1980b). Rather than despair over such situations or write off small areas as useless for conservation, it is important to bear in mind not only that they may have some worth in themselves, but also that there is every reason to entertain the notion of protecting a peripheral area in which the natural community might be allowed to regenerate itself so as to achieve a larger fragment.

Much of what has been written about the problem of reserve size has been based on an analogy between habitat remnants and actual islands. While there are inevitably differences between isolation by water and isolation by severe modification of terrestrial habitat, the analogy has been useful in a number of ways. From the general observation that large islands usually have more species than islands that are smaller yet are of similar ecology, it is easy to deduce the almost elementary rule that large reserves will generally hold more species than small ones.

Perhaps a more important conclusion emerges from two other observations about islands:

i)   continental islands (those with previous connections to a mainland obliterated by rise in sealevel) have more species than oceanic islands (those never connected to a mainland) of similar size and ecology.

ii) continental islands in turn have fewer species than a comparable mainland area.

This is interpreted as indicating that continental islands had a number of species equivalent to a typical comparable area of mainland at the time of isolation and that they have been subsequently losing species, and will probably continue to do so until they reach a number equivalent to that of an oceanic island. This would indicate that reserves will also lose species after isolation; this has been documented primarily for birds in places as disparate as Brazil (Willis, 1979; Lovejoy et al.,1983), Ecuador (Leck, 1979) and the Eastern United States (Forman et al., 1976; Whitcomb et al., 1981).

There has been much recent discussion of the theory of island biogeography. This theory considers number of species on an island to be a function of a rate of immigration and a rate of extinction, each rate being a function both of the size of the island and the distance from a colonizing source such as a mainland. Many (Diamond, 1975; Terborgh, 1975; Wilson and Willis, 1975) draw the conclusion depending of course on the goal, large areas are preferable for conservation because they will protect more species and minimize the rate of species loss. Others (Simberloff and Abele, 1976) contend that island biogeography is neutral on this point and could equally well indicate that a series of small reserves will hold more species than a single large one having an area equivalent to their total. Discussion continues on this debate as there is a lack of direct evidence. The argument swings on the extent to which natural communities consist merely of a random collection of species, as opposed to the extent to which they are structured. Proponents of large reserves recognize that there must be some structure because some species, such as large predators, are very area sensitive and would survive only in large natural areas. However, there may be other communities such as invertebrates on mangroves, which are less structured and may approach the several-small-reserves model.

Some of the most interesting island data is derived from a Panamanian hilltop (15.6 km^2) which became Barro Colorado Island (BCI) when Gatun Lake was formed during the construction of the Panama Canal. While not studied before its isolation, it has been studied fairly continuously since, during which time a large number of species of birds have disappeared. Some of those undoubtedly disappeared because of vegetation maturing from early second growth, but about 15 - the very number predicted on the basis of island biogeography - are attributable to the so-called "area effect" (Terborgh, 1974). Attempts at reintroduction of two of these species (Morton, 1978) would indicate that the attribution is probably correct for at least one species because BCI is too small to have stream habitat. The area effect is probably also the reason for the loss of the second species. It returned to the only patch of second growth which soon matured and as BCI is too small to have enough natural patches of second growth generated by windfalls to maintain a population it disappeared. When a species is still plentiful and available elsewhere, reintroduction is a powerful tool to determine why the species was lost, and thus how management might maintain it in the ecosystem.

If one approaches the problem of reserve size from the species/area relationship, the larger the area sampled the larger the species number (Figure 2) is the rule generally known to hold in nature (Watson, 1835). However, it becomes obvious that species will also be lost subsequent to isolation. Whatever the size of the reserve, that is, wherever the species/area curve is cut, the very last species added to the curve is likely to consist of single individuals or tiny, non viable populations (Lovejoy and Oren, 1981).

Each biological community has a characteristic species/area curve. The curve for an Amazon forest with, for example, 300 species of trees in 10 ha rises much faster than one for a temperate woodland, which may have only 10 or 20 species of trees both locally and spread over a huge area. This intrinsic diversity of an association is one of the reasons why the species/area curve continues to rise; another is the increasing chance of encountering patches of environmental variants (e.g. different soil moisture) with increasing area. Yet another source of additional species may be those added by geographical replacement; however, loss of those replaced is of

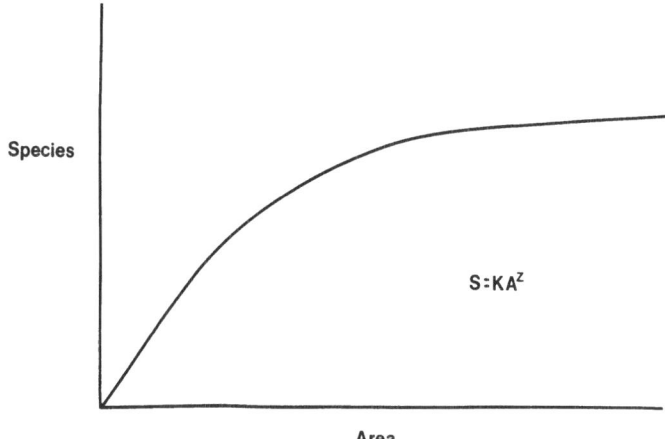

Figure 2: Typical species/area curve of the form S = KA: S = number of species, A = Area size.

less immediate concern in conservation, for they will presumably be protected in reserves elsewhere. Ignoring this particular class of species, a major goal of conservation becomes to protect not only the full array of plant and animal species, but to protect them as they occur in ecosystems - to protect a representative series of ecosystems that can be viewed as an international base of ecological standards against which ecosystem management in general can be measured. This is tantamount to establishing a series of protected species/area curves, each characteristic of the particular type of ecosystem.

A species/area curve constructed without biogeographical replacement species - a task far more easily stated than done - would in theory produce an asymptotic curve (Figure 3). Where it approaches the asymptote is equivalent to the minimum area of the phytosociologists, i.e. the minimum area necessary to encounter most of the species (often 95 %) and characteristic of the biological association (given with upper and lower limits). However, this would not be an ecologically viable unit in the sense of one which would maintain its full species richness if the surrounding natural area were to be destroyed. Such a point would be farther out on the curve and is defined as the minimum critical size of the ecosystem (Lovejoy and Oren, 1981). Functionally the curve could be cut at that point, i.e. the size at which the reserve could be set, and the biological association should maintain its characteristic species/area relationship over indefinite periods of time. For example, if it is indeed characteristic of Central Amazonian forest that ten ha has about 300 species of trees of 10 cm dbh (diameter at breast height) or greater, then the reserve should be large enough so that hundreds of years hence any ten ha sample would yield a count of some 300 species.

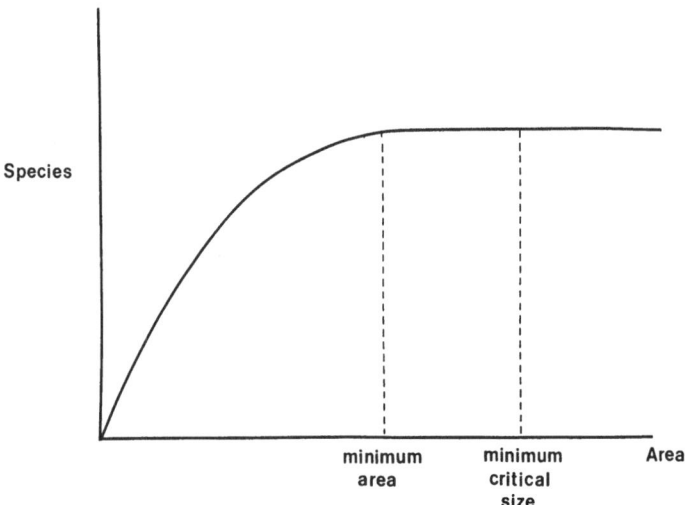

Figure 3:   Species/area curve following removal of those species
            added by biogeographial replacement, assuming an unreal-
            istically flat world. S = number of species. A = area.
            The minimum area is that necessary to encounter most of
            the species. The minimal critical size is that necessary
            to maintain full species richness if its surrounding
            natural area was destroyed.

     Defining this point of minimum critical size is, of course, very
difficult but will inevitably derive from what can be learned about
the species loss function and how it varies with the area of the
isolated habitat fragment. Another important point is the extent to
which species are lost in a predictable order. Another related
question that bears upon the controversy over large versus small
reserves is the extent to which remnant patches of the same size end
up with similar species composition. The joint World Wildlife Fund
(WWF) - Instituto Nacional de Pesquisas da Amazonia (INPA) project
at Manaus is designed to approach this problem - the Minimum Critical
Size of Ecosystems. Taking advantage of Brazilian law requiring
50 % of each development project to remain in forest, the project
has been able to study a selected size series of reserves (with rep-
licates) while the forest is still continuous, and then follow them
after isolation as they go through the species loss process (Figure
4). Ambitious as this project is, it is not able to test the effects
of differing reserve shape; nor will the results be transferable in
every detail to all types of ecosystems or to all continents - there
will inevitably be ecosystem-specific differences. Consequently a
number of similar projects around the world, even if involving only
a single habitat patch, a single group of animals or plants, and
even if not able to include pre-isolation data, will be of importance.

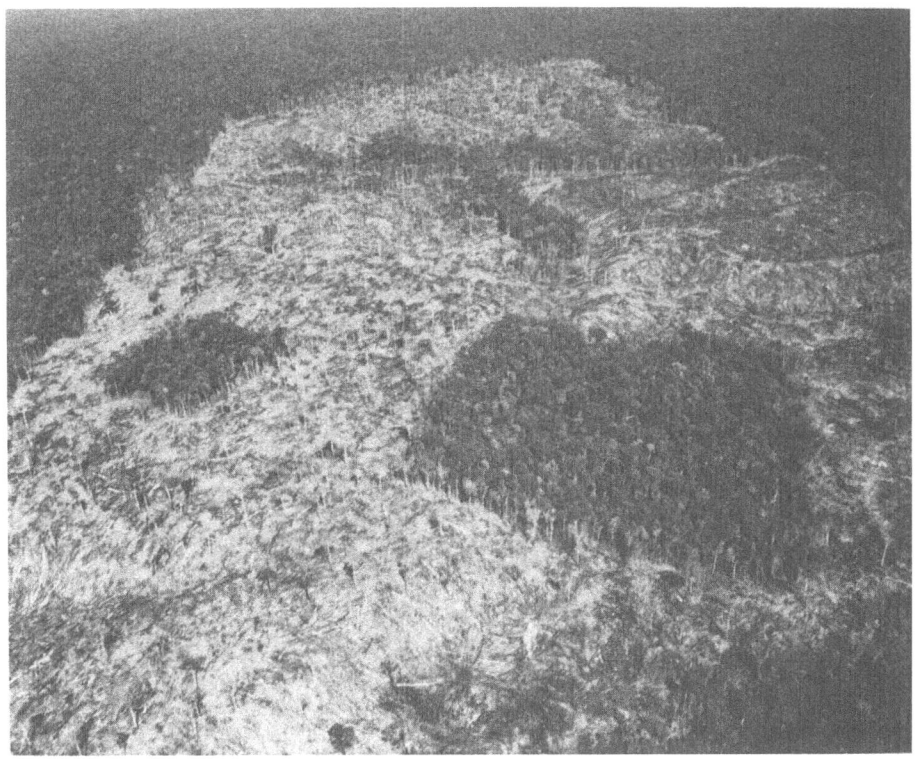

Figure 4: Aerial view of two isolated rain forest reserves, one
hectare on left and ten hectare on right. The reserves
are square but distorted in this perspective. The thin
strips of forest in the background are required, by
Brazilian law, to be retained along stream courses. The
forest is being cleared to create cattle pasture.

The history of an isolate is probably based on a founding species
composition derived from two sources (Figure 5). One is essentially
a sample taken from a more extensive and more species-rich community
when a fragment is saved. The second is whatever species may have
taken refuge in the fragment when surrounding habitat is destroyed.
This founding community is then subject to the species loss process,
but it is supplemented by whatever manages to colonize (or recolon-
ize, if it was lost subsequent to founding). Out of this will eventually come an ecosystem composed of fewer species, as subject to
species turnover as any ecosystem, but in a relatively stable state.

At the time of writing (September 1982), the Manaus project only
had a single one ha and a single ten ha patch isolated out of perhaps as many as 50 patches, which will include ones of 100 and
1000 ha, and one possibly in excess of 10 000 ha that will serve as

History of a Forest Fragment

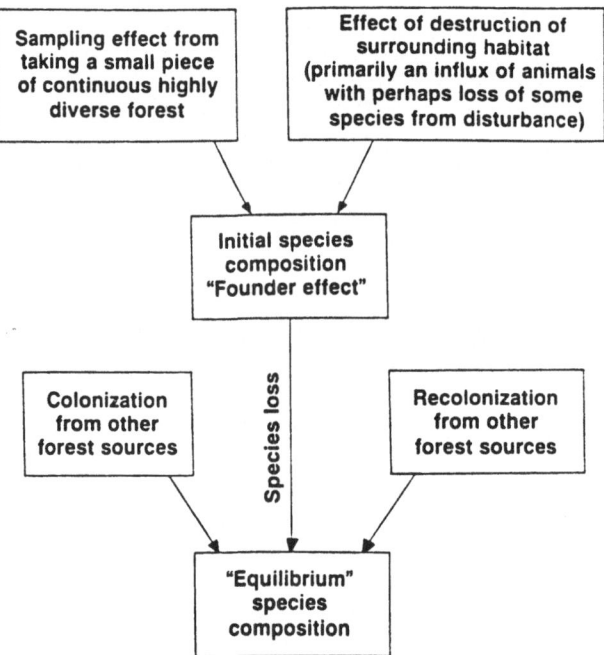

Figure 5:   The initial hypothetical model of the consequences of
            fragmentation at the species level (Lovejoy, 1980). This
            model does not take into account the effects of altered
            temperature and relative humidity conditions observed in
            the isolated reserves in the Central Amazon.

a baseline or "control" plot for comparison. Both plots showed signs
of species loss <u>within months</u> of isolation. An interesting occurrence was an immediate increase in rate of capture of birds in mist
nets once the reserves were isolated (Figure 6). This indicates
that there was an influx of birds from the surrounding destroyed
habitat, and shows that an initial over-population problem (at least
for mobile organisms such as birds) is superimposed on the area
effect. This will be a larger problem the smaller the reserve, as
it is a function of the ratio of the area of habitat destroyed to
the area left intact.

Another aspect that is magnified in a small reserve is the effect
of second growth species on the perimeter - a major difference between terrestrial "islands" and real ones with an aquatic surround.
This may also help to raise the total number of species in a recently
isolated reserve, at least for a while, until the species loss
process becomes obvious. This could easily lead to false encouragement if only species number is considered. In reality, second

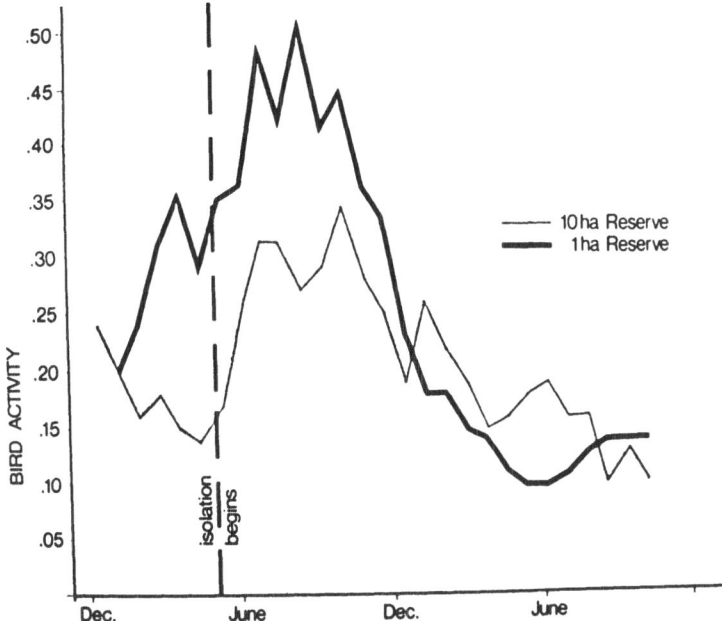

Figure 6: Bird activity as measured by capture rate (birds/net/day) from December 1979 to June 1981 in one and ten hectare reserves of Central Amazon forest near Manaus, Brazil. Capture rate increased markedly after the reserves were isolated.

growth species are not necessarily ones that conservation needs to worry about, since their high dispersal and high reproductive rates tend to make them able to take care of themselves. The edge effect in the Manaus project is particularly dramatic because it involves a major transition from the almost constant humidity and temperature of the forest interior to large daily fluctuations in both. In addition, the shallow-rooted rain forest trees that are now exposed to the wind have sharply increased rates of windthrow. These kinds of edge related effects are more important the smaller the reserve, essentially being a function of the ratio of perimeter to the area of the reserve.

The actual amount of primary forest effectively protected is thus less than the plot initially isolated. Further, it is important to bear in mind that the minimum size here defined makes no allowance for environmental fluctuations of the sort that occur predictably, but only rarely - for example the once-in-a-century dry spell experienced during the El Niño effects in Borneo in 1983. These can impose a great deal of pressure on an ecosystem and its species, and an additional increment of areas should be added to compensate for these effects, resulting in areas perhaps double the minimum size.

Animals at the end of long food chains often need large areas,
but whether these are less than or equal to minimum critical size is
unknown. This is particularly so in the case of larger animals.
What can be fairly certain is that many nature reserves will not be
large enough for long term support of such species. Not enough is
known of the biology of almost any of these species to be able to
define what may be a minimum population size. While it is known
that inbreeding can be a serious problem in some (but not all) small
populations (Lovejoy, 1978; Ralls et al., 1979), the estimates made
of minimum population size have not been tested in practice (Soule,
1981; Frankel and Soule, 1981). Complications concerning genetic
management of small populations including captive ones were explored
at a workshop entitled "Species Propagation Plans", held by the
Chicago Zoological Society, 5-6 May 1984 (Parker, Pers. comm.); a
publication will be forthcoming. Such species are likely to need
occasional or continuous assistance to remain viable as part of an

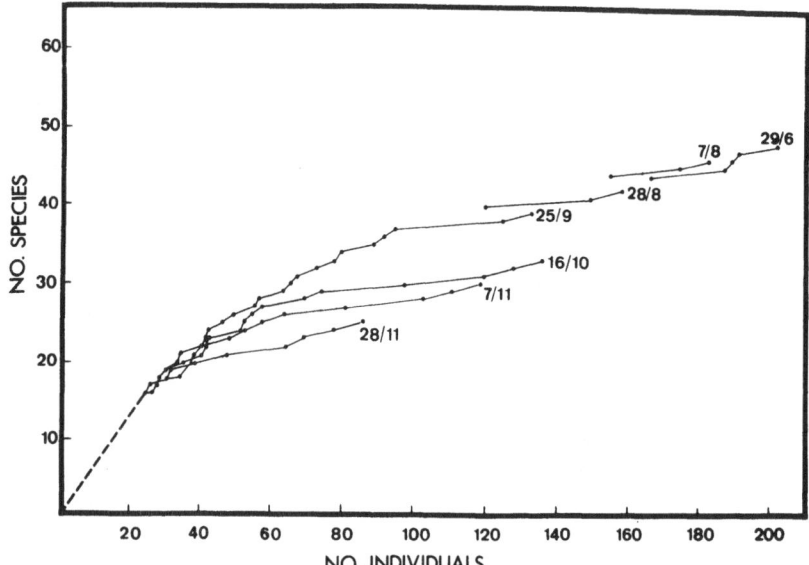

Figure 7: Species encounter functions (number of species with
           increasing number of individuals sampled) for understory
           birds in a 10 ha reserve near Manaus, Brazil after isol-
           ation in May 1980. The later the initial date (day/month)
           used to calculate a curve, the lower the rate of species
           encountered with increasing sample size. The earliest
           curves are not plotted all the way to the origin for ease
           of interpretation. These data indicate the understory
           bird community was declining in species number.

ecosystem, such as transplanting to mix blood lines, or captive propagation and re-introduction.

One approach to determining what minimum area may be necessary for a particular species is to observe its incidence over a series of islands or habitat patches of different size (Diamond, 1978) (Fig. 7). Some species will occur on all islands regardless of size, others are likely to show a threshold size below which they do not occur at all, and yet others occur 100 % of the time in all islands above a certain size, thus producing an S-shaped curve. The upper end of this threshold should provide a reliable minimum area for a species, <u>provided</u> that the island or habitat patches are known to be comparable, as well as to be in a relatively stable state and not still in the species loss process.

It is important to recognize that the extent of isolation will vary with a particular species. It may be close to zero for an edge species or close to unity (i.e. complete) for some birds of the rain forest understory. It is now recognized that many canopy species have been less affected by the isolation of Barro Colorado (Karr, 1982); certainly macaws currently use the project's minimum size isolated reserves as stepping stones. Some such species may be able to do quite well in a checkerboard landscape, but the actual requirements are not understood for any single species. In yet other situations, the effective size of a reserve may be far larger than an actual national park needs to be. Such is the case in Kenya, where first at Amboseli, as well as now for other parks, there are dispersal areas where people and wildlife co-exist during periods of the year when wildlife moves beyond the park boundaries. This model need not be restricted to animals which periodically leave a park area: it can also serve for situations in which some elements of the ecosystem being protected are always beyond the park boundaries. Migratory animals obviously present yet another set of problems.

Lastly, the concern over the size of a protected area should not become such a preoccupation as to obscure the importance of areas that are less than minimum critical size. Some environmental features can be protected with considerably less; indeed Britain's Nature Conservancy Council has determined that a great variety of species of herbs indigenous to ancient woods can be protected by a series of small reserves (Game and Peterken, 1981). When a reserve of minimum size is no longer possible for a given type of ecosystem, there may still be value in a reserve of lesser extent. Granted it will lose species and become a simplified version of what was previously present, but surely if an ecosystem in its entirety is of greater value than an impoverished version, the latter in its turn, is better than nothing.

REFERENCES

Burgess, R.L., and Sharpe, D.M., eds. 1981, "Forest Island Dynamics
    in Man-Dominated Landscapes", Springer Verlag, New York.
Diamond, J.M., 1975, The island dilemma: Lessons of modern bio-
    geographic studies for the design of natural reserves, Biol.
    Conserv., 7: 129.
Diamond, J.M., 1978, Critical areas for maintaining viable popula-
    tions of species, in: "The Breakdown and Restoration of Eco-
    systems", J. W. Holdgate and M. J. Woodman, eds., Plenum Press,
    New York.
*Ehrlich, P., and Ehrlich, A., 1981, "Extinction", Random House,
    New York.
Forman, R.T.T., Galli, A.E., and Leck, C.F., 1976, Forest size and
    avian diversity in New Jersey woodlots with some land use impli-
    cations, Oecologia (Berl.), 26:1.
*Frankel, O.H., and Soule, M.E., 1981, "Conservation and Evolution",
    Cambridge University Press, Cambridge, U.K.
Game, M., and Peterken, G., 1981, Nature reserve selection in
    central Lincolnshire woodlands, NCC (Nature Conservancy Council)
    Internal Report, London.
Karr, J.R., 1982, Avian extinction on Barro Colorado island, Panama:
    A reassessment, Am. Nat., 119:220.
Leck, C.F., 1979, Avian extinctions in an isolated tropical wet-
    forest preserve, Ecuador, Auk, 96:343.
Lovejoy, T.E., 1978, Genetic aspects of dwindling populations, in:
    "Endangered Birds", S.A. Temple, ed., University of Wisconsin
    Press, Madison.
Lovejoy, T.E., 1980a, A projection of species extinction. p.328 to
    332. In: The Global 2000 Report to the President. GPO,
    Washington.
Lovejoy, T.E., 1980b, Discontinuous wilderness: Minimum areas for
    conservation, Parks, 5:13.
Lovejoy, T.E., and Oren, D.C., 1981, The minimum critical size of
    ecosystems, in: "Forest Island Dynamics in Man-Dominated Land-
    scapes", R. L. Burgess, and D. M. Sharpe, eds., Springer-
    Verlag, New York.
*Lovejoy, T.E., Bierregaard, R.O., Rankin, J.M., and Schubart H.O.R.,
    1983, Ecological dynamics of forest fragments, in: "Ecology
    and Management", S. L. Sutton, T. C. Whitmore, and A. C. Chadwick,
    eds., Blackwell Scientific Publications, Oxford, U.K.
Morton, E.S., 1978, Reintroducing recently extirpated birds into a
    tropical forest preserve, in: "Endangered birds: Management
    Techniques for Preserving Threatened Species", S. A. Temple, ed.,
    University of Wisconsin Press, Madison.
Myers, N., 1980, "Conversion of Tropical Moist Forest", Report for
    Committee on Research Priorities in Tropical Biology of the
    National Research Council, National Academy of Science,
    Washington, D.C.
Parker, P. Pers. comm.

Ralls, K., Brugger, K., and Ballou, J., 1979, Inbreeding and juvenile
    mortality in small populations of ungulates, Science, 206:1101.
Raven, P.H., (Chairman of Committee) 1980, "Research Priorities in
    Tropical Biology", Committee on Research Priorities in Tropical
    Biology, National Academy of Science, Washington, D.C.
Simberloff, D.S., and Abele, L.G., 1976, Island biogeography theory
    and conservation practice, Science, 191:285.
*Soule, M.E., and Wilcox, B.A., eds., 1980, "Conservation Biology",
    Sinauer Associates, Inc., Sunderland, Massachusetts.
Soule, M.E., 1981, Thresholds for survival: Maintaining fitness
    and evolutionary potential, in: "Conservation Biology", M. E.
    Soule and B. A. Wilcox, eds., Sinauer Associates, Inc.,
    Sunderland, Massachusetts.
Terborgh, J., 1974, Preservation of natural diversity: The problem
    of extinction prone species. BioScience 24: 715.
Terborgh, J., 1975, Faunal equilibria and the design of wildlife
    preserves, in: "Tropical Ecological Systems: Trends in Terres-
    trial and Aquatic Research", F. B. Golley, and E. Medina, eds.,
    Springer-Verlag, New York.
Watson, H.C., 1835, Remarks on the geographical distribution of
    British plants, n.p., London.
Whitcomb, R.F., Robbins, C.S., Lynch, J.F., Whitcomb, B.L.,
    Klimkiewicz, M.K., and Bystrak, D., 1981, Effects of forest
    fragmentation on avifauna of the eastern deciduous forest, in:
    "Forest Island Dynamics in Man-Dominated Landscapes", R. L.
    Burgess and D. M. Sharpe, eds., Springer-Verlag, New York.
Willis, E.O., 1979, The composition of avian communities in reman-
    escent woodlots in Southern Brazil, Papeis Avulsos Zool., 33:1.
Wilson, E.O., and Willis, E.O., 1975, Applied biogeography, in:
    "Ecology and Evolution of Communities", M. L. Cody, and J. M.
    Diamond, eds., The Belknap Press of Harvard University Press,
    Cambridge, MA.

---

*Recommended additional reading

17

IMPACTS OF CHANGING LAND-USE PARTICULARLY IN CENTRAL EUROPE

K.-F. Schreiber

Institut fur Geographie
Westfalische Wilhelms-Universitat Munster
Robert-Koch-Strasse 26
44 Munster, Federal Republic of Germany

HISTORICAL BACKGROUND

Numerous regions in Europe show the signs of human intervention which are both the cause and the result of changes in land-use. For the present I am confining my attention to effects related to the cultivation of food, so setting aside those effects attributable to e.g. mining, construction of industrial and residential complexes etc.

Since man began to settle, towards the end of the mesolithic and in the neolithic eras, and started to cultivate and graze domesticated animals on managed pastures, he has profoundly modified the landscape of Europe (Ellenberg, 1978). The development of the "cultivated landscape" has, however, been periodically interrupted. On the one hand, population pressures led to the rapid development of agriculture in former forest regions, while, on the other, wars, epidemics and famines have led to its decline with the abandonment of fields and the reafforestation of clearings. The exposed ridges "Wolbacker", in many forests of Europe, are testimony to former agrarian land-uses. Woodland clearings and improved agricultural techniques, equipment and methods (e.g. the transition from wooden to iron ploughs, and the institution of crop rotation) have led to substantially increased arable areas, and regrettably to increased amounts of soil erosion. In the recent past, the so-called 'agrarian revolution', starting around 1800 AD with the transition from an extensive alternating arable-pasture economy to intensive continuous arable cropping, resulted in, within a few decades, the impairment of arable land by gully erosion, these areas subsequently being relinquished as 'bad lands' (see Hard, 1970) for an account of

303

events in the Westpfalz and Zweibruecker Land). Gully erosion was in
due course minimized by afforestation, changes in cropping practices,
together with purposeful soil conservation. However gully erosion
was succeeded by sheet erosion which, because it is slower posed less
of a problem. But nonetheless sheet erosion frequently leads to a
substantial lowering of the value of arable land. It seems that the
history of erosion is cyclic, the different phases being linked to
the history of man and his activities.

The European landscape has been changed by extensive pasture and
forest management. In the different climatic regions of Europe,
typical landscapes have come into existence e.g. the park-like high
pastures and Alpine mountain pastures with cattle terraces, the
Mediterranean Karst landscape and the treeless blanket bogs of N.W.
Europe. While these are frequently thought to be the result of
natural evolutionary processes - particularly those controlled by
climate (e.g. Hempel, 1979; Pennington, 1974) the influence of man
should not be overlooked. The predilection of sheep and goats for
the buds, leaves and twigs of trees, put the regenerating stands of
saplings at risk, and together with timber utilization and fire
hazards has led in some instances to the loss of forests. In the
Mediterranean climate zone damage to vegetation by browsing and
trampling linked with intense rainstorms, has led to intensified
erosion in overgrazed 'forests'. Even where there were/are cultivated terraces, erosion is still a problem, leading to denudation
(Schreiber, 1982). In contrast, in very humid but cooler regions of
N.W.Europe, organic matter accumulates in afforested areas to form
low permeable, thick peat, so adding to the underlying soil.

With the need to provide winter feed for domestic animals, attention turned to the use of forest duff, or litter and the so-called
heather sods (Plaggen) which are fed to cattle wintered in barns,
when hay and foliage (of deciduous trees) were no longer available
(Ellenberg, 1978). Subsequently the foodstuff residues, enriched
with manure, were spread to enhance soil fertility. Since the iron
age, the removal of sods has played an important role in land-use
and cultivation. Sods were cut, with a special hoe, mostly from
thinned, grazing woodlands, with the removal of the raw humus layer
and the top soil permeated with roots. With a 15-20 year rotation
of sod removal allied to grazing, the regeneration of trees was
prevented and as a result, areas of heath fit only for undemanding
breeds of sheep, expanded, for example, the large heaths of N.W.
Central Europe (Ellenberg, 1978). In the event the loss of nutrients
by sod removal from already deficient areas benefitted the fields
which received the manure enriched residues after feeding. Thus,
over the years, a type of soil, of manmade origin, "Plaggenesch",
was formed, sometimes ground-level being raised 1.5 m. The most
fertile fields, the "Eschfluren", in infertile sandy soil regions,
and areas with loessic loam soils, often owe their origin to these
enriched residues (Eckelmann, 1980). Thus, the removal of sods

decreased the fertility of their sites of origin, while benefitting the locations receiving the residues after being fed to animals - a switch of nutrients having a profound effect on landscape.

If this is what happened in the past, what is happening at present ? Can our acquired knowledge help prevent the repetition of earlier mistakes ?

THE ABANDONMENT OF AGRICULTURAL LAND, AND ITS CONSEQUENCES

The abandonment of many fields, during the last twenty years, has triggered lively debates and a considerable amount of research. As already pointed out abandonment is not a new phenomenon. It has recurred periodically leaving its mark on the landscape. For the purpose of the ensuing discussion the causes of abandonment are immaterial - no attempt is made to distinguish between social reasons or those attributable to low productivity. (In some languages 'abandoned' is sometimes regarded as synonymous with fallow. In these instances long term fallow is distinguished from a 1-year break in cropping by referring to the latter as 'black fallow' describing the colour of the weed free soil).

Recent changes in the appearances and structure of landscapes attributable to the abandonment of agricultural fields

The big increase in the areas of abandoned and derelict land within the countries of the European Common Market, especially those in the central and southerly parts of Central Europe, has aroused concern among farmers and landscape architects, not to mention Society at large. To begin with, the mainly urban populations were 'offended' by the neglected appearance of the landscape - but within fifteen years social attitudes, particularly those related to conservation, have changed. Initially, people missed the seasonal changes in colour associated with ploughing and cropping, with the continuous green of managed pastures being superseded in summer and autumn by shades of brown - the brown of withering and withered parts of plants. Additionally farmers were apprehensive about a possibly increased incidence of pests, of augmented supplies of weed seeds and general site deterioration. In contrast, landscape architects and many plant conservationists were concerned that large sections of highland landscapes, which have remained uncultivated, might be transformed by afforestation (Buchwald, 1970).

It was assumed that abandoned fields would transform themselves into grasslands, which - like abandoned grasslands - would develop into thickets en route to forest climax vegetation (Stählin et al., 1972; Schreiber 1974); a sequence supported by observations, made at one time, of different developmental stages separated by space (geographically)(Büring, 1970; Hard, 1976). However, too little

regard was given to a model described by Egler (1954) which stressed
that the initial assemblage of plants has a profound effect on subsequent succession, a concept subsequently confirmed by field experiments done by Schmidt, (1975) and Schreiber (1977).

The quick growth of woody plants, at an early stage, can only be
expected on abandoned agricultural land. Species with wind blown
seeds or with vegetatively reproducing underground structures tend
to predominate with the latter being commonplace more often in former
grasslands (Hard, 1976; Schmidt, 1981). Assemblages of this sort
comprise the scrubby vegetation of hedgerows, the bushes of forest
'belts' and glades (Prunetalia and Epilobietea-communities). In due
course thicket-type vegetation develops, sometimes forming thick
copses but development towards a natural forest is unlikely. Hard
(1976) indicated that even wind-dispersed species have only a limited
ability to scatter up to a few hundred metres. Developing plants of
this sort are therefore likely to be restricted to small border
zones, - the distance effect (Hard, 1976). Hard reckons that the
possibility of establishing woody vegetation on abandoned fields
declines after the first 3 - 4 years as the cover afforded by annuals
and biennials becomes increasingly dense with an intensification of
competition. This early ruderal stage of annuals/biennuals, rich in
species of grasses, is attained rapidly and can remain relatively
stable over many decades.

The abandonment of orchards should be regarded as a special case.
In the temperate, humid conditions of Central Europe, cherries,
plums, apples and pears, and to a lesser extent peaches, sprout root
suckers which occasionally develop into a thick almost impenetrable
forest of poles (Schreiber, 1980b). However, in Mediterranean
climates this does not happen; here abandoned fruit trees favour a
succession, with species whose seeds or fruits are dispersed by
animals, notably birds, playing a much stronger part as has happened
on the abandoned terraced orchards of Mt. Carmel and the Judean hills,
Israel. The early developments after abandonment seem more likely to
lead to climax forest, but a development to this end may be interrupted and a pattern of herbs and grasses established which leads
to a stable shrub stage on the terrace slopes.

In contrast to that of abandoned arable fields, the development
of vegetation on abandoned grasslands is remarkably constant with
the exception of the effect of grazing and mowing. Grasslands become
invaded by plants with underground stolons, rhizomes or other storage
organs (Schiefer, 1981), species that are not strongly influenced by
the mulching effects of layers of litter. With the development of
these species at the expense of rosette plants, the establishment of
wind-dispersed woody vegetation is impeded. Thus it is not surprising
to find that, even after 20 years, former pastures and meadows seem to
resist the growth of scrub; this is particularly true of unproductive
pastures (lowland, upland and mountain) on acid soils, which develop
matgrass stands (Nardetalia-communities).

In humid, wet conditions, abandoned grasslands develop high growing perennials (Hochstaudenflur), frequently rich in large sedges,
expecially when soils are water-logged (Schiefer, 1981). The assemblages of plants are also influenced by the chemical nature of bedrock.

The post-abandonment picture of development will be greatly
affected by the presence of trees that form root suckers. From his
observations of sparse sub-Mediterranean White Oak woodlands (*Quercus
pubescens*) Jakucs (1972) developed a theory, regarding the phases of
a root suckering succession. Mother plants of woody species growing
in hedges and wind breaks thrust their roots into abandoned fields,
with the subsequent development of emergent terminal shoots. This
type of development can be observed where former fields, meadows,
pastures and vineyards are bordered by heaped stones (e.g. in the
South German limestone areas); the stone borders having been formed
manually. While the fields are still cultivated the intruding tender
and tasty shoots are regularly lost as a result of ploughing and
browsing. In their absence Blackthorn (*Prunus spinosa*) can, on
fertile soils, advance by as much as 1 m a year. On shallower limestone soils blackthorn can advance, mostly in dense stands, by about
0.5 m per year, but by only a few cm on dry soils which restrict
root penetration (Schreiber, 1980b). Jakucs (1972) and Hard (1976)
have observed the disintegration of the older parts of root suckering
trees, which become replaced by faster growing trees. The speed of
this change seems to be highly variable.

Abandoned fields do not fit a simple grouping: structural successional changes can lead to increased landscape diversity but change
can be minimized by mowing, mulching, or grazing. Interventions of
this sort are only needed at infrequent intervals (Schmidt, 1981),
they are needed to minimize the growth of woody incomers.

In summary: after abandoning arable fields there is an immediate
increase in species diversity (Schmidt, 1981), reaching in due course
"saturation", after which recruitment virtually ceases: there could
be a net loss. But this is not to say that short term changes
can be induced by weather, particularly dry weather (Schiefer,
1981; Schmidt, 1981). After lying fallow over a long period, grasslands, however, show a gradual decrease in species (Schreiber, 1980b;
Schiefer, 1981). The net loss in species diversity after prolonged
abandonment is not unexpected. It probably relates to the elimination of colonists that are not adapted to withstand later competition.

How does abandonment, one form of change, affect the availability of nutrients, energy and water ?

Whereas plant biomass, with its content of nutrients, is partially or completely removed during cropping, after abandonment,
these resources are retained - a major difference.

Biomass/production and turnover.   The annual production of foliage
by abandoned grasslands ranges, according to site, between <500 and
>10 000 kg/ha dry matter (Schreiber 1980b):   a similar range was
given by Schmidt (1981) for abandoned arable fields. Furthermore
in moderately moist to moderately damp or fresh (frisch) sites these
amounts of biomass can be decomposed by soil fauna (e.g. earthworms)
and soil microbes within a year. Nevertheless, sufficient remains
to increase the depth of litter. Unlike the shoots, a proportion
of the roots of grasses remain active for long periods. Thus only
a fraction of the annual root production is decomposed within the
next 12 months.

   Drought has the same limiting effect as water-logging, it inter-
rupts litter decomposition, with consequent increases in soil organic
matter. Whereas these temporary increases improve moisture retention
in dry conditions they lead in wet anaerobic conditions to the accumulation of undecomposed or partially decomposed organic material,
leading to the formation of 'mor', humic or bog soils. The rates of
these changes tend to change from season to season. Together the
seasonal and episodic events combine to influence production in a
way that cannot yet be predicted.

Changes in soil properties. In permanent experiments, no mentionable
change in the pH count of fallow soils has yet been observed. While
a small decrease in pH has been observed in the topsoil of a field
after being abandoned (Ellenberg, 1978), corroborative experimental
evidence is not available (Schreiber, 1980b, c; Schmidt, 1981). In
contrast many substantial increases in carbon and organic substances
have been recorded (Borstel, 1974), the increases tending to reach
a plateau which is thereafter maintained. This course of events
confirms the idea that soils develop and maintain optimal humus concentrations (Kohnlein, 1964; Klapp, 1971).

   The initial increases in humus would be linked with enhanced
water holding capacity and ability to neutralize - features constantly being exploited when cropped intensively. Strangely, prescribed burning does not seem to decrease total amounts of carbon in
abandoned soils (Schreiber, 1981).

   Studies of abandoned arable fields and grassland by Schmidt
(1981), Schreiber (1980b,c) and other unpublished data of grassland
fallow, show that concentrations of P_2O_5 and K_2O increase slightly
with the probable transfer of nutrients from moribund foliage to
roots and underground storage organs (e.g. rhizomes). At this stage
it should be remembered that plants with storage organs characteristically play a decisive role following abandonment only to be
replaced at a later stage, when an internal nitrogen cycle has been
developed, by tall growing grasses and shrubs (Schreiber, 1980b).

The increased amounts of P_2O_5 and K_2O in abandoned soils do not equate with the total amounts of nutrients transferred to biomass and therefore potentially available for release during decomposition. As yet, however, little is known about the nutrients that are lost, (Schmid, 1974).

Changes in water economy.   It was feared that the well maintained open drains on intensively managed agricultural holdings would deteriorate after abandonment with consequent water-logging. This fear has, in some instances materialised, e.g. the former irrigated meadows of the Central European highlands.

Undoubtedly land drainage by pipes buried to depths of 40 to > 100 cm are at risk particularly to penetration by tree roots. But, for the present, the problem is more apparent than real.

Biomass production is usually greater before, than after, abandoning arable fields if for no other reason than the application of optimal amounts of artificial fertilisers. On the assumption that soil water controls the optimal applications of fertilisers, would there be an excess of soil water after abandonment and would the excess, if any, be damaging (Schreiber, 1977, 1980b,c; Schiefer, 1981)?   When making comparisons with the effects of mulching, there seems to be no evidence of a deleterious effect. Even a change in the species formation - to species with a higher humidity index (Ellenberg, 1979), is by no means evident in permanent experiments. On the contrary there is a trend instead towards the substitution of early colonists by species with larger nitrogen requirements, hardly an indication of site impairment (Schiefer, 1981). These deductions have been made from abandoned meadows and pastures, and not from abandoned intensively managed grasslands on nutrient-rich sites. When this happens grass cultivars, selected or bred for their exceptional abilities to use nutrients will be replaced by plants with appreciably smaller nutrient demands. In these instances there may be real changes in soil water status which may be influenced as a result of competition between plants.

In summary it seems that the accumulation of biomass after abandoning dry sites will enrich soils with the mulching effect minimizing evaporation losses. As a result fluctuations in water content will be minimized and periods of soil drying decreased. At the same time organic matter accumulations will increase sorption capacity and nutrient supplies, changes that may cause an undesired modification of plant assemblages in dry meadows (with calcareous soils (mesobrometum)) which have many protected species.

RECULTIVATION OF ABANDONED FIELDS

What are likely to be the problems encountered if it were to be desirable to recultivate abandoned fields ? If the branches of the developing scrub are not thicker than 3 cm they can be cut with purpose-built machines. The trunks and roots of large trees can be extracted with a cable winch. But to bring abandoned areas back into production, cutting and stump removal need to be complemented by ploughing usually with a heavy plough/high powered tractor combination.

To be prudent, and possibly because it may be necessary to bring abandoned land back into production, preferably with the minimal use of costly energy, can steps be taken to ensure that major problems are avoided ? Thus to avoid the problem created by trees, which mainly develops in the first two years after abandonment, a case could be argued for occasional mowing or mulching, without removing the debris, during the first two years, knowing that, thereafter, the ingress of trees will be slower. Extensive grazing with sheep, goats, horses or cattle on suitable sites, and in many cases prescribed burning, can have the same effect. However, these treatments should be considered in combination: none is wholly effective. These measures would not be adequate for the control of rhizome forming plants, geophytes, which seem to be particularly able to establish themselves on abandoned sites.

On balance it would seem that the fertility of reclaimed 'abandoned-land' is likely to be better than that of land in continuous cultivation, attributable to an extended pasture phase in an alternating arable-pasture economy. The extended pasture phase would provide a soil with enhanced organic matter and nutrients which could be utilized in the arable phase to give a 'fertilizer effect'.

The ability of commercial cultivars of pasture grasses to compete is variable. However, if pastures are abandoned for short periods, a commercially worthwhile sward can be obtained by applying fertilizers and repeatedly mowing. These measures would not produce the desired results if applied to grasslands abandoned for a considerable number of years. However if the abandoned grassland were to be mown in the autumn of every third year then the sward before abandonment could be recreated.

Should fields become wetter after abandonment either because of differences arising from competition, or due to the 'breakdown' of open drains the consequent increased biomass production can, by the demand for transpiration, lead to "biological drainage", so improving turf condition. To re-establish cropping on these sites it would be essential to repair or replace defective pipe drainage. Particularly when the grassland is to be used as a pasture, the turf should be tread resistant.

IMPACTS OF THE "MAIZE-REVOLUTION"

For centuries maize was restricted to moderately rich sites devoid
of water-logging and free of frost. With 'improvement' present-day
hybrids have the tolerance allowing maize to be grown where it used
to fail. Thus up to the early fifties, maize was restricted to the
warmest locations of Central Europe - the lowlands of Upper Rhine
and France: now its triumphal march extends not only to the cool,
temperate sites of North Germany, but also to large expanses in the
plains and hill regions on the northern border of the Alps. The
acreage of maize in the Federal Republic of Germany has increased
from 55 000 ha in 1961 to 862 000 ha in 1981 (126 000 ha of grain
and 736 000 ha for silage).

This rapid expansion has been accompanied by a proportionate
increase in erosion particularly in loess areas, an increase attributable to the lack of contour ploughing found to be desirable in
N. America. In the absence of contour ploughing gully erosion
occurred during spring while the ground was without plant cover.
In Southern Bavaria erosion gullys, up to 50 cm deep could be seen,
with the movement on sloping sites of >50 t of soil per hectare per
yr (Schwertmann, 1980, 1981): amounts between 70 - 100 $tha^{-1} yr^{-1}$
have not been uncommon. This loss should not be considered only in
terms of silt, clay .... but also of plant nutrients, especially
phosphate (e.g. Schwertmann, 1981). These losses jeopardize the
growth of terrestrial plants but in doing so give an undesirable
boost to the growth of aquatic plants as a result of eutrophication.

The successful cultivation of maize requires large amounts of
nitrogenous fertilizers, up to 200 kg N $ha^{-1} yr^{-1}$, have been recommended. But that was before potentially alarming reports of groundwater contamination had been documented. Nitrate concentrations of
90 mg/l and more have since been reported (Obermann, 1981) so exceeding the legal limit in the Federal Republic of Germany. As with all
plant crops it is absolutely necessary to apply more fertilizers to
maize than are actually 'used' to get the desired effect. Presumably
the 'excess' poses the threat to ground water. Because of their
sensitivity to competition, crops of maize must be kept weed free.
This used to be done by tilling but now herbicides tend to be used
instead. We are assured that herbicides in current use, and their
metabolic products, are decomposed relatively speedily but do we
know sufficient about the sensitivity and tolerance of non-target
species ?

The examples that I chose to discuss illustrate some of the
hazards of land-use change, particularly abandonment. However, there

are many more e.g. monocultures, the construction of ski runs with
the risks of exacerbating soil erosion etc. These problems are not
insoluble, very often we can hazard a guess about the most appropriate
action to take particularly if we take the trouble to learn from our
earlier mistakes. But above all, an integrated approach is essential.

REFERENCES

Borstel, U.-O. von 1974, "Untersuchungen zur Vegetationsentwicklung
    auf ökologisch verschiedenen Grünland- und Ackerbrachen hessis-
    cher Mittelgebirge (Westerwald, Rhon, Vogelsberg)", Diss. Univ.
    Giessen.
Buchwald, R., 1970, Auswirkungen des Agrarstrukturwandels auf
    Struktur und Bild der Landschaft, Schr. r. Forsch.rat.Ernährung,
    Landw. Forsten, 4:29.
Büring, H., 1970, "Sozialbrache auf Äckern und Wiesen in pflanzen-
    soziologischer und ökologischer Sicht", Diss. Univ. Giessen.
Eckelmann, W., 1980, "Plaggenesche aus Sanden, Schluffen und Lehmen
    sowie Oberflächenveränderungen als Folge der Plaggenwirtschaft
    in den Landschaften des Landkreises Osnabrück", Geol. Jb., R.F.
    10, Schweizerbart'sche Verlagsbuchhandlg., Hannover.
Egler, F.E., 1954, Vegetation science concepts. I. Initial floristic
    composition, a factor in old-field vegetation development,
    Vegetatio, 4:412.
Ellenberg, H., 1978, "Vegetation Mitteleuropas mit den Alpen in
    ökologischer Sicht", Ulmer, Stuttgart.
Ellenberg, H. 1979, "Zeigerwerte der Gefässpflanzen Mitteleuropas",
    Scripta Geobotanica, 9, Göttingen.
Hard, G., 1970, Exzessive Bodenerosion um und nach 1800, Erdkunde,
    24:290.
Hard, G., 1976, "Vegetationsentwicklung auf Brachflächen", KTBL-
    Schr., 195, Landwirtschaftsverlag, Münster-Hiltrup.
Hempel, L., 1979, Wenn der Boden zum Skelett abmagert, Umschau,
    79:405.
Jakucs, P. 1972, "Dynamische Verbindung der Wälder und Rasen",
    Budapest.
Klapp, E. 1971, "Wiesen und Weiden", Parey, Berlin.
Köhnlein, J. 1964, Ober die Beziehungen zwischen Ertragsbildung,
    Bodenfruchtbarkeit und Humus, Schr.r. Landw. Fak., Univ. Kiel,
    37:5-40.
Obermann, P. 1981, "Hydrochemisch/hydromechanische Untersuchungen
    zum Stoffgehalt von Grundwasser unter dem Einfluss landwirt-
    schaftlicher Nutzung", Ruhruniversität, Bochum.
Schiefer, J. 1981, "Bracheversuche in Baden-Württemberg", Beih.
    Veröff. Naturschutz Landschaftspflege Bad.-Wurtt. 22, Inst.
    für Naturschutz u.Ökologie, Karlsruhe.
Schmid, G. 1974, Umweltprobleme durch Brachflächen ? ("Wenn Brach-
    land zum Landschaftsproblem wird..."), Arbeiten der Deutschen
    Landwirtschafts gesellschaft, 141:24.
Schmidt, W., 1975, Vegetationsentwicklung auf Brachland - Ergeb-
    nisse eines fünfjährigen Sukzessions-Versuches, in: "Sukzessions-
    forschung", ed. W. Schmidt, Cramer, Vaduz.

Schmidt, W., 1981, "Ungestörte und gelenkte Sukzession auf Brach-
äckern", Scripta Geobotanica, 15, Göttingen.

Schreiber, K.-F., 1974, Landschaftspflege mit oder ohne Landbewirt-
schaftung - wie sieht es der Landschaftsökologe ?, ("Wenn Brachland zum Landschaftsproblem wird...") Arbeiten der Deutschen
Landwirtschafts gesellschaft, 141:7.

Schreiber, K.-F. 1977, Zur Sukzession und Flächenfreihaltung auf
Brachland in Baden-Württemberg, Verh. Ges. Ökol, 5:251.

Schreiber, K.-F., 1980b, Brachflächen in der Kulturlandschaft,
("Ökologische Probleme in Agrarlandschaften") Dat. Dok. Umwelt-
schutz, Sonderr. Umwelttag. Univ. Hohenheim 30:61.

Schreiber, K.-F. 1980c, Entwicklung von Brachflächen in Baden-
Württemberg unter dem Einfluss verschiedener Landschaftspflegema-
ssnahmen, in: Verh. Ges. Ökol., 8:185.

Schreiber, K.-F. 1981, Das kontrollierte Brennen von Brachland -
Belastungen, Einsatzmöglichkeiten und Grenzen. Eine Zwischen-
bilanz über feuerökologische Untersuchungen, Angew. Botanik,
55:255-275.

Schreiber, K.-F., 1982, The Origins of Ecosystems and the Effects
of Human Intervention, Appl. Geogr. Development, 19:126.

Schwertmann, U. 1980, Stand der Erosionsforschung in Bayern,
("Ökologische Probleme in Agrarlandschaften") Dat. Dok. Umwelt-
schutz, Sonderr. Umwelttag. Univ. Hohenheim 30:96.

Schwertmann, U. 1981, Grundlagen und Problematik der Bodenerosion,
Bayer. Landw. Jb., Sonderh. 1, 58:75-79.

Stählin, A., Stählin, L., und Schäfer, K., 1972, Ober den Einfluss
des Alters der Sozialbrache auf Pflanzenbestand, Boden und
Landschaft, Z. Acker-und Pflanzenbau, 136:177.

18

NATURE AND LANDSCAPE PLANNING IN A RAPIDLY CHANGING REGION

A. J. Beenhakker

Provinciale Planologische Dienst, Zeeland
Postbus 153, 4330 AD Middleburg
The Netherlands

INTRODUCTION

In 1906, a storm flood swept over the shorelands of the
Zuiderzee, a shallow, almost landlocked bay of the North Sea in the
centre of the Netherlands. Although plans for land reclamation of
the bay had been prepared since 1886 it took 12 more years of discussion before the Dutch people could pluck up the courage to decide
that the Zuiderzee was to be enclosed (dammed), so virtually pre-
cluding the possibility of similar future disasters. At the same
time and because the Zuiderzee was shallow with fertile deposits it
was possible to reclaim part of the Zuiderzee by means of artificial
drainage. Since the 1930s, considerable areas have been reclaimed;
many farms, villages, even cities, and nature reserves have been
founded 4 or 5 metres below sea level. The work is still continuing.

In 1953 a much more severe storm hit the South-Western part of
the Netherlands, which is commonly known as the Delta Region and
includes the mouths of the rivers Rhine, Meuse and Scheldt (Figure 1).
On this occasion it took only two years to decide to dam the different estuaries excepting the most important shipping lanes to Rotterdam
and Antwerp. In this instance land reclamation was of little importance: the waters are deep and the shallow flats are generally sandy.
Instead, safety against inundations and the possibilities of storing
fresh water were considered of paramount importance. Work is still
in progress.

The difference in the speed with which decisions were made to
dam the Zuiderzee and the Delta Region reflects the change in the
Netherlands during a period of 50 years from a rustic and rural
country into a bustling and rich industrial nation. In neither

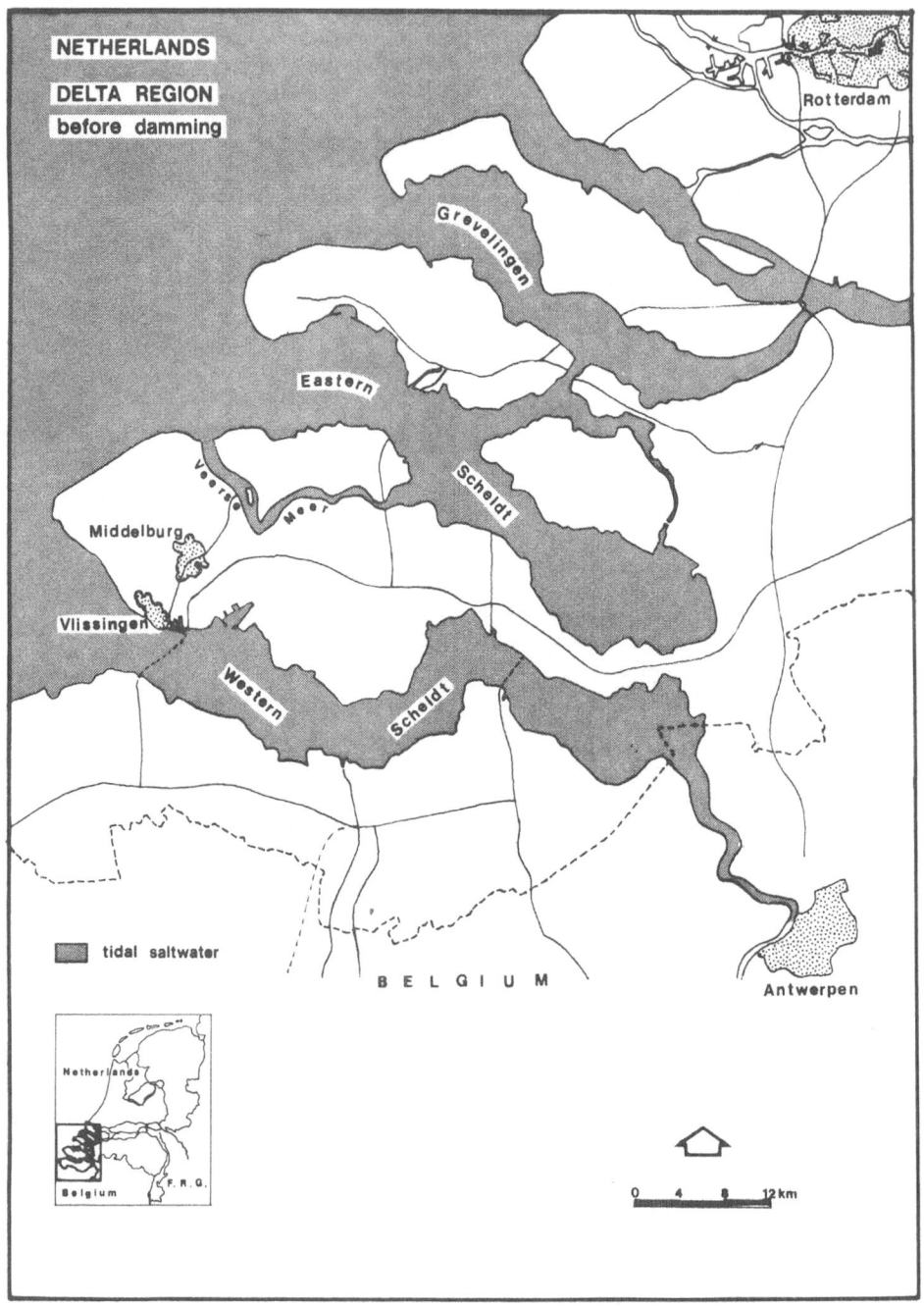

Figure 1:   The Delta region of the Netherlands before damming
            commenced after the 1953 floods, showing the system of
            four tidal inlets.

instance was prior consideration given to nature conservation and landscape values, aspects of planning which did not gain political significance until the 1960's, when the Zuiderzee scheme was 75 % complete and when the Delta works were in full progress. However, discussions of the advisability of enacting the last reclamation project in the Zuiderzee area are still in progress, although the national government recently decided to adhere to the original plan. In the Delta region the original plan has been subject to a major change: instead of being constructed as a closed dam, the most important dam, the Eastern Scheldt enclosure, will be installed as a storm flood barrier. As a result most of the estuary will continue to be tidal so ensuring the conservation of (a) the natural environment and (b) the fisheries which, together, provided the major consideration for this important change of policy.

The introduction of nature conservation considerations into planning has posed the necessity to find nature and landscape criteria which are suited and fitted to the planning process, remembering that conservation and preservation are not synonymous. The changes caused by the construction of dams have been so large that the earlier ecological systems were bound to be changed. The government had to decide the way in which management of nature and landscape in the rapidly changing areas should be directed - a decision with technical and political implications. The ecologist, as a technical expert, has to advise which values should be preserved or developed. The engineer must provide guidance on topics within his/her sphere of competence. The policy maker, i.e. the government, must integrate the different pieces of advice (economic, social and ecological) and decide on future land-use.

In the present chapter the criteria that may be used to formulate a policy for landscape and nature management are discussed using examples mainly taken from the development of the Delta region.

NATURE PLANNING AND MANAGEMENT

A great many of man's planned activities have been developed with little thought for their effects on the natural environment, effects which have too often been considered as side-effects.

Very often, the possibilities for conserving nature are not great, for example, in housing projects and the development of industrial estates. In these instances the original natural environment will disappear completely with the subsequent need to create a new environment. Sometimes these newly created situations provide enhanced opportunities for the management of the environment and in these instances it is increasingly important to take premeditated decisions about what should be done. This is what happened in the Delta region: an estuary region which was 'partitioned' into a

complex of basins with or without tides, having salt, brackish or
fresh water (Figure 2). The different basins provided a set of new
starting points for environmental management remembering that the
interests of 'nature' and 'fisheries' were second only to protection
against floods when the scheme was sanctioned. How should the
ecologist meet this challenge ?

In the 'nature management' case studies in the different Delta
basins, four steps were identified when providing a framework for
decision-taking:

(i)   the preparation of an <u>inventory</u> of existing natural
        features and circumstances;
(ii)   the <u>evaluation</u> of the inventory to assign priorities to
        natural features and circumstances;
(iii)   the <u>analysis of projected changes</u> and their likely
        impacts;
(iv)   the <u>agreement</u> on a <u>strategy</u> for future development and
        conservation.

(i)   <u>Inventory</u>

The indispensable starting point for all wildlife impact studies
is an inventory. In addition to a list of species it is desirable
to survey their inter-relations one with another, and with their
habitats. It is usually impossible to be completely comprehensive:
instead it is necessary to be selective.

When considering landscape ecology, the natural environment can
be divided into a number of 'spheres' (atmosphere, lithosphere, pedosphere, hydrosphere....), which can be ordered in a hierarchy where
the next 'sphere' is dependent on what happens to the preceeding
'sphere'. On the other hand, the preceeding 'sphere' will not be
influenced to any great extent by changes in the next 'sphere' in
the hierarchy with one very important exception : man has the ability
to strongly influence events in all spheres. Zonneveld (1982) provided the following scheme of 'spheres':

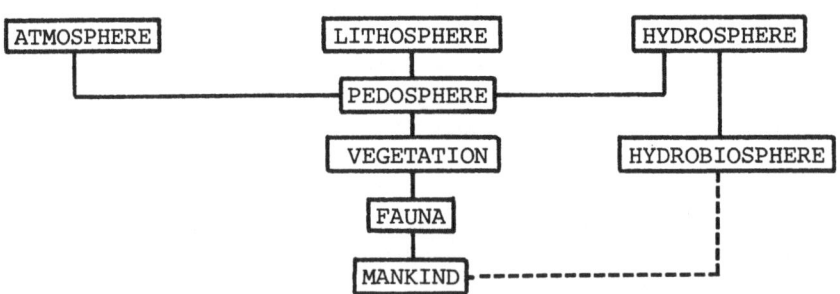

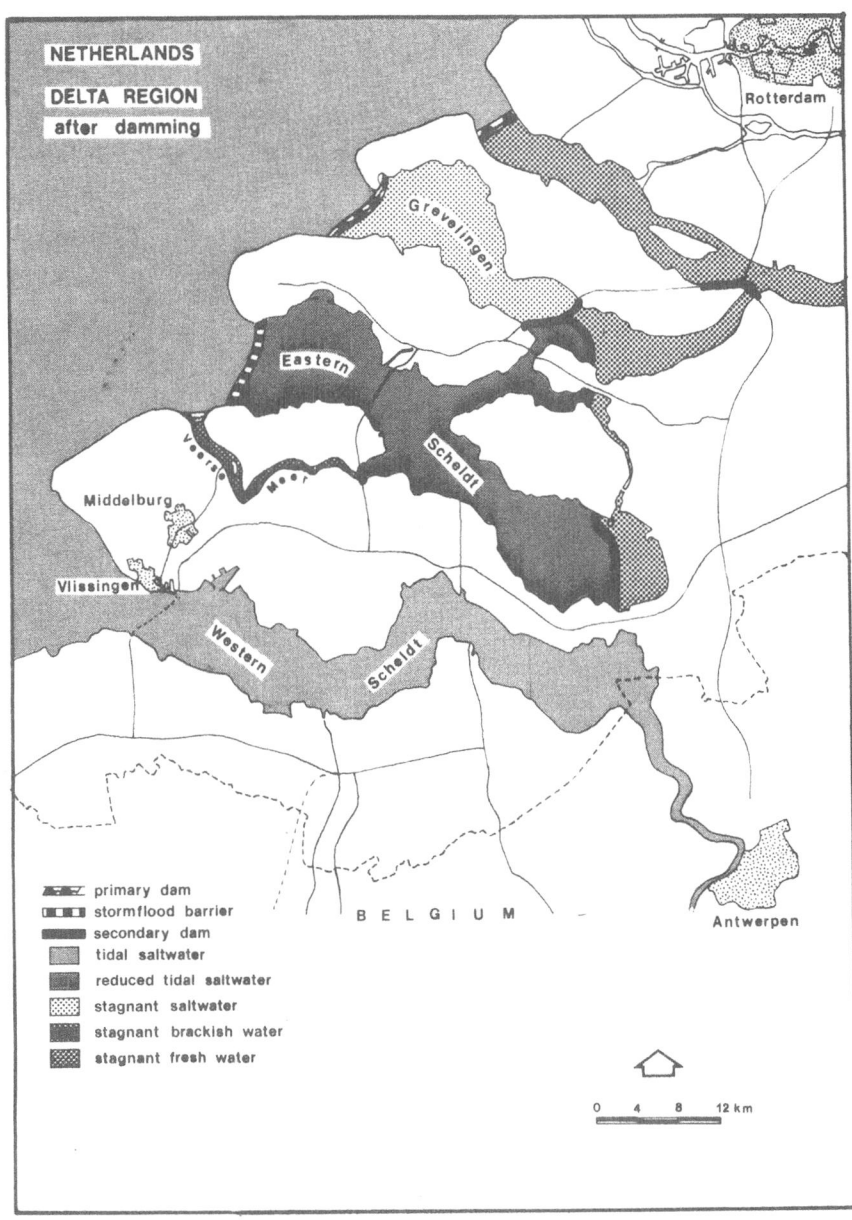

Figure 2: The Delta region, following the construction of 4 primary
           and 6 secondary dams, showing the resultant range of water
           type in a series of basins, each providing different ecol-
           ogical conditions. The storm flood barrier at the western
           end of the Eastern Scheldt is normally open allowing a
           somewhat restricted tidal entry, it is closed if storm
           floods are threatened.

From this scheme it is clear that if man's activities change
the lithosphere then all the 'higher' spheres will be influenced.
For instance, the conversion of an estuary into a stagnant freshwater
basin (change of hydrosphere) will affect its flora and fauna.
Changes of vegetation (e.g. afforestation or deforestation) will
affect the assemblages of animals with at the same time a secondary
influence on the pedosphere (erosion).

It is essential to know which 'sphere' is likely to be the first
directly influenced by the implementation of the planning decisions.
This sphere must be carefully surveyed - an absolute necessity - while
sufficient should be known of succeeding spheres to enable reliable
predictions of effects to be made. In practice, the first sphere to
be influenced will usually be the lithosphere (pedosphere), hydrosphere or both of them. This being so the inventory must be based on
geological, geomorphological and pedological maps with accompanying
hydrological information (Figures 3, 4). Additionally it is desirable to know about the assemblages of plants including their distribution. In practice it is easier to concentrate on plant species
rather than the occurrence of assemblages of plants which are sometimes difficult to define. The occurrence of animals usually depends
on the availability of plants. If sufficient is known about the
latter it is often possible to predict the range of animals, with
the exception of strongly mobile animals, which can exploit vegetation over relatively large areas. For this reason separate studies
are usually required for mammals and birds (especially migratory
birds). Experience suggests that hydrobiological studies should be
specially focused on benthos and plankton.

Thus an "inventory" to be useful should include data about :

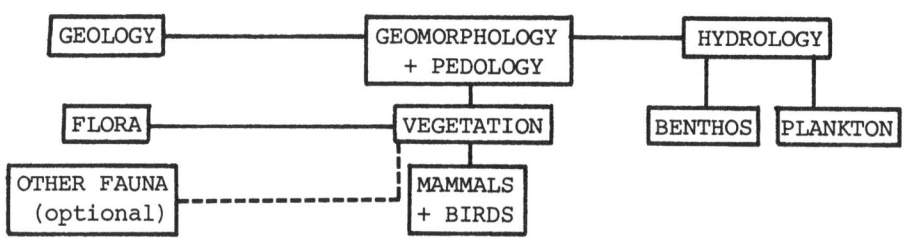

(ii) Evaluation

What is it particularly important to pay special attention to ?
Which plants and animals demand special care ? Man is inclined to
categorise plants into the beautiful and/or useful, with a third
group of awkward, nuisance weeds. He will favour horses, seals and
young mammals while stigmatizing animals such as scorpions, snails
and snakes. However, ecologists must recognise that all species
have their own niches and are equally important. Objective ecol-

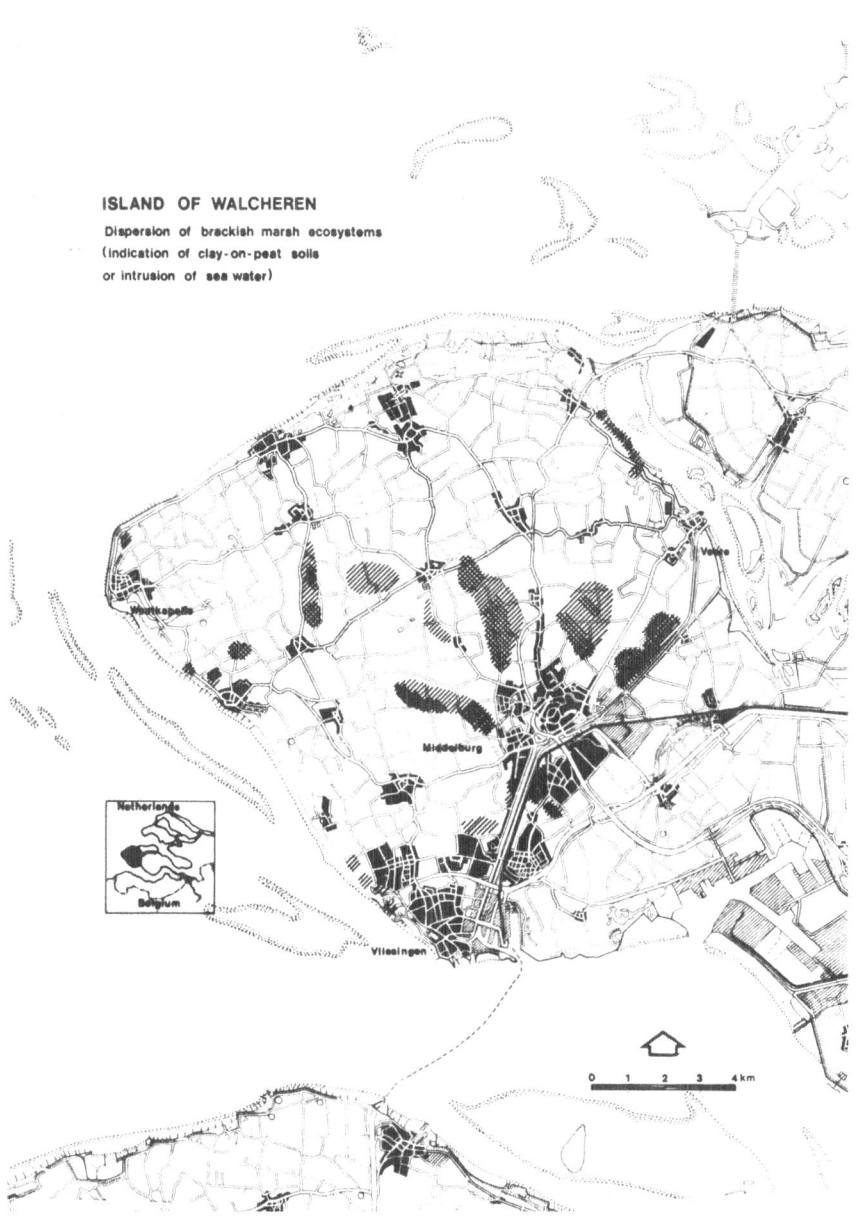

Figure 3: An example of the information gathered for an inventory
of factors changes in which will affect wildlife - in
this case the distribution of brackish marsh ecosystems
(part of the hydrosphere) on the island of Walcheren
(see inset for its position in the Delta Region).

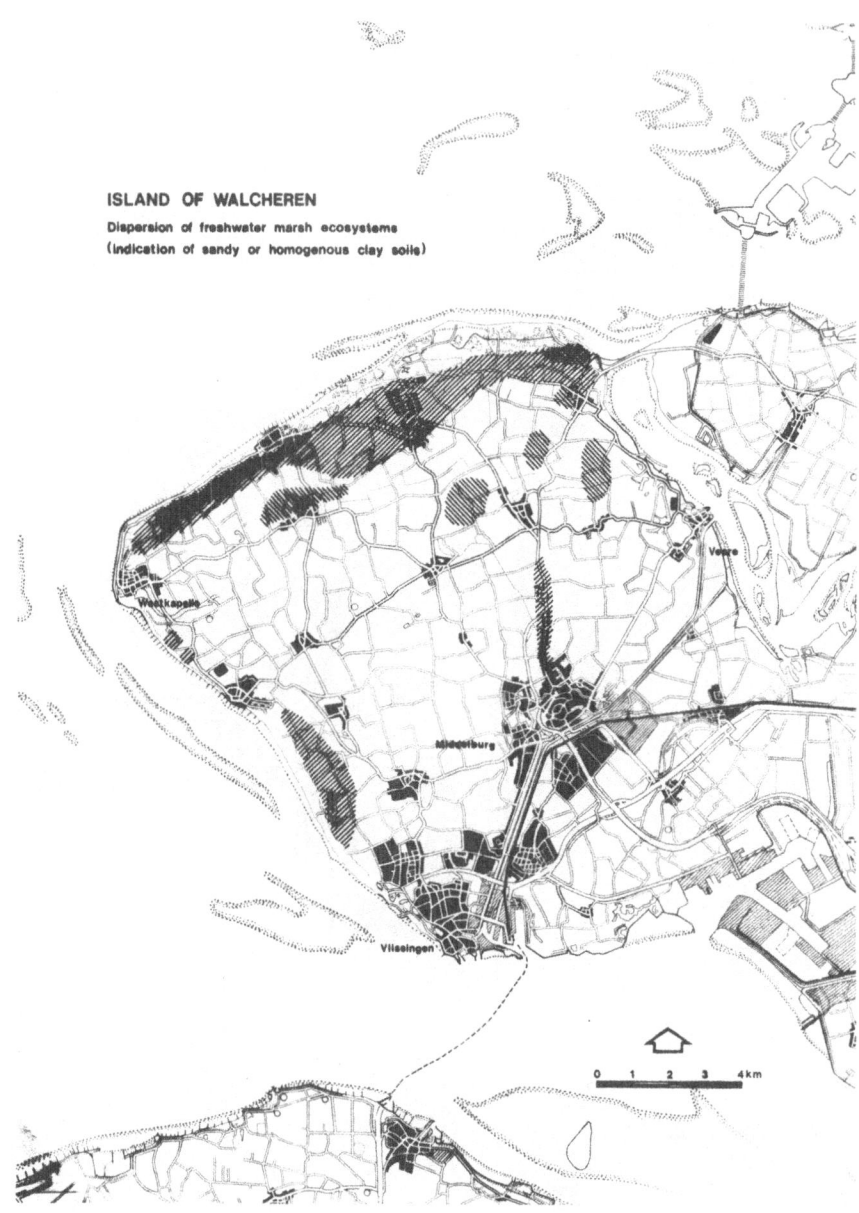

ISLAND OF WALCHEREN
Dispersion of freshwater marsh ecosystems
(indication of sandy or homogenous clay soils)

Figure 4: An example of the information gathered for an inventory of
          factors changes in which will affect wildlife - in this
          case the distribution of freshwater marsh ecosystems (part
          of the hydrosphere) on the island of Walcheren (see inset
          on Figure 3 for its position in the Delta Region).

ogical planning must focus on the conservation of all species by
ensuring the survival of their habitats. This requires a knowledge
of species/habitat interactions. Nonetheless to be realistic it
should be recognised that some species and communities are more
vulnerable than others with highly specialised species being particularly at risk. How then can assemblages of plants and animals be
judged ? By,

    a. species diversity or richness;
    b. the occurrence of rare species;
    c. the occurrence of rare communities;
    d. the completeness of the assemblages (how many of the
        potential species are present);
    e. the potential for introducing species which are found
        elsewhere in comparable habitats.

In general it is usual to put special values on complex and
stable communities that have remained undisturbed for long periods
and which may contain rare species. However, the arbiters (a-e) are
sometimes contradictory. Thus mud flats may have few species, but
are important because of their rarity. Many important feeding grounds
for migratory birds are monotonous and uninteresting grasslands.
There is obviously no one formula to define 'value'. The evaluation
may also depend upon the amount known of a particular location thus
making it universally, nationally, regionally or locally important.
As far as the Delta region is concerned, it has a rich diversity of
species and communities, and is also one of the focal points of
Western European bird migration; it is therefore of international
importance.

(iii) <u>Analysis of changes</u>

Having prepared inventories and assessed the importance of the
locations subject to planning decisions it would then be necessary
to consider the impact of those decisions. Thus by damming a tidal
inlet, and so creating a new habitat it is necessary to explore the
possible ways in which this new habitat may be managed to foster
wildlife.

Initially the losses, caused by the interference, and especially
those considered to be grievous, must be taken into account. The
possibilities for compensation have to be investigated including the
introduction of an alternative plan or the implementation of possibly
minor changes to the original plan, which would lessen the losses.
Sometimes the creation of a nature reserve with controlled conditions
is possible. In the Delta region the construction of a storm flood
barrier in the Eastern Scheldt instead of a complete barrier was
devised to reduce unacceptable losses (Figure 2). However, its
construction decreased tidal amplitudes with persistent effects concentrated on the intertidal zones. Inevitably these zones will be

smaller and as a result, the benthic species that live between ebb
and flood, will be harmed; also the birds, Oystercatcher (*Haematopus
ostralegus*) and different Stints (*Calidris* spec.) that feed on the
benthic species. However, had the complete barrier been constructed
the enclosed water would have become stagnant with the total loss of
species whose habitats are between ebb and flood. Another approach
involves the possibility of creating new habitats (biotopes) for
threatened species. On the sandy islands of the Grevelingen basin
(Figure 4) it is possible to create new breeding grounds for Little
Terns (*Sterna albifrons*) and Sandwich Terns (*Sterna sandvicensis*),
whose breeding places elsewhere in the Netherlands were disappearing
(Beyersbergen & Van den Berg, 1980). The enormous fields of Eelgrass
(*Zostera marina*), that developed spontaneously in the Grevelingen
basin now provide ideal feeding grounds for the Brent Goose (*Branta
bernicla*).

In addition to these measures consideration is being given to
the use of the newly reclaimed lands for breeding and conserving non-
indigenous threatened mammals, such as Tarpan (*Equus caballus gmelini*)
and Przewalski's Horse (*Equus przewalskii*) which may be able to share
the reclaimed land with breeding colonies of grassland birds. Already
a flock of Heck Rinds (a German-bred race resembling Aurochs) has been
installed on the Grevelingen flats.

Having considered the steps necessary for habitat conservation
and the protection of rare species it is then appropriate to consider
what else is desirable. The potential of the new environment, however, will be largely influenced by its geomorphology. In the
Grevelingen and the Veerse Meer Basins gravel dams were constructed
to screen flat sandy islands from wave erosion.

(iv) Strategies for conservation and management

The final step is the preparation of a conservation plan, based
on the "evaluation" of the original site and the potential impact of
changes. It should be remembered that priorities are decided arbitrarily to a greater or lesser extent and are therefore always subject
to change.

Generally natural assemblages develop best when an area is left
alone; nevertheless their development sometimes has to be helped.
In the Netherlands the climax vegetation, the end stage of natural
succession, is likely to be some type of woodland, but the landscape
is prized because of the variety provided by grasslands with great
botanical and ornithological value. To halt succession and sustain
the continuing presence of grassland, management has to include
grazing by sheep and cattle. Thus it is essential to maintain sheep
farming to conserve grasslands thus retaining their botanical diversity and ornithological usefulness.

Until now I have dealt with aspects of nature conservation.
But the development of a comprehensive plan, even where nature conservation is the prime concern, necessitates a consideration of recreation, agriculture, infrastructure and other land-uses. The Delta
region in the Netherlands is one of the outstanding water recreation
areas of Western Europe; it is an important region for the collection
of mussels and the breeding of oysters. Clearly it is essential to
consider the inter-relations between the terrestrial and aquatic
environments and the interplay between their different methods of
management - the task of the councils governing the different basins
of the Delta region. Further, those relationships are likely to alter
as time proceeds and this being so it is essential to periodically
reassess earlier planning and nature conservation decisions, and to
seek, where necessary, modifications.

LANDSCAPE PLANNING AND MANAGEMENT

'Landscape' is wider than a consideration of wildlife; it
includes human activities, past and present, and the inter-relations
between wildlife and man. Landscape planning, as a planning activity,
entails cohesion, for landscape as an entity is more important than
the component parts. There is another difference between nature conservation and landscape planning: man's task towards nature is that
of a guardian, a superintendent, a steward of existing plants and
animals, whereas landscape is his own creation, and is continually
evolving, but landscape planning is not a goal in its own right.
Landscape is the result of landscaping, planning and a whole variety
of other activities, done for other reasons: housing, agriculture,
industry, infrastructure, recreation. It is a by-product of these
activities, a durable and often highly regarded by-product.

Why should we concern ourselves about landscape ? There is one
reason above all others: 'landscape' is synonymous with our own
environment. We need it - whether planned or not - landscape
(including townscape) is part of our life. We identify with landscape. Continuity is the most important quality of landscape, both
in space and time. Thus the goal of landscape planning should be
the provision of coherence and continuity in our environment and in
achieving this goal there are 4 aspects to be considered:

      (i)   inventory;
      (ii) evaluation;
      (iii) analysis of changes;
      (iv) strategy and planning.

(i)   Inventory

As for nature conservation, landscape planning is dependent
upon geological, geomorphological, pedogical and hydrological

inventories - accepting that human activities, past and present, can
have overwhelming effects. In a flat and geologically speaking,
rather uniform country, such as the Netherlands, variety is mainly
attributable to the activities of man over the centuries. This
being so the second facet of inventory is concerned with human
settlement, history of occupation, and the visible effects of that
occupancy including the pattern and types of settlement, the
apportionment of land, the location of roads and waterways and the
occurrence of intensive cultivation. Geology, pedology, hydrology
and human settlement are the main determinants of landscape and an
inventory may consist of :

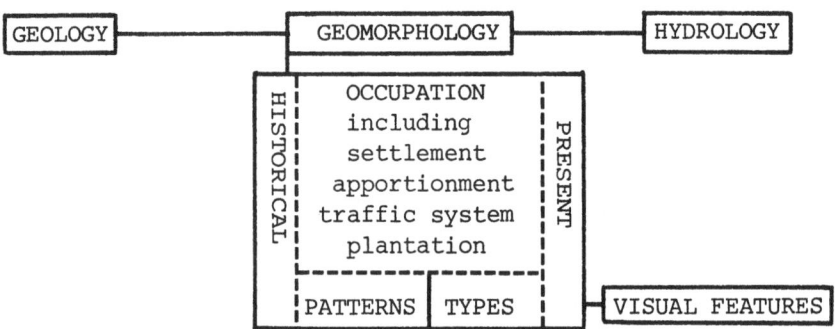

While the comprehensive mapping of landscape might appear to
be a desirable objective, is it really essential to do this to
expedite landscape planning ?

(ii) Evaluation

The main quality of any landscape is its uniqueness. No two
landscapes are identical. Therefore, landscape planning cannot be
approached in terms of rarity or complexity. Landscape is as a
human being: every man, woman or child is an individual, and unique,
and thus they are all equal in value. The same is applicable to
landscapes: the combination of features makes each unique. A landscape should have a recognisable coherence, its qualities being
judged by the beholder. The ideal landscape is a highly personal
matter; it is usually the landscape in which an individual or a
group, tribe, of individuals was born and brought up, irrespective
of how barren or unfriendly the region may seem to strangers. In
evaluating landscape it is necessary to take account of its coherence
and the structure and its component parts some of which may be
regarded as being particularly characteristic. Needless to say it
is essential to ensure that details of these characteristics are
included in the inventories.

(iii) Analysis of changes

As already indicated, continuity of landscape is very important; changes should not be made too rapidly. Development should be
continuous and harmonious with what already exists but in reality
this is not always possible. Large scale housing and industrial
development will cause rapid changes, and consequent feelings of
discomfort. Not all of the characteristics of the landscape can
be conserved but attempts should be made to avoid obscuring the
dominant regional features.

Landscape design must be attuned to the functions pertaining to
the newly planned activities. On the other hand the designer must
care for the 'continuity' of the development in a regional context
taking note of the features and patterns that made the pre-existing
landscape unique.

(iv) Strategies for landscape planning

'Landscape' is the habitat of man who needs a niche in which to
feel at home. Therefore the human dimension must always prevail in
landscape planning, both in space and time. In space, this implies
that local and small-scale changes may be readily accepted, if the
main regional characteristics are left intact. To avoid discontinuities in scale and function large scale changes are only acceptable
in large Regions. Changes in time should, ideally,be gradual recognising that sudden changes in man's surroundings are nearly always
regarded, rightly or wrongly, as undesirable.

In the South-Western part of the Netherlands the Delta scheme
has involved enormous landscape changes. Fortunately space allows
these to be developed on a large scale. From the point of view of
the inhabitants, it was, and is, fortunate that the Delta works are
scheduled over at least 30 years; much longer than originally
intended. Thus many of the inhabitants will grow up with the development of the new landscapes; they will become accustomed to it.

In planning the landscape of newly reclaimed areas of the Delta
Region, planners have tried to harmonize their plans with the scale
and character of the neighbouring islands. Nevertheless it will
take more than a lifetime before the trees make a significant landscape impact, giving stability.

Landscape planners are not artists who give vent to free
expression. With detailed knowledge of geology, hydrology, ecology,
demography and history they attempt to conserve unique and signif-

icant features while at the same time attempting to assist man's
wellbeing. In so doing the emphasis is on conservation not preservation recognising that the world does not stand still.

REFERENCES

Beyersbergen, J. and Berg, A. van den, 1980, "De Grevelingen, de
    vogels van een afgedamde zeearm", Kerckebosch, Zeist (The
    Grevelingen, the birds of a dammed estuary).
Zonneveld, J.I.S., 1982, The ecological backdrop of regional ecology,
    in: "Perspectives in Landscape Ecology", S.P.Tjallingii and A.A.
    de Veer, eds., Pudoc, Wageningen.

SUGGESTED READING

Anon., 1982, "Streekplan Midden-Zeeland", Provincie Zeeland,
    Middelburg (Regional plan for Central Zeeland).
Beenhakker, A.J., 1975, Goed dat er grenzen zijn, Zeeuwsch-Vlaanderen
    in ruimte en tijd, in: "De Gouden Delta 2", H. Gysels, ed.,
    Pudoc, Wageningen. (It is well that there are limits. The
    region of Zeeland Flanders in space and time).
Haperen, A.M.M. van, 1983, "De vegetatie van Midden-Zeeland",
    Provinciale Planologische Dienst voor Zeeland, Middelburg
    (The vegetation of Central Zeeland).
Loenen, M., 1982, Planning and research for new nature areas, in:
    "Perspectives in Landscape Ecology", S.P.Tjallingii and
    A.A.de Veer, ed., Pudoc, Wageningen.
Loenen, M. and Pinkers, M., 1980, "The development of the Grevelingen
    area for nature and recreation", Rijksdienst voor de IJsselmeer-
    polders, Lelystad.
Stroband, A.G. and Poel, K.R. de, 1981, "Een beeld van het Zuid-
    Hollandse landschap, deel 1", Buro Maas, Utrecht (A view of
    the landscape of the province of Zuid-Holland).
Tjallingii, S.P. and Veeer, A.A. de, 1982, "Perspectives in Land-
    scape Ecology", Pudoc, Wageningen.

19

TOWARDS BETTER LAND USE - THE ROLE OF GEOGRAPHERS

J. Bonnamour

Ecole Normale Supérieure
5 Rue Boucicaut
92260 Fontenay-aux-Roses
France

How can we best use our land resources ? This apparently simple
question confronts everyone concerned with land use, and it raises a
complex set of problems that have to be solved within the constraints
posed by the physical environment and the need to use investment
capital in the most profitable way, the two constraints being balanced
so as to create uses that will support lifestyles conducive to the
ultimate user's well-being. In spite of its modern expression, it is
indeed a very old question, and its answer requires the introduction
of parameters that complicate the planning process, particularly in
industrial societies with high living standards. These parameters,
by their very existence, tend to give an impression that perfection
is possible, even though they continue to require evaluation in
relative terms. Given this situation, geographers can help specialists in other disciplines to avoid the pitfalls of what may prove to
be a blinkered, narrow, approach. They can help evaluate probable
impacts of proposed developments within an extensive, but nevertheless circumscribed framework taking full cognisance of complexities
inherent in time and space. Probably the best way of elaborating
these points is by using examples, drawing attention to successes
and failures.

THE SCIENTIFIC APPROACH

The fundamental question is whether scientific precision is
desirable and/or feasible. Indeed, the problem may well be amenable
to a scientific approach if it involves a set of didactic processes
that assess the contributions of a variety of disciplinary approaches
within a given geographic and historical context. Thus it is not

possible to give an acceptable definition of the quality of arable
land without including value judgements that incorporate the financial
base and technical capability of the corporate enterprise that seeks
to exploit the resource. Many case histories show clearly that development options that favour one soil type over another are often
followed more on the basis of social considerations than on the intrinsic land values. Thus the arid Champagne region of France (la
Champagne pouilleuse), was arbitrarily (and mistakenly) designated
as barren by the physiocrats (founded by Quesnay in France in the
18th century who maintained that society should be governed according
to an inherent natural order). But today it yields magnificent
harvests and is known as 'la Champagne crayeuse'. These lands have
become cultivable as a result of an integrated set of human and social
circumstances that have also determined the extent of the usable area.
Thus the most dangerous concept of all must be a view of "Land
Potential" that ignores social and economic dimensions, particularly
the potential for introducing and assimilating new technologies. An
excellent illustration of this is the failure of so many agricultural
aid programmes in tropical countries, such as the painfully slow penetration of more intensive systems of cultivation in Africa.

Years ago the concept of climax was applied to much of the world's
natural vegetation, the assemblages reflecting an equilibrium with
their environment. But nowadays ecologists recognize that plant associations are ceaselessly changing, often as a result of soil changes
(evolution). Despite the great increase of knowledge, it seems that
we often spend more time lamenting its deficiencies than using what
exists in constructive ways. In these circumstances, the geographer
can continually reassess the multiplicity of conflicting 'solutions'
presented by neighbouring communities in formally identical environments. Viewed by satellite, the frontier between Canada and the
United States, across the western prairies is strikingly visible.
Two types of land use in regions with similar environments (edaphic
and atmospheric) but contrasting population densities appear in
marked contrast along a frontier seemingly traced in chalk along
latitude 49 $^{\circ}$N, separating the patchwork of cultivated fields on one
side from the broad expanses of grazing pastures on the other. In
western Europe the Belgian Ardennes is seen to be merely a controlled
buffer zone between the sprawling wooded Ardennes in France and the
garden-like Ardennes in Luxembourg.

In summary, land use studies need, above all, humility when considering the interests and views of those affected, and scientific
rigour when integrating information derived from a variety of disciplines.

THE REQUIREMENTS OF THE SCIENTIFIC APPROACH

Problems must usually be attacked in three dimensions - on different scales, at different times, and from different viewpoints.

Scale: at a very large scale, the conflicts and harmonious relationships between land-uses, including the effects of rivers etc. and the changes associated with minor climatic variations can be seen as in Morand's studies of the Laonnais region.

But the terrain is only one aspect of the larger whole. On the medium scale, studies of slope and aspect or the analysis of cleared areas come into their own. But there is no doubt, the examination of changes at a small-scale has been neglected, and their significance often overlooked. Charvet (Chapter 26) has shown the extent to which a series of land-use maps of different scales can contribute to achieving a better understanding of existing conditions and problems. His work has stressed the importance and desirability of considering analyses at a variety of levels, none more so than when considering the consequences of the Aswan Dam, which has impacts extending from the shores of Lake Nasser, the lake created in southern Egypt, to the Mediterranean Sea.

The dimension of time: the importance of historical sequences and the fundamental significance of long term observations have already been discussed, but by themselves, these aspects do not provide a satisfactory overall understanding of temporal effects. Sites must be evaluated at different seasons of the year, and not only must the effects of normal climatic cycles be well understood, but knowledge of predictable less frequent influences is essential. Fortunately, climatologists are continually improving their ability to provide data and information about such events. Recent climatic events, such as the 1956 frost and the 1976 drought were responsible for major long-term modifications to the vegetative cover of certain regions of France. The desertification and consequent decrease in population in the Sahel is an even more thought-provoking and painful example of the need for appropriate scales of temporal analysis and their logical integration with spatial evaluation.

Viewpoints: Land use in a region traditionally flows from its history, and each period leaves a more or less visible mark. It represents a projection in space. Thus the great estates of the Seville region of Spain which belong to the few are now deserted by a constant stream of migrants bound for the factories of northwest Europe. The region's geography and population dynamics reflect

a social system in which the masses were dominated by the few - a
knowledge of land tenure is fundamental to an understanding of the
Seville region and, ultimately, to proposed changes of land use
planning.

When considering the impacts of time and space on land use it
is important not to undervalue the social significance of landscape.
Some kinds of land use have symbolic value, with undesirable psychological connotations for societies such as ours, so fraught with
anxieties that we have abandoned attempts to create anything new or
imaginative. Instead, whenever it is economically feasible we tend
to copy or restore old models, both in urban centres and rural areas.
The sociological significance of landscape is an essential feature
that must always be considered, (Naveh, Chapter 18). The complexity
of the problems that confront planners and the dangers of accepting
schematic "solutions" show clearly that there is no alternative to
a preliminary interdisciplinary study that is free from preconceived
hypotheses. This is the only foundation on which it is possible to
build models combining social, economic and institutional dimensions
of real situations. Without simulation, we do not have a basis for
rational decisions concerned with planning.

20

LANDSCAPE ECOLOGY AS THE SCIENTIFIC BASIS FOR HOLISTIC LAND

APPRAISAL, PLANNING AND MANAGEMENT

Z. Naveh

Lowdermilk Faculty of Agricultural Engineering
Israel Institute of Technology
Technion
Haifa 32 000
Israel

INTRODUCTION

Rational land use - is it achievable using ecological principles of integrated land appraisal, planning and management ? Many of the contributions to this book outline principles and techniques but at the same time they have reminded us, again and again, that there are severe constraints on a fully-integrated holistic, approach to the real world of decision making on land uses. In September 1981 I had the opportunity to participate in the International MAB-UNESCO Scientific Conference "Ecology in Practice, Establishing A Scientific Basis for Land Management" in Paris, marking the anniversary of the Man and the Biosphere Programme (Di Castri et al, 1984). This contributed greatly to the international and interdisciplinary basis of research and training aimed at rational land use, but how should the communication gaps between scientists, the land use planners, managers, sociologists, economists and decision makers be overcome ?

In my opinion, one of the main reasons for these gaps is the fact that most of us have been exposed to discipline-oriented, reductionistic education. In fact, the higher the academic and/or professional level of this education, the more pronounced are these tendencies of specialization. As a result many of us lack the proper cognitive tools for synthesising, planning, studying and managing complex, interrelated systems. To confront the problems now facing us we need to adopt a holistic approach. We need to interrelate rather than separate. In the words of Waddington (1977) we urgently require "new tools of thought" to deal efficiently with

these unfamiliarly complex situations. One tool is the system
approach, now commonly applied to the two major basic integrative
land appraisal sciences, namely ecology and geography. With a
common pool of concepts and methods from these sciences, it should
be possible to formulate a unified theory of integrated land use
bridging the gaps between different disciplines.

At present however, there is greater emphasis on system analysis
than on system synthesis. Furthermore existing system concepts,
especially those dealing with natural and man-made ecosystems, are
too vaguely defined to bridge the communication gaps.

I believe, however, that landscape ecology as a holistic and
multidisciplinary planning- and management-oriented science, drawing
its paradigms not only from ecology and geography but also from
systems theory and bio-cybernetics (i.e. the science of communication
and control of biosystems and ecosystems), can provide the essential
platform to bridge the gap between scientific and humanistic attitudes related to rational land use.

THE EVOLUTION OF LANDSCAPE ECOLOGY

Landscape ecology is a young branch of modern ecology, dealing
with the interrelationship between man and his rural and urban environments. The concept evolved in central Europe as a result of the
fully integrated holistic approach adopted by geographers, ecologists,
landscape planners, designers and managers in their attempts to
bridge the gap between natural and man-made environments including
agricultural, silvicultural (plantation forests) and urban.

The earliest reference to 'landscape' seems to be the Book of
Psalms (48.2) where Landscape ("noff" in Hebrew, probably etymolog-
ically related to "yafe" (beautiful)) refers to the beautiful view
over Jerusalem from Mt. Zion, where King Solomon's temple, castles
and palaces can be seen. This visual-aesthetic connotation of
'landscape' is usually equated, in the English language, to "scenery".

As Whyte (1976) has shown in his important book on land appraisal, the meaning of landscape has undergone great changes, but the
original visual-perceptual and aesthetic connotation has been
adopted in literature and art, and is still used in landscape planning and designing, and by gardeners who are frequently more concerned with aesthetic landscape perceptions than with ecological evaluations. This attitude is reflected in the great amount of English
literature devoted to landscape assessment (see reviews by Arthur
et al (1977), Zube et al (1975) and others). Unfortunately, the
epistemological (scientific study of knowledge) development of the
landscape concept, leading to the discipline of Landscape Ecology
is almost unknown outside Europe, and even there, it has been mainly
centred on Central and East Europe.

In German, Landscape, and its etymological equivalent "Landschaft", include a geographical-spatial connotation. Since the Renaissance, and especially in the 18th and 19th centuries, this spatial connotation has acquired a more comprehensive meaning in which the landscape is envisaged as applying to the total environment.

The term 'Landscape' was introduced as a scientific-geographical concept in the early 19th century by A. von Humboldt, the great pioneer of plant and physical geography. He defined it as "Der Totalcharakter einer Erdgegend" - the total character of an earth region. However, with the rise of classical Western geography, geology and earth science, its meaning was narrowed to the characterization of physiographic, geological and geomorphological features of the earth's crust; it becomes a synonym of "landform". However, Russian geographers have again given 'landscape' a much broader meaning by including biological and non-biological phenomena - "landscape geography". These semantic and developmental changes have been described in detail by Troll (1971), a leading German biogeographer, who defined Landscape as the total spatial and visual entity of human living space, integrating the geosphere with the biosphere and its man-made artifacts. He regarded Landscape as a single fully integrated entity, where the "whole" is more than the sum of its parts; it should, therefore, be studied in its totality. He recognized that aerial photographic interpretation of landscapes by remote sensing would have great potential. By 1939, while studying problems of land use and development in East Africa, Troll coined the term Landscape Ecology. In doing so he hoped for closer collaboration between geographers and ecologists so that earth and life sciences might evolve with the development of "ecoscience"- as contrasted with "geoscience", the study of inanimate rocks. In practice, Landscape Ecology combined the "horizontal" approach of geographers when examining the spatial interplay of natural phenomena with the "vertical" approach of ecologists who study processes and mechanisms in a variety of habitats - "ecotypes".

The Division of Land Research (now named Land-Use Research) of the Commonwealth Scientific and Industrial Research Organization (CSIRO) in Australia, was one of the first major organizations to adopt a holistic approach to land survey and evaluation (for development). Under the leadership of C.S. Christian and G. A. Stewart, the concepts of "Land Units" and "Land Systems" (Christian, 1958) were applied with great success in large scale multidisciplinary integrated surveys (Christian and Stewart, 1968). Since then, and with the advent of the new Division for Land Use Research, the holistic approach has been further broadened to include socioeconomic and ecological parameters when producing land-use plans (Division of Land Use Research, 1978).

Another important agency the Netherlands-based Institute for Aerial Survey and Earth Science (ITC) working chiefly in developing

subtropical and tropical countries has also developed the holistic
Landscape concept. As described by Zonneveld (1972) in a comprehensive textbook about photo-interpretation, Landscape Ecology is
a crucially important subdivision of "Land(scape) science" concerned
with the interrelations of the different elements that determine
landscape (Figure 1).

Zonneveld proposed a number of landscape units that logically
fit a hierarchy judged by geographical (spatial) extent:

i)   The ecotope (or site) is the smallest; it is characterized
      by its homogeneity; and lack of excessive variation.
ii)   Land facets (or micro-chores) are combinations of eco-
      topes, which owe their juxtaposition to the same dominant
      land attribute (e.g. landform).
iii) Land systems (or meso-chores) are combinations of land
      facets, together forming convenient mapping units.
iv)   Main landscapes (or macro-chores) are combinations of
      land systems together delimiting geographical regions.

The holistic perception of the landscape is being increasingly
adopted by ecologists who are usually well-versed in the management
of natural and man-made assemblages of plants and animals - they have
moved from the narrow examination of individual species to a concern
for mixed assemblages and the factors determining their existence
including the mosaics of different habitats. The development of
landscape ecology in Central Europe which has been reviewed elsewhere
(Naveh, 1982) can be summarized as follows:

    Landscape ecology is viewed presently in Europe as the
    scientific basis of land, and landscape, planning, management,
    conservation, development and reclamation, and as such it
    extends beyond the realm of classical ecological sciences
    which were concerned with natural phenomena to include socio-
    economic and other human-centred aspects.

This development is reflected in the adoption of a broad range of
landscape-ecological (environmental), planning-oriented and other,
studies of the interrelations between man and his open, cultural and
built-up landscape which together aim to reconcile cultural and
socio-economic aspirations/demands while at the same time, ensuring
the enrichment of man's biotic environment. Many similar approaches
are being made to the problem especially in the English speaking
world, where the terms "environmental conservation" or "environmental management" are sometimes used. Because of the challenge to
conserve and enhance man's total living space (urban and rural) there
is an urgent need for the consolidation and convergence of all these
aims and methods towards a holistic human ecosystem science of landscape ecology (Naveh, 1982). For this reason sections of this
chapter will be concerned with definitions used in General Systems

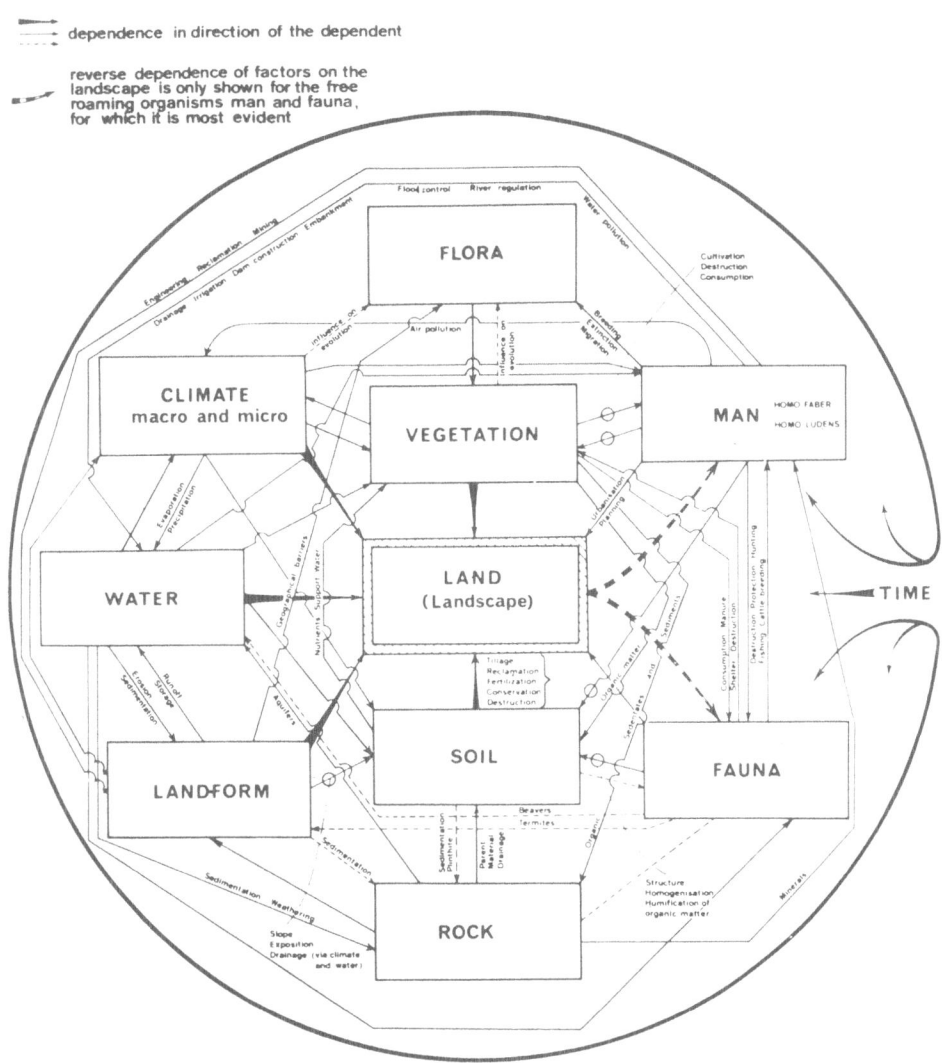

Figure 1: The interrelations of biotic and abiotic factors that together influence land (or landscape)(Zonneveld, 1979).

Theory, non-equilibrium thermodynamics and bio-cybernetics, disciplines which have prepared the ground for a new general bio-systems theory and the study of human ecology.

TOWARDS A GENERAL BIOSYSTEMS THEORY

New Insights into the holistic axiom

The relations between ecology and General Systems Theory have
been discussed from the point of view of (i) natural resource management by Schultz (1967) and (ii) human ecology by Young (1974). For
Landscape ecology both approaches are equally important.

The basis of a general systems theory is the identification of a
hierarchical organisation of nature as a series of open systems with
increasing complexity through the appearance of new qualities that
evolve at each higher level of organisation (Von Bertalanffy, 1968;
Weiss, 1969). Egler (1942) was the first ecologist to recognize this
approach, which is now generally accepted in ecology; it is central
to the study of plant associations (Rowe, 1951; Dansereau, 1957;
Büchner, 1971). Before proceeding it is necessary to define certain
terms:

A system in the most general sense, is a set of elements whose
relations, one with another, are closer than those with their
environment. Because of these relations, a system is always
more than the sum of its elements - it is a whole. This
concept of wholeness encompassing qualities that emerge from
the systems own behaviour, is at the heart of the holistic
approach.

An interaction system (in German, "Wirkungsgefüge") is a special
class of system whose elements are connected, or coupled, with
each other in such a way that if one element of the system is
altered then all the others will be, irrespective of the nature
of the perturbation (Sachsse, 1971). In living systems these
perturbations can be biotic and abiotic - in a flock of geese,
they may relate to social factors; in rural communities they
may be cultural.

A concrete system is a non-random accumulation of matter/energy
in a region, in physical space/time, which is organized into
interacting subsystems or components (Miller, 1975). The units
of 'concrete-systems' and their interrelationships (spatial,
temporal etc.) are usually determined empirically. Most concrete
systems are open systems with boundaries usually permitting the
transfer of energy-matter and information.

Abstracted systems are sets of relationships abstracted or
related to the interest of the observer concerned. The boundaries of abstracted systems may be conceptually established so
that they do not conform to recognised units with established
positions in physical space. All definitions of ecosystems
emphasize the interactive nature of the relation between living

organisms and their non-living environment. Tansley (1935),
when introducing the term, stated that ecosystems comprise
"not only the organismic (plant/animal/microbe) complex but
also the mix of physical factors forming what we call the
environment". Whittaker (1975) in one of the most lucid
textbooks on community and ecosystem ecology, defined an
ecosystem as "a community and its environment treated together
as a functional system of complementary relationships, a transfer
and circulation of energy and matter". According to Ellenberg
(1973) ecosystems are "interaction systems (Wirkungsgefüge) of
living organisms and their non-living surroundings, regulating
themselves to a great extent". The term "ecosystem" can be
applied to abstracted systems also to concrete systems, for
example, 'aquatic ecosystem' would be an abstracted system and
the "Lake of Galilee-aquatic system" an example of a concrete
system. Excepting well-defined waterbodies, the borders of
'concrete' aquatic ecosystems - as open, functional systems -
are hard to define. This difficulty and the failure to distinguish between abstracted and concrete, space/time defined
ecosystems pose problems of terminology as does the distinction
between natural and human ecosystems.

As already mentioned, Egler (1942) was one of the first to realize
the holistic nature of ecology; he criticized the conceptual and
methodological weaknesses of contemporary plant ecology in failing to
take account of this. More than 20 years later, Egler (1964) used
the failure to cope with the pesticide problem to demonstrate the
urgent need for holistic human-ecosystem ecology where "man-plus-his-
total-environment" form a 'whole' that should be studied in its
totality - "total human ecosystem". Egler (1970) wrote :-

"The chief goal of human ecosystem science is a knowledge of,
and man-oriented technology towards, a permanent balance
between man and his total environment, both operating as part
of a single whole, that will afford a life of the highest
quality".

In an important attempt to bridge the gap between division and
integration Koestler (1969) introduced the term "holon" to charact-
erise the unique nature of systems at each integrative level of the
hierarchical order: they are intermediary entities, functioning as
self-contained units relative to their subordinates, but at the same
time depend on entities at the next higher hierarchical level of
integration. The term "holon" helps encapsulate the complementary
aspects of (i) being-part-of, and wholeness and (ii) differentiation
and integration.

The holistic axiom that the whole is more than its parts, has
been restated in quantitative terms by Mesarovic et al (1970). Using
a quantitative description, Weiss (1969) showed, because of the

constraints on the degrees of freedom of component parts as a result
of system behaviour, that the variance of the total system V_s is less
than the sum of the variances of its elements V_a, V_b, V_c, ...V_n:

i.e. $V_s << (V_a + V_b + V_c ......V_n)$.

Thus, coordination and control become the emergent qualities of
a new holon. Lorenz (1973) referred to the work of Hassenstein (1970)
who showed, using a simple electronic model, how the coupling of two
independent circuits causes the sudden, flash-like emergence of a new
system with its own qualities and behaviour. Lorenz called the sudden
birth of a system with new properties, which cannot be expected or
predicted from the properties of each holon, fulguration, (fulguratio
= lightning flash) and attached great importance to its occurrence in
evolution. He stated that "cybernetics and system theory have freed
this sudden creation of a system with new properties and new function
from the odium of a miracle".

Vester's basic bio-cybernetic rules of biosystems and human systems

Cybernetics, the scientific discipline dealing with regulation
of ecological systems, was developed by Wiener (1948) for system engineering and communication and extended to biology by Ashby (1964).
It is now also widely used in the rapidly growing fields of mathematical ecology and system analysis (e.g. Dale, 1970; Patten, 1972);
it has been applied also to environmental management (Bennet and
Chorley, 1978) including landscape ecology (Langer, 1970; Van Leeuwen,
1966). But of greatest relevance in this respect for our discussion,
is the theory of bio-cybernetic regulation, developed by Vester (1976,
1980) and its application in regional planning and environmental
education.

In his book, "Urban Systems in Crisis", Vester (1976) has recently pointed to the high technological efficiency of bio-systems in
utilizing energy/material, also of the strong degree of regulation,
both contrasting with the low efficiency of human technology. He
explained the efficiency of bio-systems by referring to basic biocybernetic rules including negative feedback coupling, to ensure
stable equilibria; the circulatory process of recycling; the highly
efficient and economical utilization of energy - particularly if
sources other than direct solar energy are involved (for example, in
energy cascades, chains and coupling); the principle of "judo" (the
Asian method of self-defence) in which force is not combated with
counterforce, but the force applied by the opponent is merely diverted
and controlled cybernetically, thus utilizing it for one's own purpose. These laws also take account of symbiosis (the co-existence
of basically different forms of life to their mutual benefit), and
the principles of multiple use. Finally cognisance should be taken
of basic biological design, bringing organizational cybernetics
together with creative "bionics".

The rules of bio-cybernetics imply that bio-systems have the ability to alter their behaviour and "jump to a higher organization level", as when populations reach critical densities. These stages have been reached now. In Vester's view a critical stage has been reached in the world with the exponential increase in human population linked with a multitude of chemical, physical, energetic and social interactions. He stressed that new tools for decision making are needed to cope with this situation, but first it is necessary to comprehend the ground rules. His book shows, with the help of an environmental simulation game, how the new tools could be applied to integrated land-use planning as in the comprehensive sensitivity model for the Regional Planning Association, Lower Main, prepared as part of the West German Man and the Biosphere project 11 (Vester and Hesler, 1980). Similar principles were applied to a sensitivity matrix concerned with the land use factors and multiple-use strategies of Mediterranean uplands (Naveh, 1979).

Non-equilibrium thermodynamics: Prigogine's Theory of self-organisation and human systems

Prigogine (1976) and Nicolis and Prigogine (1977) have recently expounded the principle of "order through fluctuation", where systems, partly open to the inflow of energy/material and information and having many degrees of freedom, are in a state of non-equilibrium. They tend to move through a sequence of mutating transitions to new regimes which in each case generate conditions of renewal within a new and higher organization, by high entropy production, and thus create possibilities for continuing life-support. The recognition of such a principle has opened the way for a new theory concerning the self-organization of physical systems which Jantsch (1975), has attempted to apply. He suggested that "order through fluctuations" seems to be a basic mechanism, operating at all levels and in all organizations, institutions and cultures, including the dynamic behaviour of mankind with its evolution from the involvement of individuals in hunting and primitive agriculture and all the way to our present sophisticated highly interacting global systems of involvement. He distinguished three basic types of internal self-organizing behaviour:

i)   "Mechanistic systems" which do not change their internal organization
ii) "Adaptive (or organismic) systems" which adapt to changes in the environment by changing their structures in accordance with pre-programmed information (e.g. engineered or genetic templates) including non-human bio-systems depending on "bio-physical information"
iii) "Inventive (or human action) systems", which change their structure by invention (internal generation of information).

In this instance the information is generated within the system,
with the possibility of iterative responses with the environmental
'forces' of change. The evolutionary time scale for adaptive systems,
in the biological domain, corresponds to what Teilhard de Chardin
called the "unfolding of biogenesis", whereas for inventive systems,
in the human domain, it would correspond to noogenesis ("cultural
information"). In the realm of planning this implies, according to
Jantsch, that man, who is intimately involved, acts as the regulator
in the systems to be regulated.

THE ROLE OF LANDSCAPE ECOLOGY AS A HUMAN ECOSYSTEM SCIENCE

Attention has been focused on some conceptual developments which
have arisen almost simultaneously and which have contributed to the
consolidation of a general biosystems theory. They serve as a basis
for a unified ecological theory enabling trans-disciplinary, multidisciplinary, concepts of human ecology. This theory is holistic,
with a hierarchy of holons (open, living and ecological systems)
being subject to biocybernetic self-regulation and feedback control.
The feasibility of this theory is supported by the thermodynamic
findings of Prigogine concerned with "order through fluctuation of
dissipative structures" recognised by Lorenz as a 'fulguration' in
human cultural evolution, as noogenesis in "inventory human systems"
in Jantsch's socio-ecological terms, and as a "jump to a higher
system level of organisation" in Vester's biocybernetic language.

With this systems background there is now a need to elucidate the
role of landscape ecology, a task that can be approached using
Ellenberg's (1973) recently proposed ecosystem classification as a
starting point.

Ellengberg's ecosystem classification: Bio-ecosystems and Techno-ecosystems

Contrary to the widely held belief, Ellenberg claims that the
biosphere should be regarded as the most extensive and diverse
'concrete global ecosystem' and not as a higher level of integration
on its own. Ellenberg subdivided it into two major groups, namely
natural ecosystems, which depend more or less on solar energy, and
urban-industrial ecosystems, which depend on fossil energy (and
increasingly on nuclear energy). The biosphere can be regarded as
the largest natural or, more precisely, biological ecosystem (bio-
ecosystem) whereas the urban-industrial or technological ecosystems,
techno-ecosystems, which are not directly controlled by the photo-
synthetic conversion of solar energy, contribute to the technosphere.
Together the biosphere and technosphere, plus relevant parts of the
geosphere, form the ecosphere. In the technosphere, man with his
technological skills, has gone beyond the ecological and spatial
limits set to life in the biosphere and thus the ecosphere of "homo

industrialis" includes all those parts of the geosphere, the atmosphere, lithosphere and hydrosphere, which are directly or indirectly
influenced by him. He has also influenced the stratosphere with far-
reaching ecological consequences for life on earth. Thus no consideration of the ecosphere can be complete without reference to the
stratosphere.

All natural or man-modified ecosystems should be regarded as bio-
ecosystems in which autotrophic organisms i.e. those capable of manufacturing complex foods from simple inorganic substances, are the
primary producers. These ecosystems are maintained by solar energy,
biotic and abiotic resources; they are regulated in the main by biophysical information. On the other hand, techno-ecosystems are designed, made, maintained and controlled by man using fossil energy,
materials made or converted by man from the geosphere and biosphere;
they are regulated by cultural-noospheric (scientific, political,
spiritual, technological etc.) information, processed by industrial
man.

Biosphere Technosphere and Man as holons of the Total Human Ecosystem

As already mentioned biocybernetic feedback control has evolved
as an essential concomitant of increasing complexity during the long
evolution of natural bio-ecosystems. On the other hand techno-ecosystems, the recent creation of man, depend on positive feedback
loops of energy/matter and cultural information, but lack the biocybernetic feedback control (sensu Jantsch) inherent in "adaptive
biological systems". Since the second industrial revolution techno-
ecosystems have grown rapidly into progressively larger urban-industrial complexes, with symptoms of what Vester (1976) called the
"urban crisis" syndrome, while at the same time threatening bio-
ecosystems by substitution, environmental pollution and what Naveh
(1973) chose to call the "syndrome of neo-technological landscape
degradation". However, as man depends for his existence on the viability of natural and agricultural bio-ecosystems, his techno-ecosystems cannot survive without the biosphere. Thus the exponential
growth of technospheres endangers not only the biosphere, but in the
long term the very existence of technospheres themselves.

Modern man occupies a dual position serving as a receiver of
vital inputs from the biosphere and geosphere and by using the outputs when developing technospheres he modifies the biosphere and
geosphere. He is thus affecting, and being affected by, these modifications (Figure 2). This dichotomy in man's position, of dependence and independence, has probably been the major cause of the
undesirable separation of 'human' and 'biological' ecosystems. Until
his cultural evolution added unique psycho-sociological and techno-
economical dimensions to his biophysical nature, primeval man was an
integral part of natural bio-ecosystems.

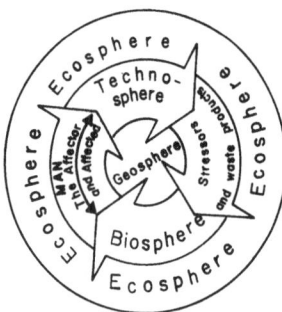

Figure 2: Modern man - the "affector" and "affected" (Naveh, 1980).

   The complex, cybernetic nature of human systems has been
discussed by Laszlo (1972) and Jantsch (1975), the latter inter-
relating the basic dimensions of these systems with a hierarchy of
natural (physical, biological, social and spiritual) or "cognitive"
factors stressing their dynamism and ability to interact.

   From the ecological point of view the dual nature of man's
position can only be resolved by recognizing the holon properties
of man, biosphere and technosphere as autonomous entities, while
at the same time being dependent on the controlling and integrating
influence of man and his environment, the biosphere, technosphere
and geosphere.

Re-definition of landscape and landscape ecology

   The landscape is the visual and spatial integration of biosphere,
technosphere and geosphere. It encompasses the concrete, space/
time entities of the 'total' human ecosystem, with the ecotope as
the smallest, and the ecosphere the largest units.

   In a model showing the relationships between different, major
ecosystems 'holons' and their landscape units, different types of
open, cultural and built-up landscapes are identified according
to the kind and size of energy, material and information inputs and
the increasing dominance of man-made artifacts (Figure 3). The
landscape units can be viewed as a continuum of increasing modification, conversion and replacement of natural bio-ecosystems. At
the present exponential rate of urban-industrial expansion, there
is an alarming tendency towards the lower left corner of this
ordination and towards the creation of more and more monotonous
cultural landscapes. It is now obvious that this "neo-technological
landscape degradation" is characterized by an increasing input of
cultural information, fossil energy and man-made waste material
from urban-industrial techno-ecosystems, together with the loss of
natural and spontaneously occurring organisms and the natural
negative feedback loops which ensure environmental stability and

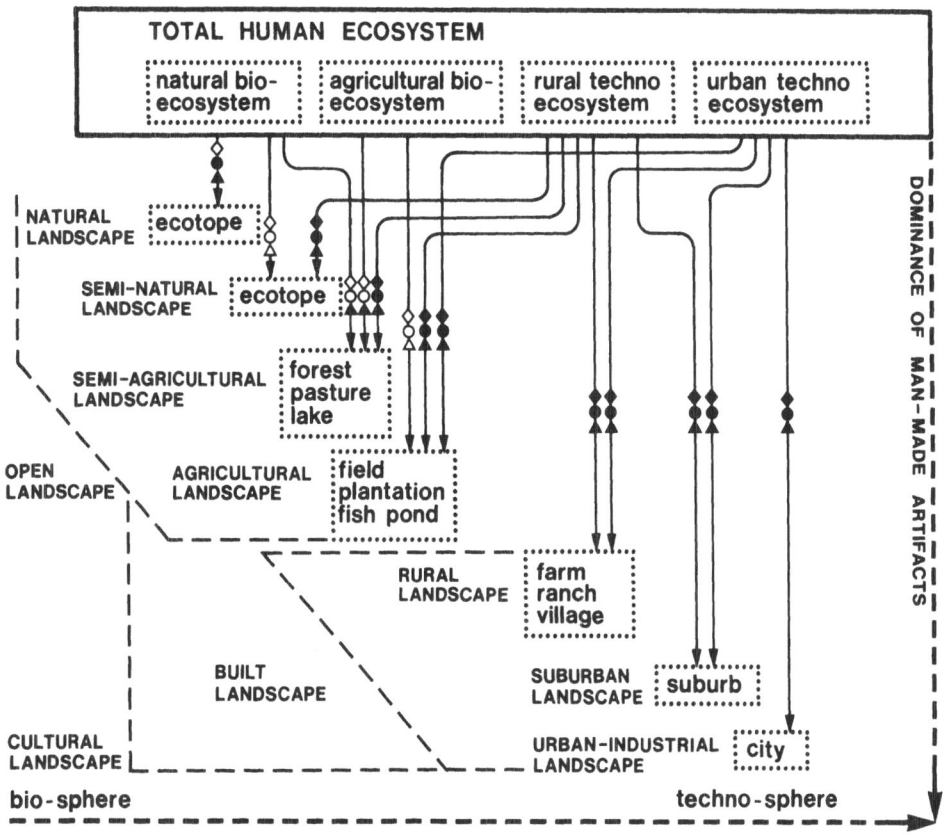

Figure 3:   Ordination model of landscapes along a gradient of change
            from natural landscapes to urban ecosystems (Naveh, 1980).

resilience. These are substituted by agro-technical, chemical and
engineering feedbacks which trigger "side effects" e.g. attributable
to environmental pollution. At the same time, the vital role of
natural vegetation acting as a "living sponge" and ensuring the
recycling of nutrients and other products of bio-geochemical cycling
is being diminished.

     To ensure 'man's' survival and that of our techno-ecosystems,
it is essential to enmesh visual and spatial integration into the
landscape, a new dynamic system is required at a higher level of
organisation/complexity.   As Prigogine (1976) has shown for dissi-

pative structures such as open ecological systems, the new equilibrium
should be considered as continually evolving: it should not be considered as a stationary state but as a flow process. For its preservation Waddington (1975) coined the term 'homeorhesis'. There should
be feedback in the decision-making process with landscape ecology
playing an important role in determining land use. This role should
be the chief goal of landscape ecologists.

RECAPITULATION: THE ROLE OF LANDSCAPE ECOLOGY IN THE NEW CYBERNETIC
SYMBIOSIS BETWEEN MAN AND NATURE

   According to the Holon concept, a dynamic equilibrium will be
created if man can find a balance between his wish for self-assertion
and the need to integrate as part of a larger whole, the Total Human
Super-ecosystem.

   In primitive societies the integrative, self-transcendent, tendencies lead to a symbiotic I/Thou relation of man with nature,
sanctified in a process of cybernetic adaptation. But the development of science and technology has disrupted this cybernetic adaptation and has replaced man's respect for nature and its natural ecosystems, as reflected in the overwhelming self-assertive tendencies,
responsible for the I/It relationship.

   Proponents of radical environmental/ecological movements (Sessions,
1979) insist that man's survival can only be ensured by substituting
a biocentric egalitarian attitude to nature for our present short-
sighted, anthropocentric attitudes. We shouldn't continue the disastrous exponential expansion of our industrially determined landscapes but can we abandon the present superior assertive status which
we have developed as inventive systems in the hierarchy of nature?
The reconciliation between our self-assertive, and self-transcendent,
tendencies will be expressed in our attitudes towards landscape.
Positive self-organizing and negative self-stabilizing feedback loops,
steered by bio-physical, cultural, scientific and technological,
political and educational information can ensure the creation of the
new post-industrial landscapes. This "jump to a higher level of
complexity" - a further step in human evolution - will yield a higher
level of ecosphere organization with emerging qualities of diversity,
stability, productivity, utility and beauty.

   This new development cannot be achieved by technocratic control
models of "intervention" or by linear, deterministic, global models
of "synthesis", described recently by Bennet and Chorley (1978). It
goes far beyond "the interfacing of socio-economic and physico-
chemical systems" envisaged by these authors. Instead it requires
the fully integrated, holistic, approach of a creative Human Eco-

system determined to ensure that land use decisions are taken in the
context of an ever responsive, evolving system subject to continual
reconsideration - landscapes are not static; like everything else
they are subject to evolutionary change. There are encouraging signs
that this realisation is already finding favour. The distinguished
biologist, environmental educator and policy maker Lord Ashby (1978)
has outlined the evolution of new social 'norms' and ethics in
regard to 'nature' as reflected in the decision making process with
respect to land use planning and management. However, time is not
on our side - these trends need to be strengthened and broadened
urgently.

In practical terms, landscape ecology should contribute to the
reconciliation between the need for conserving (not preserving) and
evolving open and rural landscapes and their biotic and cultural
resources on the one hand the socio-economic needs on the other hand.
To do this it is necessary to develop new tools for holistic analysis
and assessment, planning and designing, management and conservation.
In this respect, it is desirable to transform intangible, ecological,
scenic and recreational "non-economic benefits" into quantitative
parameters enabling the optimization of landscape values (Naveh,
1978). It is here that cybernetic sensitivity matrices can be valuable tools in helping to appraise the compatibility of different
land-use factors, and determining the sensitivity and tolerance of
key variables; examples of their use have been described by Vester
(1976), Vester and Hesler (1980), Naveh (1979) and Naveh and
Lieberman (1984). For the present the most pressing needs of landscape ecology are :

i)   The intensification of integrated, multidisciplinary long-term
     ecosystem projects studying the direct and indirect effects of
     different land uses: the results should be targeted on the
     ability of existing ecosystems to 'absorb' these changes; they
     should be expressed quantitatively so that they can be used in
     predictive models guiding, planning, future management and
     conservation decisions.

ii) The conversion of degraded and destroyed natural bio-ecosystems
     into thriving systems with many attributes of value to indus-
     trial man - ecological/environmental, aesthetic etc.

iii) The need to recognize the important role of landscape ecology
     in the development of 'society' - it is a form of interdiscipl-
     inary task-oriented environmental education where decisions
     have to be continually reviewed and re-assessed. What may be
     thought to be final or correct today may prove to be incorrect
     tomorrow: flexibility and the need to be iterative are of the
     essence.

REFERENCES

Arthur, L.M., Daniel, T.C. and Boster, R.S., 1977. Scenic assessment - an overview, Landscape Planning 4: 109.

Ashby, W.R., 1964, "An Introduction to Cybernetics", Chapman and Hall Ltd., University Paperbacks, London.

Ashby, E., 1978, "Reconciling Man with the Environment", Stanford University Press, Stanford.

Bennet, R.J. and Chorley, R.J., 1978, "Environmental Systems: Philosophy, Analysis and Control", Methuen and Co. Ltd., London.

Bertalanffy, L., von, 1968, "General System Theory Foundations, Development and Applications", George Braziller, New York.

Buchner, H.K., 1971, The ecosystem level of organization, in: "Hierarchically Organized Systems in Theory and Practice", P. A. Weiss, ed., Haffner, New York.

Christian, C.S., 1958, The concept of land units and land systems, Proceedings of Ninth Pacific Science Congress, 20: 74.

Christian, C.S. and Stewart, G.A., 1968, Methodology of integrated surveys, in: "Aerial Surveys and Integrated Studies", Proc. Toulouse Conf., UNESCO, Paris.

Dansereau, P., 1957, "Biogeography - an Ecological Perspective", Ronald, New York.

Dale, M.B., 1970, "System analysis and ecology", Ecology 51: 2.

Di Castri, F., Baker, F.W.G., and Hadley, M. (eds.), 1984, "Ecology in Practice", Vol. 1, Ecosystem Management, Vol. 2, The Social Response, Tycooly International Publishing Ltd., Dublin, UNESCO, Paris.

Division of Land Use Research, 1978, "Research Report 1973" CSIRO, Canberra.

Egler, F.E., 1942, Vegetation as an object of study, Philos. Sci., 9:245.

Egler, F.E., 1964, Pesticides in our ecosystem, American Scientist, 52:110.

Egler, F.E., 1970, "The Way of Science, A Philosophy of Ecology for the Layman", Hafner Pub. Co., New York.

Ellenberg, H. ed., 1973, "Ökosystemforschung", Springer-Verlag, Berlin; Heidelberg, New York.

Hassenstein, B., 1970, "Biologische Kybernetik", Quelle and Meyer, Heidelberg.

Jantsch, E., 1975, "Design for Evolution, Self-organization and Planning in the Life of Human Systems", George Braziller, New York.

Koestler, A., 1969, Beyond atomism and holism - the concept of the holon, in: "Beyond Reductionism - New Perspectives in the Life Sciences", A. Koestler and J. R. Smithies, eds., Hutchinson, London.

Langer, H., 1970, Die Ökologische Gliederung der Landschaft und ihre Bedeutung für die Fragestellung der Landschaft-Pflege, Landschaft - Stadt., 3:2.

Laszlo, E., 1972, "Introduction to Systems Philosophy - Toward a New Paradigm of Contemporary Thought", Harper and Row Pub., New York.

Leeuwen, C.G. van., 1966, A relation theoretical approach to pattern and process in vegetation, *Wentia*, 15:25.

Lorenz, K., 1973, "Die Rückseite des Spiegels, Versuch einer Naturgeschichte menschlichen Erkennens", R. Piper and Co., Munchen-Zurich.

Mesarovic, M.D., Macko, D. and Takahara, Y., 1970, "Theory of Hierarchical Systems", Academic Press, New York and London.

Miller, J.G., 1975, The nature of living systems, *Behavioural Sci.*, 20:343.

Naveh, Z., 1973, The neo-technological landscape degradation and its ecological restoration, in: "Pollution - Engineering and Scientific Solutions", E.S.Barrekette, ed., Plenum Press, New York, London.

Naveh, Z., 1978, The role of landscape ecology in development, *Environmental Conservation*, 5:57.

Naveh, Z., 1979, A model of multiple-use management strategies of marginal and untillable Mediterranean upland ecosystems, *in*: "Environmental Biomonitoring, Assessment, Prediction and Management - Certain Case Studies and Related Quantitative Issues" J. Cairns, G. P. Patil and W.E.Waters eds., Int'l. Co-op. Pub. House, Fairland, Md.

Naveh, Z., 1980, Landscape ecology as a scientific and educational tool for teaching the Total Human Ecosystem, *in* "Environmental Education: Principles, Methods and Applications", T.S.Backshi and Z. Naveh, eds., Plenum Press, New York, London.

Naveh, Z., 1982, Landscape Ecology as an Emerging Branch of Human Ecosystem Science, *in*: "Advances in Ecological Research", A. Macfadyen, ed., London-New York.

Naveh, Z., and Lieberman, A.S. 1984, "Landscape Ecology - Theory and Applications", Springer Series on Environmental Management. R.S. de Santo, Series editor, Springer Verlag, New York.

Nicolis, G., and Prigogine, I., 1977, "Self-organization in Non-equilibrium Systems", John Wiley and Sons, New York.

Patten, B.C., (ed.), 1972, "Systems Analysis and Simulation in Ecology", Vol. 2. Academic Press, New York, London.

Prigogine, I., 1976, Order through fluctuations: self-organization and social systems, *in*: "Evolution and Consciousness", E. Jantsch and C. H. Waddington, eds. Addison-Wesley, London.

Rowe, J.S., 1951, The level of integration concept and ecology, *Ecology* 42: 420.

Sachsse, H., 1971, "Einführung in die Kybernetik", Vieweg and Sons, Braunschweig.

Schultz, A.M., 1967, The ecosystem as a conceptual tool in the management of natural resources, in: "Natural Resources: Quality and Quantity", S.V.Cirancy-Wantrup and J.J.Parsons, eds., Univ. of Calif. Press, Berkeley, Calif.

Sessions, G., 1979, "Ecophilosophy (No. 2)", Philos. Dept., Sierra
    College, Rocklin, California.
Tansley, A.G., 1935, The use and abuse of vegetational concepts and
    terms, Ecology, 16:284.
Troll, C., 1971, Landscape ecology (Geo-ecology) and Bio-eoceonology-
    a terminology study, Geoforum, 8:43.
Vester, F., 1976, "Urban Systems in Crisis", Deutsche Verlags -
    Anstalt GmbH, Stuttgart.
Vester, F., 1980, "Neuland des Denkens, Vom technokratischen
    Aeitalter Zum kybernetischen Zeitalter", Deutsche Verlags -
    Anstalt GmbH, Stuttgart.
Vester, F., and von Hesler, A., 1980, "Sensitivity Model, Ecology
    and Planning in Metropolitan Areas", Regionale Planungsgemein-
    schaft Untermain, 6000 Frankfurt am Main 1.
Waddington, C.H., 1975, A catastrophe theory of evolution, in:
    "The Evolution of an Evolutionist", C. H. Waddington, ed.,
    Cornell Univ. Press, Ithaca, New York.
Waddington, C.H., 1977, "Tools for thought". Granada Publishing
    Limited, Paladin, Frogmore.
Weiss, P.A., 1969, The living system, determinism stratified, in:
    "Beyond Resuctionism, New Perspectives in the Life Sciences",
    A Koestler and J. R. Smithies, eds., Hutchinson, London.
Whittaker, R.H., 1975, "Communities and Ecosystems" (2nd Ed.)
    McMillan Pub. Co., New York.
Whyte, R.O., 1976, "Land and Land Appraisal", Junk, The Hague.
Wiener, N., 1948, "Cybernetics", John Wiley and Sons, New York.
Young, G.L., 1974, Human ecology as an interdisciplinary concept:
    A critical inquiry, in: "Advances in Ecological Research",
    A. Macfayden, ed., Academic Press, New york.
Zonneveld, I.S., 1972, "Textbook of Photo-Interpretation", Vol.VII,
    Use of aerial photo-interpretation in Geography and Geomorphology,
    ITC. Enschede.
Zube, E.H., Brush, R.O., and Fabos, J.G., (eds.), 1975, "Landscape
    Assessment: Values, Perception and Resources", Strondsburg,
    Pennsylvania.

COMMENTARY : CONSERVATION/LANDSCAPE

'Landscape' is a concept that encompasses the entire human experience. It refers to the environment in which we live, and as such expresses our continuity with the past: it reflects the interactions between the environment and the past and present activities of man in some instances leaving the resource unchanged, in others, bringing about change through massive energy inputs that produce environments/habitats that bear little resemblance to what went before. The three, very different, concepts of preservation, conservation and direct management are fundamental to the decisions that have to be made so as to derive the maximum 'benefit' from our resources. However, their differences are not widely and explicitly recognised by planners and their agencies.

The concept of 'landscape ecology' was evolved in an attempt to develop a basis for conservation that can then be applied in a planning and management orientation. Despite its origin, the principles upon which 'landscape ecology' rest can be derived thermodynamically by considering the energy inputs required to maintain systems in elevated states. This derivation leads directly to an appreciation that ecosystems collapse if their energy sources are withheld. It is therefore necessary to recognize the nature of the collapsed ecosystem or 'environmental absolute'.

There is no doubt that man, the technocrat, has been trying to overlook the bounds imposed upon us by our environment, and, in doing so, is unlike his primitive ancestors who were well aware that they had no option but to live within its constraints, whether or not they understood them. The core of conservation is centred upon the maintenance of life support systems; conservation is therefore a fundamental principle to be adopted by planners. The sustainable management of the environment and its resources, necessitates that we live in harmony with it.

Conservation is not restricted to the maintenance of individual species like the white rhinoceros; it is about life on earth. While being concerned with the 'rare' or 'special' it is more importantly concerned with the commonplace. While not widely recognised as part of the conservation debate, land use is pivotal. In Britain the strategic rationale for self sufficiency is being questioned, as is the policy for maximizing food production when large surpluses

are accumulating in the other countries of the European Community.
To an extent, the policy of maximum food and timber production has
been responsible for damage, which some consider unnecessary, to the
environment, with the 'reclamation' of peatlands and moorlands by
drainage schemes and afforestation. Conservation, amenity, and
recreation are claimed by an increasing number of people to be more
in the public interest, with the different groups of people making
charges and countercharges. Within the European Community the environmentally aware citizens of the Netherlands, Denmark and Germany
deprecate the Commission's seeming willingness to downgrade conservation to a hobby, albeit an increasingly important one.

In the U.S.A. the more negative environmental attitudes of
government have more than once been overridden by the pressures of
an environmentally-aware electorate. Institutionally, it is not a
matter of sympathetic treasuries providing resources; the problem
is more deepseated. Governments and planning authorities have been
slow to embrace the concept of conservation. Through legislation
and regulation, they have tended to discriminate against conservation
which, like Science, is seen as a part of a group of cultural
activities - moderately desirable, but unlike engineering, neither
useful nor productive, an attitude particularly prevalent in developing countries, where the emphasis is on technological development.

However, in Brazil there is now a growing conviction that the
conservation of the Amazon forests is not only important for the
conservation of rare species and assemblages of species, but also for
the conservation of Man. It amounts to a holistic approach to the
management of resources. Half of the world's increase in atmospheric
carbon dioxide, not to speak of changes in the oxygen/nitrogen
balance, are believed to be attributable to the destruction of the
Amazon forests. It is not often realised that, in geological times
changes of this sort have been tantamount to the creation of new
habitats with their own distinctive assemblages of plants and animals.
How much do we really know about the effects of urbanisation on the
island biogeography of plants and animals, or the impacts of burning,
ploughing etc. recognizing that ecosystems are dynamic and therefore
continually changing and responding.

Planners like to have access to 'norms'. What is the minimum
area, 'biogeographic island', for sustaining populations and/or
assemblages of plants and animals. For a plant growing in arctic
tundra environments it might be only 20 square metres, increasing
to many square kilometres for the maintenance of caribou. As with
the conservation of anything else, there is safety in numbers
which provide security against disasters: many rather than one
individual, many rather than one representative sample of each and
every assemblage of plants and animals, of ecosystems etc. While
some species will cross from one 'island' to another, the effective
sizes of the different islands can be greatly decreased by the

treatment of the intervening areas which may exacerbate temperature fluctuations, humidities, wind exposures and changes in the amounts and chemical composition of groundwater etc. Much more needs to be known about the spatial factors determining the survival (sustained management) of species and assemblages recognizing that the structure of the latter is highly ordered - there is a strong element of inter-dependence.

B

LAND CLASSIFICATION

21

MONITORING, OPTIMIZING AND PREDICTING IMPACTS FROM MULTI-SOURCE

SPATIAL DATA

S. W. Bie

Norwegian Computing Centre
P.O.Box 335, Blindern
Oslo 3, Norway

INTRODUCTION

Monitoring implies a capability to predict the uses, and
changes of use, of land, whereas mapping involves the precise assessment of all facets of land use, monitoring is not necessarily dependent upon complete enumerations/inventories. In the event, land-
information-systems depending upon completeness often overshadow
simpler, and sometimes more reproducible, systems.

In this chapter methods are suggested to produce working
solutions for problems where it is desirable to observe and quantify
land use changes suggesting at the same time procedures whereby
accepted optimization tools may be linked to distribution data.

GEOGRAPHIC SPACE - PROPERTY SPACE

The continuous and complete monitoring of geographic space by
direct observations is not attainable in practice. Instead, it is
necessary to rely on samples and to be equipped with a framework
for extrapolating from samples.

Environmental classifications are almost always multi-dimensional with classes being defined by sets of attributes. All
attributes are not necessarily equally discriminatory or valuable;
the classes may not provide the optimal separations for particular
attributes. Classifications constitute a subdivision of property
space into compartments. Unless there are distinct discontinuities
in property space, the derived classes will be heterogeneous with
'noise' possibly obscuring small, subtle or slow but nevertheless
important, changes as when monitoring.

The monitoring of land may involve the observation of variable attributes, some only adequately depicted on maps at the largest scale (e.g. the garden of a smallholder, the networks of paths, roads and railways); others such as the vast expanses of grain fields, forests, desert and ice, in contrast being easily portrayed on small-scale maps. This being so we have a need of a flexible approach to monitoring, with access to detailed information when needed, while allowing the exploitation of more generalized information particularly in less variable locations.

POLYGONS VERSUS RASTER

There are two digital ways of storing geographical information:

- by polygons
- by cells

Polygon systems rely on the boundaries (edges) of the polygons being defined by sets of absolute or relative x, y-coordinate pairs. The accuracy of polygon delineation is a function of :

- density of x, y-coordinates
- the relevance of the 'interpolator' linking sets of x,y-coordinates.

Polygon-based systems facilitate the representation of smooth, "natural" boundaries. Polygon data-files are usually relatively small, and may be condensed by suitable combinations of .(i) co-ordinate data reduction algorithms ("weeding" or "sieving") and (ii) interpolation algorithms (e.g. linear, arcs, polynomials). Polygon data-files are usually complex with 'pointers' allowing the shared use of common lines, and enabling attribute values to be associated with the different polygons.

Cell-based systems (or raster cell systems, grid systems or picture-element = pixel systems) depend upon the use of discrete geometric objects of defined form. Quadrate or rectangular cells are commonest; triangles and hexagons, although having attractive geometric properties, have not been widely used (triangles are commonly used in some digital terrain models). Unlike polygons, cells are usually of equal size. Raster data files are usually simple, the position of a cell being its position in an array or matrix with little or no built-in reference to its neighbours - they are independent.

In practice, the size of a cell is determined by the spatial variability of relevant attributes and by the number of cells that can be cost effectively processed whether by computers or manually.

Polygon, and raster cell, systems have both been widely used in computer-aided cartography, the value of which has been restricted by the coarseness of commonly used output-devices (line printers, elect- rostatic plotters and TV-screens).

Raster cell systems have been favoured when considering data analyses largely because of the lack of efficient algorithms for processing polygon data, in particular when overlaying several inde- pendent sets of polygons. Foreseeably, but not imminently, this is likely to change, either by the arrival of new and more powerful algorithms, or by a hardware solution to the overlay problem. Exist- ing computer software and hardware are particularly suited to the handling of raster-based matrices. Currently there are many devel- opments associated with computers and their output devices (screens, laser plotters); they will improve the performance of raster systems and reduce the objections related to their display.

In practice it is possible to convert a polygon system to a grid system either by using computer hardware and/or software, or by hand encoding onto an overlaid grid. The amount of information lost depends on cell size in relation to the irregularity of polygon boundaries and polygon size distributions. It is similarly possible to convert a raster cell system to a polygon system using computer software or manual interpolation. The computer conversion is not without its problems, the degree of information loss depending on (i) the sizes of the grid cells and (ii) the characteristics of the interpolator used to construct polygon boundaries: manual inter- polation is highly subjective.

CELL SHAPE AND SIZE

Ideally a 'cell' should fit accurately the area of land about which a set of accurate and precise statements can be made. They may be small in complex landscapes and large in uniform landscapes. Cells may be convoluted, e.g. in areas of complex geomorphology or simple, rectangular e.g. in a reclaimed polder area. In reality the world can be divided into a set of more or less irregular polygons of different sizes.

Because of the difficulties of handling polygons by computers it has been argued that square cells are the most convenient. However, with manual (analogue) data-handling polygons are probably as convenient as squares and in terms of accuracy probably prefer- able. But because of the volume of data to be handled, manual methods should rarely be countenanced.

The size of a cell should in the first instance be determined by scientific, not technological, factors. Knowing that soil is variable but that various forms of land use (plants, fields, houses)

frequently integrate short-range variability, it seems that cells
rarely need to be smaller than 10 m x 10 m - the expected resolution
of satellite imagery expected in the 1980's. World-wide this would
yield 1.34×10^{12} cells, a large number, but technologically not
impossible to handle, being equivalent to the optical (video) storage
disks expected in the second half of the 1980's.

If 10 m x 10 m is the size of the smallest cells, few need to be
larger than 10 km x 10 km: in most environments the large degree of
variability could make larger cells worthless.

Perhaps a nest of cell sizes catering for at least four levels
of generalization is the best approach i.e.

Level	Cell Size	No. cells globally
1	10 m x 10 m	1.34×10^{12}
2	100 m x 100 m	1.34×10^{10}
3	1 km x 1 km	1.34×10^{8}
4	10 km x 10 km	1.34×10^{6}

SAMPLING STRATEGIES

It is necessary to minimize uncertainty (variance), associated
with measurement errors and at the same time obtain estimates of this
uncertainty. Further, regional and local planners are interested
in where and under which circumstances change takes place in addition
to if, and how many changes occur. In addition to these considerations a monitoring strategy should take note of technical, personnel
and financial constraints.

Sampling procedures for the compilation of inventories

Provided appropriate measuring devices are used with a discip-
lined approach to recording, direct observations are often the most
accurate. Different groups of scientists have developed different
methods for sampling climate and socio-economic factors and biological characteristics: this chapter will concentrate on the sampling
of soil, and land, attributes.

When soils are sampled, direct observations involve the examination of a profile or the collection of samples; as far as land
use is concerned, visits must be made. How should the location
of samples be determined

    - by subjective positioning ?
    - by probability sampling ?

Subjective positioning.   Subjective positioning has been the
rule when choosing where to survey soils and land uses. Surveyors
decide locations in the light of their perception of the landscape.
The success of this sampling strategy depends on the ability of the
soil scientist to obtain a set of ground truths reflecting the range
and frequencies of values occurring for each attribute. Because the
perception of the landscape relies intuitively on assumed attribute
covariance and spatial autocorrelation it is likely that success will
depend both on scientist and landscape. Subjective positioning may
yield data of uncertain value for which it is impossible to ascribe
probability.

Probability sampling. "Probability sampling" has been widely
used when surveying land uses. Surveyors select sets of locations
to be recorded based on a strategy where all locations have the same
chance of being selected. If the sample is of reasonable size,
estimates made from the sample will have an accuracy of known
certainty/uncertainty. Probability samples are becoming increasingly
used in other surveys: grid surveys and regular or random transects
exemplify this approach. Even though the boundaries of the locations
may be constructed from other sources (visual inspection, air photo
interpretation), probability samples should yield good estimates.
They are being used increasingly in a wide variety of surveys.

There are many ways of designing a probability sample for area
estimates: random, aligned stratified random, unaligned stratified
random, regular grid, random or regular transects with random or
regular points etc., regular grid and aligned stratified random
procedures are widely favoured.

Evaluation of sampling procedures

When creating inventories of soil and land use it is essential
to have estimates of the degree of uncertainty. It is therefore
recommended that inventories are based on surveys determined by
probability sampling, or that subjective surveys are supplemented as
a probability sample, describing the variability of the mapping
units.

Monitoring by repeated direct observations

Changes may be estimated by returning to the same locations
and making repeated assessments as happens when censusing populations.
Land use is an attribute that can be readily monitored provided that
due consideration is given to seasonal factors e.g. as they affect
cropping.

When changes in time are being measured, per cent before or
after, there is no absolute need to visit exactly the same locations
although the error will be less if they were to be revisited.

Increasingly there is evidence of large short-range variations in
soil and other environmental attributes - in N.W. Europe, parts of
the USA, USSR and Australia and possibly elsewhere. There may be
major differences within a few meters or decimeters. In this
instance the probability of finding the original location is not
high; it is suggested that the effects of heterogeneity may be minimized by using bulked samples.

METHODS OF ESTIMATING THE VALUE OF AN ASSESSMENT MADE OF AN
ATTRIBUTE WITHIN A GRID CELL

(i)   <u>Given value</u> - in this instance the value cannot be given with
        confidence limits.

(ii)   <u>When a supercell or polygon is subdivided</u> - When the area of
        an original (super) cell or polygon is larger than required
        one of two courses can be taken: (a) the supercell contains
        exactly 'n' desired cells and can therefore be easily sub-
        divided or (b) the supercell cannot be easily subdivided
        without the production of part-cells where the peripheral
        cells are bisected. In the latter instance, the allocation
        of values to border cells must be controlled by a decision
        algorithm. There are many possible strategies. Some are
        geometric, either minimizing misfit, preserving area or
        retaining shape, locally or globally; others are thematic
        minimizing statistical measures of heterogeneity, either
        locally or globally. Polygon to cell conversion is a special
        case of (b).

(iii)   <u>Realignment of cells</u> - If the areas of a set of cells (matrix)
        are aligned differently from another set of cells and a con-
        version is necessary, one of a number of techniques can be
        employed. Geometric or thematic misfits may be minimized or
        if the cells are unequal size, it may be easier to make a
        Dirichlet tesselation using the midpoint of each cell as a
        (pseudo-) observation point.

(iv)   <u>Interpolation from a point observation - discrete variables.</u>
        If variables are measured at discrete points interpolation
        strategies are required to estimate the value (class) to be
        allocated to each cell. This is normally done by Thiessen
        polygons through a Dirichlet tesselation process involving
        the use of Delauney triangulation to calculate midlines
        between points of dissimilar values (classes). Thiessen
        polygons are subsequently gridded into cells of desired size
        usually using a geometric decision algorithm. If discontin-
        uities (fault lines, barriers) are present, the intermediary
        triangulation process is modified so providing modified
        Thiessen polygons respecting the discontinuities.

(v)   <u>Interpolation from a point observation - continuous variables.</u>
         If a variable is continuous, e.g. altitude - it is necessary
         to use <u>digital surface modelling,</u> a powerful interpolation
         technique that is not necessary for discrete variables. A
         digital surface can be displayed as a <u>contour</u> joining points
         of equal value in the digital surface model. "Contouring
         packages" have digital surface model algorithms as an essen-
         tial component.

         There are essentially two approaches to digital-surface model-
         ling: (a) using the geometric properties of the sample on
         which the digital surface model is to be based and (b) (in
         addition to a) using known or estimated spatial autocorrela-
         tions (as in kriging). Simple geometric approaches are
         commonest. Several fast algorithms now exist to create good
         digital surface models. The other approach, b (geostatistical
         procedures), is more resource-demanding but provides a better
         estimate of the 'surface', together with estimates of the
         errors of the estimated surface. Both methods are normally
         able to handle discontinuous data.

THE RASTER OVERLAY PROCEDURE

     This is the traditional procedure of overlaying transparent maps,
often coloured or cross-hatched to achieve a synthesis map. This
overlay-procedure is essentially Boolean (stack = layer 1 and layer
2...and layer n) with each layer being similarly important. With the
careful use of colours it is sometimes possible to obtain integrated
quantitative interpretations.

     The digital <u>raster overlay</u> procedure is an extension of the
traditional overlay procedure, a matrix of values of a variable
corresponds to "a layer" and a set of matrices constitutes the
"stack". The matrices are not visible; they are numbers stored in
the memory of a computer (each matrix may of course be displayed by
a plotter or on a screen if desired). Instead of being inspected
visually, the overlays can be challenged by a mathematical procedure.

$$\text{Synthesis map} = \sum_{\text{cell } n}^{\text{cell } 1} \text{function (variable 1, variable 2, variable n)}$$

PREDICTING IMPACT

     Some operational research tools are now being used for planning,
e.g. non-linear and linear programming. These require estimates of
basic attribute-value pairs in order to simulate the values of other
attributes. It is preferable to record individual attributes separ-

ately. The raster overlay procedure allows the storage of both
original, and simulated, estimates. By overlaying, new alternative
or preferred solutions can be generated.

NEW RESOURCE SATELLITES

The launch of LANDSAT-4 with the high-resolution Thematic Mapper
during the summer of 1982 provides 30 m x 30 m pixel information;
data will not be received routinely until 1985. In the meantime the
French/Belgian/Swedish SPOT-satellite due in late 1985, although
with only 3 wavebands will provide imagery with a resolution of 20 m
x 20 m in colour or 10 m x 10 m in monochrome, so enabling land use,
even in complex areas, to be monitored.

CONCLUSIONS

Simple and powerful data-processing 'tools' are now becoming
available encouraging the processing of data rather than its storage.
The use of sampling procedures whether using polygons or raster cell
interpolation, classification and presentation necessitates computer
programmes which are now generally available.

With these aids planners/decision makers should attempt to
optimize the solution of limited problems, e.g. the monitoring of
land, by objective sampling procedures rather than use total enum-
erations.

ACKNOWLEDGEMENTS

The philosophy of this chapter owes much to discussions with
Jürgen Lamp, Institute of Soil Science and Plant Nutrition,
University of Kiel, West Germany, during a joint consultancy with
United Nations Environment Programme, Nairobi, Kenya in November
1981.

22

LAND CLASSIFICATION IN RELATION TO PHYSICAL PLANNING

F. D. Mathiesen

Landbrugsministeriet
Area datakintoret
Enghavevej 2
DK-7100 Vejle
Denmark

INTRODUCTION

Since 1938 the area devoted to agriculture in Denmark has decreased from 3.2 million hectares to 2.9 mha, the latter area being equivalent to 70 % of the total land area. Part of the former agriculture area, especially marginal land - has been afforested.

This notable reduction in the area of agricultural land has been one of the main reasons for incorporating provisions into agricultural legislation to protect the resource of tillable land in Denmark. Another is the increased public interest in a balanced, planned, development of rural land resources taking into account social, economic, and ecological aspects and aiming to provide directives for land-use with respect to urban development, infrastructure, agriculture, forest and areas for raw material extraction, conservation interests, and areas for recreation. With these objectives in mind, it is necessary to provide land-use planners with an effective, and flexible land data system. To meet the demand for information at different administrative levels, the Ministry of Agriculture, Bureau of Land Data (ADK), has developed (i) an extensive data base, with special emphasis on soil-, forest-, water- and raw-material parameters and (ii) an electronic data processing system. This chapter briefly describes the ADK data processing system which has been operative for five years, and outlines some of the programmes of investigation carried out by ADK in 1982.

THE ADK DATA PROCESSING SYSTEM

The system was originally designed for thematic mapping (Platou, 1971, 1975, 1982). It has been extended to handle the Danish Soil Classification and a number of other subjects. Data are organised in files which are described in a catalogue, the Common Parameter File (CPF) which is of fundamental importance for accessing the different groups of stored data. Area-related data are generally referred to the Universal Transverse Mercator Grid (UTM).

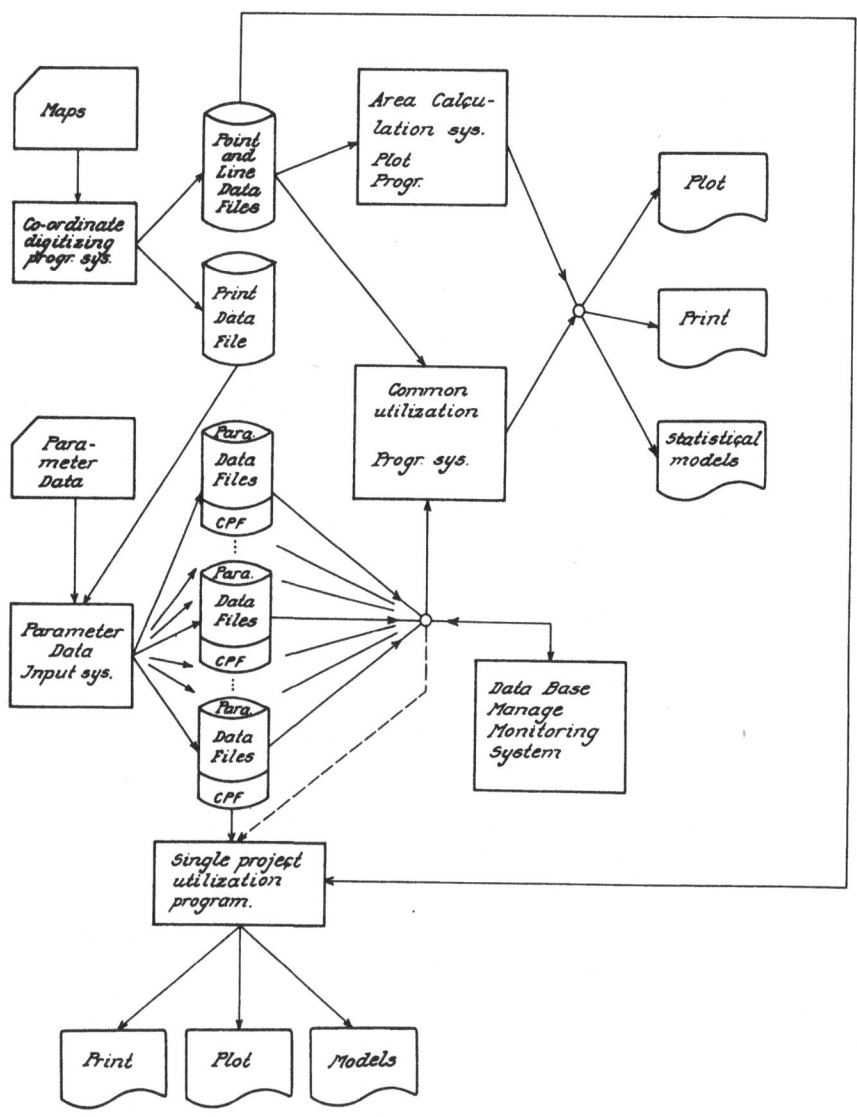

Figure 1: The Land Data system developed at the Danish Ministry of Agriculture's Bureau of Land Data.

The different groups of data can be presented graphically,
tested statistically and used for calculating areas of different
soil types, catchment areas etc. (Figure 1).

INVESTIGATIONS DONE WITH THE ADK DATA PROCESSING SYSTEM

The Danish Soil Classification - the main project (Mathieson et al.,
1979; Mathiesen, 1980).
     A new soil classification was initiated nationwide in 1975
(Commission of the European Community, 1980) (Figures 2, 3, 4). It
was required to meet the following five main requirements remembering
that it was likely to be used at different political levels - communities, counties and central government:
     - the classification had to be based on constant, or slowly
       changing parameters.
     - the parameters chosen had to be strictly and unambiguously
       defined in order to ensure nationwide comparability.
     - the classification scale should be wide enough to encompass
       the range from very good, to very poor soils.
     - the results of the classification should be capable of pres-
       entation on maps so as to facilitate their use by planners.
     - the classification project had to be completed *in toto*
       within a few years.

The other requirements included:
     - complex factors should be capable of being broken down into
       their component parts.
     - it should be possible to add new information at a later
       stage.
     - there should be sufficient flexibility to enable future re-
       evaluation of the weighting placed on different factors.

In planning the structure of the data system, the following
facets were stressed:
     - preparation and publication of information relating to soil
       types, slopes etc.
     - co-ordination of the soil classification with physical plan-
       ning in the counties.
     - the immediate availability of data for regional, and local,
       planning with a back up interpretive service.

     The nationwide soil classification, meeting the above require-
ments, was completed in 1980 with the preparation of 400 multicoloured soil maps (scale 1:50 000) covering the whole of Denmark.
These maps have subsequently been digitized in order to provide
facilities for the production of computer-drawn soil maps at different scales and with different combinations of parameters (Figure 5a,
5b). Calculations of areas occupied by different soil types etc.
can be made by computer for administrative regions, catchment areas

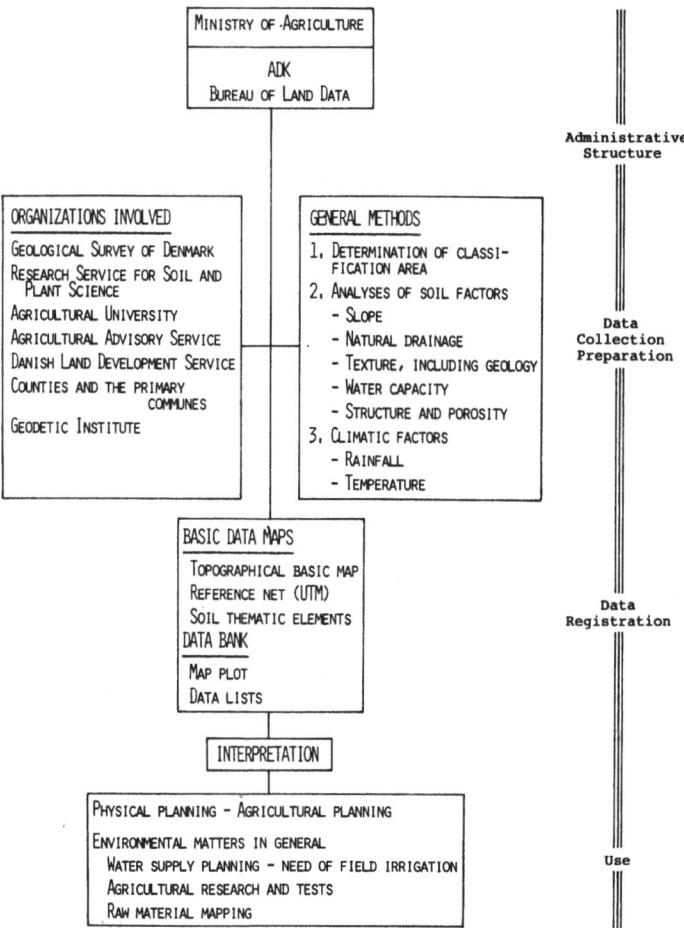

Figure 2: Structure and organization of the Danish Land Classification project in 1975-1980.

etc. This facility is at present extensively used in planning,
administration and research.

The Danish Forest Inventory

Since 1979 a Danish forest inventory programme has been made
by ADK in co-operation with (i) the State Forestry Enterprise and
(ii) the forest organizations. The purpose of the inventory is:
  - to establish a data-base which is easily updated, of Danish
    forestry.
  - to facilitate the implementation of forest legislation in
    relation to land use planning.

The programme includes maps of forests complying with the directives given by the FAO. Boundary lines are digitized so as to
facilitate map-drawings, area calculations etc. (Figure 6a, b). The
various forest areas can be described with details of e.g. tree

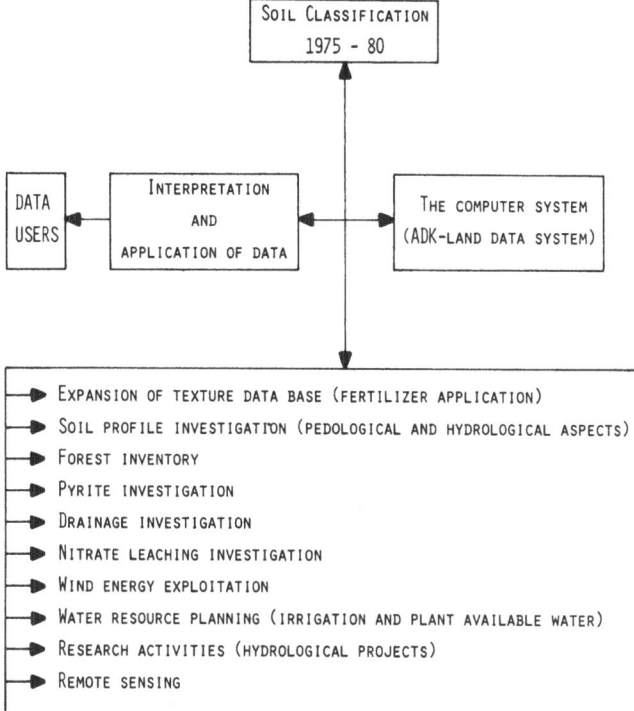

Figure 3: Land classification and sub-projects undertaken in 1980-
1985 by the Danish Ministry of Agriculture's Bureau of
Land Data.

species, their age (Figure 6c), wood production. At present the
inventory is used, with others, to calculate forest resources.

Soil Profile Investigations
    The construction of pipe-lines across Denmark and North Sea gas-
fields provides a unique opportunity to make pedological studies of
soil profiles overlying different geological deposits (Madsen, in
press). These investigations are being done by ADK in co-operation
with the Government Laboratory for Soil and Crop Research, the
Geographical Institute at the University of Copenhagen, the Chemical
Institute at the Royal Veterinary and Agricultural University and
the Geological Survey of Denmark, funds being provided by the Danish
Natural Science Research Council, the Agricultural and Veterinary
Science Research Council and the Ministry of Agriculture.

SOIL TYPE	%	JB-nr.	Percentage by weight				
			Clay < 2 µm	Silt 2-20 µm	Fine Sand 20-200 µm	Total Sand 20-2000 µm	Humus 58.7%C
Coarse Sand	19	1	0-5	0-20	0-50	75-100	
Fine Sand	8	2			50-100		
Clayey Sand	22	3	5-10	0-25	0-40	65-95	
		4			40-95		
Sandy Clay	20	5	10-15	0-30	0-40	55-90	≤ 10
		6			40-90		
Clay	5	7	15-25	0-35		40-85	
Heavy Clay or Silt	1	8	25-45	0-45		10-75	
		9	45-100	0-50		0-55	
		10	0-50	20-100		0-80	
Organic Soils	6	11					> 10
Atypic Soils	1	12					

Figure 4: Definition and distribution of the soil types used in the Danish system of Land classification. The JB-nr. is a sub-specification of the soil type, related to texture analyses.

The project includes pedological characterisation of soil profiles at intervals of 25 metres in pipe-line trenches approximately 2 m deep. Detailed studies are undertaken of 3 profiles per 1 km length of trench when soil samples are taken from each horizon in the profile. The samples are analysed for texture, organic content and various chemical and physical parameters.

The results from this project are scientifically useful in addition to aiding land use planning. They will increase knowledge of soil development in various geological deposits and additionally provide new information to soil maps already published by ADK. Soil-water retention measurements will act as basic data for evaluating the requirements for irrigation (Figure 7). Furthermore, the soil analyses contribute to a better understanding of the problems related to the leaching of nutrients, e.g. nitrate, from the soil.

Investigation of soils containing pyrite

Water quality of some Danish water courses can be adversely affected by strong acidity and a high concentration of dissolved iron (Madsen, 1980). A programme for producing detailed maps of poorly drained soils, containing pyrite, has been initiated by ADK in co-operation with the Ministry of Environment.

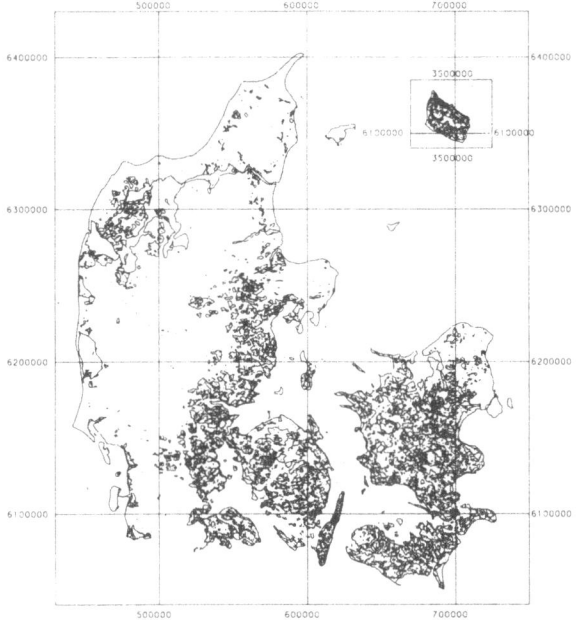

Figure 5: Computer drawn maps of the distribution in Denmark of a) soil types 1 and 2 (sandy soils) b) soil types 4, 5 and 6 (clay soils).

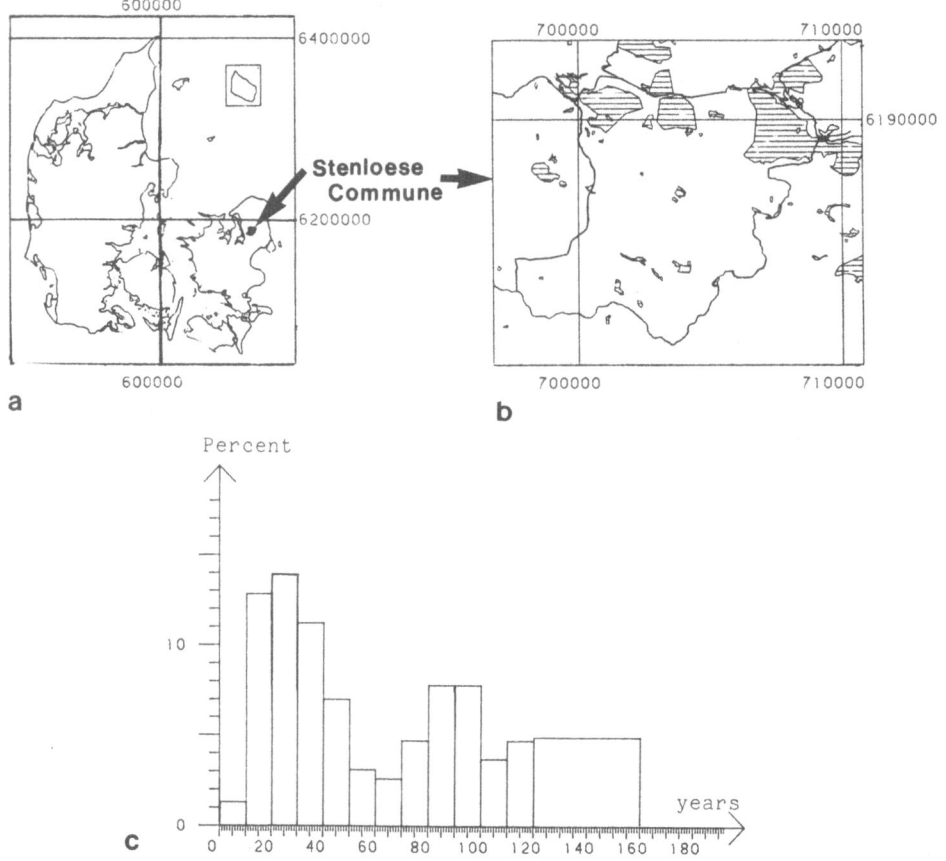

Figure 6:   The Danish national forest inventory: an example of a
            plot: (a) The location of Stenloese commune in Denmark,
            (b)   distribution of the forest area in the commune,
            (c)   Age-distribution for beech in Stenloese commune.
            Beech is the main species, covering 70 % of the forest
            area.

        The investigation is based on measurements of pyrite concent-
rations in soil profiles: more than 7 000 profiles have been investigated.   Data are stored in data files at ADK, where computer based
maps are produced depicting areas with high pyrite concentrations.
They will be used in future planning of the infiltration of water
courses by acid water enriched by dissolved iron, a problem which
in some regions may be influenced by the need for agricultural
drainage.

SOIL TYPE	ROOT ZONE (cm)	WATER CAPACITY (mm)
Coarse Sand	50	70
Fine Sand	70	125
Clayey Sand	70	100
Sandy Clay	70	125
Clay	90	160

Figure 7: Preliminary assessment of 'plant-available' water capacity in the root zone for the different soil types in Denmark. The data are based on soil-water retention measurements.

CATCHMENT AREA CODE	LENGTH OF RIVER KM	CATCHMENT AREA SIZE IN HA	WET LAND AREA		OCHRE POTENTIAL AREA	
			CASE 1 HA	CASE 2 HA	CASE 1 HA	CASE 2 HA
4047	20.2	15000	1200	300	300	250
4201	8.1	9000	200	90	50	50
4109	10.7	12000	500	300	100	80

Figure 8: Example of calculation of the size of wet-land and ochre-potential areas in different catchment areas. Case 1 and 2 represent different drainage conditions as a function of the difference in ground water level.

Investigation of drainage conditions

The investigation will provide data for a balanced evaluation of drainage requirements taking note of possible ecological impacts of drainage on water courses.

Together with the pyrite investigation, the drainage investigation provides a basis for an evaluation of the need for control of ochre in relation to drainage (Figure 8).

FUTURE ORGANISATION OF DANISH LAND DATA INFORMATION SYSTEMS

In order to co-ordinate the work done by different authorities a committee has been established under the supervision of the Ministry of Environment with representatives of the Finance, Environment, Agriculture and Housing Ministeries and also of county and municipal organizations. This committee has been given responsibility for the following:

- analysing the needs of computerized data handling in relation to land-, natural resource- and environmental parameters.
- securing interaction between different land data centres.
- investigating possibilities for standardizing methods of date collection.

For the future, emphasis will be placed on the interaction between different data-centres. This will provide an essential and improved basis for the physical and administrative planning of Danish land resources.

REFERENCES

Commission of the European Communities, 1980, "Land Resource Evaluation", (Report of seminar on land use and rural resources held at Wexford, Ireland, 1978), J. Lee and L. van der Plas, eds., Commission of the European Communities, Brussels.

Madsen, H.B., 1980, "Fast laeggelse af okkerpotentielle omrader i Himmerland", (Mapping of areas with a high potential of ochre in Himmerland), ADK, Landbrugsministeriet, Copenhagen.

Madsen, H.B., in press, "Himmerlands Jordbundsforhold", Ph.D. thesis.

Mathiesen, F.D., 1980, Soil classification in Denmark: its results and applicability in; "Land Research Evaluation", J. Lee and L. van der Plas, eds., Commission of the European Communities, Brussels.

Mathiesen, F.D., Platou, S.W., and Plum, P.M., 1979, Datasystem for det abne land (A data system for physical planning), Ugeskrift for Jordbrug, 124:29.

Platou, S.W., 1971, "An electronic data processing system for
     geological field and laboratory data", Danish Geological
     Survey, Copenhagen.
Platou, S.W., 1975, Aarhus University System II, <u>in</u>: "Computer-
     based systems for geological field data",W.W.Hutchison, ed.,
     Canadian Geological Survey, Ottawa.
Platou, S.W., 1982, "The ADK computer program system for handling
     parameter data", ADK, Landbrugsministeriet, Copenhagen.

23

LAND CLASSIFICATION SYSTEMS IN PRACTICAL PLANNING : A NORWEGIAN

PERSPECTIVE

E. Langdalen

Department of Land Use Planning
Agricultural University of Norway
P.O. Box 29, N-1432
AAS-NLH, Norway

INTRODUCTION

Following the expansive period of the 1960s, the decade of 1970-80 witnessed a general introduction of ecology in land use planning, a new conservation attitude among planners, politicians and the public, and an increasing demand for scientific knowledge and insight, which had formerly been largely overlooked. Environmental awareness, new types of survey methods and new systematic planning approaches were typical of this decade, which saw the development of suitability and capability maps, conflict analysis and environmental impact assessment.

Although these approaches are still valid in land use planning, in the 1980s we face a somewhat different situation. The forces of development have regained much of their relative dominance, due to the worsened economic situation, the will and ability to sacrifice economic gains to long range ecological considerations have been weakened, and at the same time the status of the rational planning approaches of the 1970s has been questioned by social scientists as well as by politicians.

This very brief and schematic description may fit most of the western industrial countries. It certainly reflects the situation in Norway, from where most of the experience behind this chapter is drawn.

Land use planning may be seen as an art of compromising between several interests, which may be compatible or incompatible. The general purpose of comprehensive planning is to identify the characteristics of the land area in question, the direct and

indirect interests involved, their compatibility, alternative
solutions and their possible impacts and thus prepare the base for
sound, long range political decisions, that involve public participation. The type of planning depends upon the general political
system in a particular country and the actual planning structure,
which may be identified as a set of ranges:

Strong government ◄─────────► free enterprise
Centralized authority ◄─────────► decentralized authority
Authority through representation ◄─► public participation
Strategic planning ◄─────────► reactive planning
Continuous planning with
gradual revisions ◄─────────► time limited planning
Goal oriented planning ◄─────────► process oriented planning
Sectoral planning ◄─────────► integrated planning
Development planning ◄─────────► adjustment planning

The position of planning on these scales determines the degree of
participation, comprehensiveness and effectiveness of the actual
planning process. In recent years, there has been a certain tendency
to shift from left to right on these scales.

Planning theory and planning practice differ from country to
country in accordance with political systems and the influence
exercised by the principal actors: professional planners, politicians
and the public. The entire planning process, from data collection to
decision making, is related to the value systems of these actors and
their activity in the planning procedure.

In principle, the demands of the 1970s remain unchanged: there
is still an urgent need to change a planning tradition based on
short range development goals, strong sectoral power and wide individual freedom to encroach upon the environment; need to strengthen
the long range defensive interests; need to expose conflicts instead
of subdueing them; need to involve new professional groups as well as
the public; need for research in both the natural and social
sciences.

These needs have been met by initiative and enterprise. Planning
theory and methods have been developed and pioneering projects have
been carried out, but a number of questions remain. How much has
actually reached and influenced the practical planning procedures ?
To what degree have inventories and systematic analysis strengthened
the base for sound decision making ? Have decisions been influenced
by the new knowledge, insight and awareness which have been gained
over the last decade ? There are no general answers to such
questions, but they have to be raised regularly so as to be
answered on all planning levels - national, regional, and local.

LAND USE CONFLICTS IN NORWAY

To understand the land use planning situation in Norway one has
to bear in mind some of the characteristics of the country: a total
area of 324 000 sq kilometres provides the 4 million inhabitants with
more land per person than any other European country, Iceland
excepted. A sense of having unlimited land resources has long pre-
vailed and it is still dominant when considering certain types of
land use. The fact is, however, that land resources for habitation
and exploitation are very scarce in Norway. If only the highly
productive areas are considered (the low flatlands, the fertile
valleys and the narrow coastal plains where the major population
centres are located), the population density is around 200 inhabitants per sq. kilometre, against 12 for the country as a whole.
In these areas there is strong tension between the principal comp-
etitors for land. These are primarily between proponents of development and conservation, between different development interests
and, to some extent, also between protective interests. The land
encroachment results from urban growth, industry and transport
facilities, especially roads. The most important protective
interest is concerned with productive agricultural land, which
comprises only 3 % of the land area. Protection of the arable land
gained heavy political support during the 1970's, partly reflecting
national independence and resistance against joining the European
Economic Community.

Conflicts also occur in settled areas outside the urban regions,
often due to very limited land resources in constricted topographical sites. In most of the medium and small sized urban settlements there is competition between development schemes, agricultural
potential and conservation/recreation interests. Linear developments,
especially new roads and road improvements, encroach heavily on landscape qualities and ecosystems, especially along rivers and lake
shores.

The main land use conflicts in the uninhabited mountain areas,
which represent more than 70 % of the total land area, are related
to the diversion and management of water systems for electricity
production. This dominant land use conflict, of much public concern
during the 1970s, remains an issue in the 1980s.

Outlying districts, the mountains, inland valleys and coastal
areas still face the pressures of "second home" construction, which
became popular during the growth period in the 1960s and raised
fundamental questions about land ownership and private development
rights. This was also related to the debate on membership of the
European Economic Community.

In addition to the high suitability for outdoor life, a very
special recreational asset in both Norway and Sweden is the free

public access to the countryside. This is independent of ownership
and extends year round to all but agricultural land in production,
and even to productive land in wintertime. This concept, stemming
from Medieval tradition and confirmed in recent law, is a key to the
understanding and handling of land use problems in rural areas.

No other European country has a coastline so relatively long,
dissected and broken by fjords, or sheltered by thousands of islands,
islets and skerries. The Norwegian coast is in general very rugged
and rough, so that erosion is not a major problem, and human habitation and land cultivation are concentrated in a few, limited areas.
Major land use conflicts have been limited to the attractive
recreational coast of Southern Norway and certain limited areas of
urban and industrial development. The oil age, however, raises a
new set of potential conflicts, and conflicting uses of the coastal
zone have suddenly become a prime subject for survey, research and
planning.

PLANNING LEGISLATION

Planning legislation in Norway is based on a number of laws
concerning the use of resources, conservation, land use, etc. The
most important is the Planning Act of 1965, which prescribes public
planning on communal (municipal), regional and county level.
Conservation and development are weighed and balanced in relation
to each other for final approval by the Ministry of Environment.
This means that a consistent hierarchy of plan types are developed,
ranging from <u>national</u> outlines to the formal <u>county</u> plan, <u>regional</u>
plans for groups of communes, <u>communal</u> master plans, and more
detailed plans for smaller areas.

The trends in Norwegian planning over the last four decades have
developed in response to a series of events and developments. The
destruction and stagnation of the Second World War was followed by
a period of rebuilding damaged cities, towns and districts. For
many years planning throughout the country was focused on housing,
due both to the war and to the rapid urbanization and rising standard
of living of the population.

The planning Act of 1965 was passed in response to widespread
criticism of the lack of regional thinking, local development policy
and physical planning, and it established a platform for integrated
planning. The next wave of criticism, around 1970, was directed
against the objectives and practices involved in the planning
strategies. It called for more public participation, decentralization, protection of arable land, nature conservation, protection
and rehabilitation of existing built up areas and buildings, traffic
safety and countermeasures against proliferating motor traffic.
These ideas are now to a great extent reflected in the platforms of
the main political parties.

Two issues that have been widely discussed during the last decades have particularly strong support:

(i) the defence and strengthening of the existing pattern of dispersed settlement and maintenance of the economic and physical regional balances; and

(ii) the protection of all arable land and its reservation for agricultural production.

A new planning bill worked its way slowly through the parliamentary process in the 1970s, stirring some opposite political views, particularly concerning the bill's absolute requirement of a plan for any environmental change. The Labour Government saw the bill approved by Parliament in June 1981, but before the Act could come into operation, an election in the autumn of the same year brought a Conservative majority to Parliament, and the legislation was annuled. A new committee has developed a new bill which, however, is much like its cancelled predecessor.

Elements of environmental impact assessment have been important, but informal, parts of land use planning for a long time. There are also certain formal requirements for impact assessment in the existing legal and administrative practices governing some sectors, such as the planning of hydroelectric power developments, large industrial developments, and major road constructions, as well as in traffic planning. The annuled Planning Act of 1981 had introduced a general, comprehensive system of environmental impact assessment for major projects with potential environmental effects. The Act specified 5 project categories, which covered mainly: new built up areas above 20 hectares; new buildings and plants requiring a certain labour intensity or employment level; transport terminals, roads and harbours above a certain size; stone quarries and gravel extraction greater than a stated encroachment level; new cultivation of more than 50 hectares, and forest planting covering more than 500 hectares.

LAND USE PLANNING - METHODOLOGICAL APPROACHES

The planning legislation calls for systematic, comprehensive, and continuous planning on the three levels of county, regional and municipal concerns. After the approval of the existing law of 1965, regional planning (defined as planning for two or more municipalities) had a promising introductory phase, which included a most valuable recognition of regional relationships, and saw the introduction of survey methods and land classification systems. However, regional planning on this scale proved to be of limited value as an inter-municipal planning discipline, due to lack of executive power. When the law was revised in 1973, county planning became the over-riding planning format above the municipal level. Up to now, this planning system has had little to do with land use problems, which are mainly left to the municipal master plans.

By 1982 approximately 50 % of Norways 440 municipalities had completed their first generation master plans, and many are now under revision. The Central Government is trying to ensure that the other municipalities wind up their first master plans as soon as possible.

Municipal planning has partly been carried out by contracted consultants, partly by municipal staff, or in combination, depending upon the size and type of municipality. Basic ecological data are usually provided by the central agencies responsible for surveys and inventories. Special data, or derived data, are usually a task for the local planning authorities.

The central planning authorities do provide instructions and advice, but this is mainly of a general type, and methodological approaches to land use planning are matters for local ability and resources. They are also very much dependent upon geographical and societal conditions as well as political intentions.

The basic thrust behind a methodology for land use planning should be to :-
(i)   enable the contributors develop an idea from its conceptual-
       ization through to a synthesis of the social, economic,
       technical, functional and cultural considerations that impinge
       on or form part of the plan;
(ii)   provide a general view of the planning process and establish
       a format for rational and effective work, stimulating
       professional and geographic integration;
(iii) provide a base for communication between political authorities,
       planners and public, in order to involve all those concerned
       in the planning process.

The planning documents must be accessible to all interest groups - and above all, be understandable. The scientist who provided material, knowledge and judgement has a right to be able to follow the way in which his contributions are used in a complex planning procedure. Politicians and the public confronted with a proposed plan must equally be offered the opportunity of seeing how it is derived and the ways in which it relates to the basic data, information and evaluations.

Satisfactory involvement of the natural and social sciences in planning is dependent upon a rational methodology. While rational, goal-oriented approaches may have political shortcomings as a base for sound decision making, even incrementalist approaches must be founded on a certain degree of rationality.

Figure 1 gives a condensed outline of the approach to land use planning developed by the author and colleagues at The Agricultural Univerisity of Norway around 1970, inspired partly by Ian McHargh's "Design with Nature". The schematic outline indicates the phases

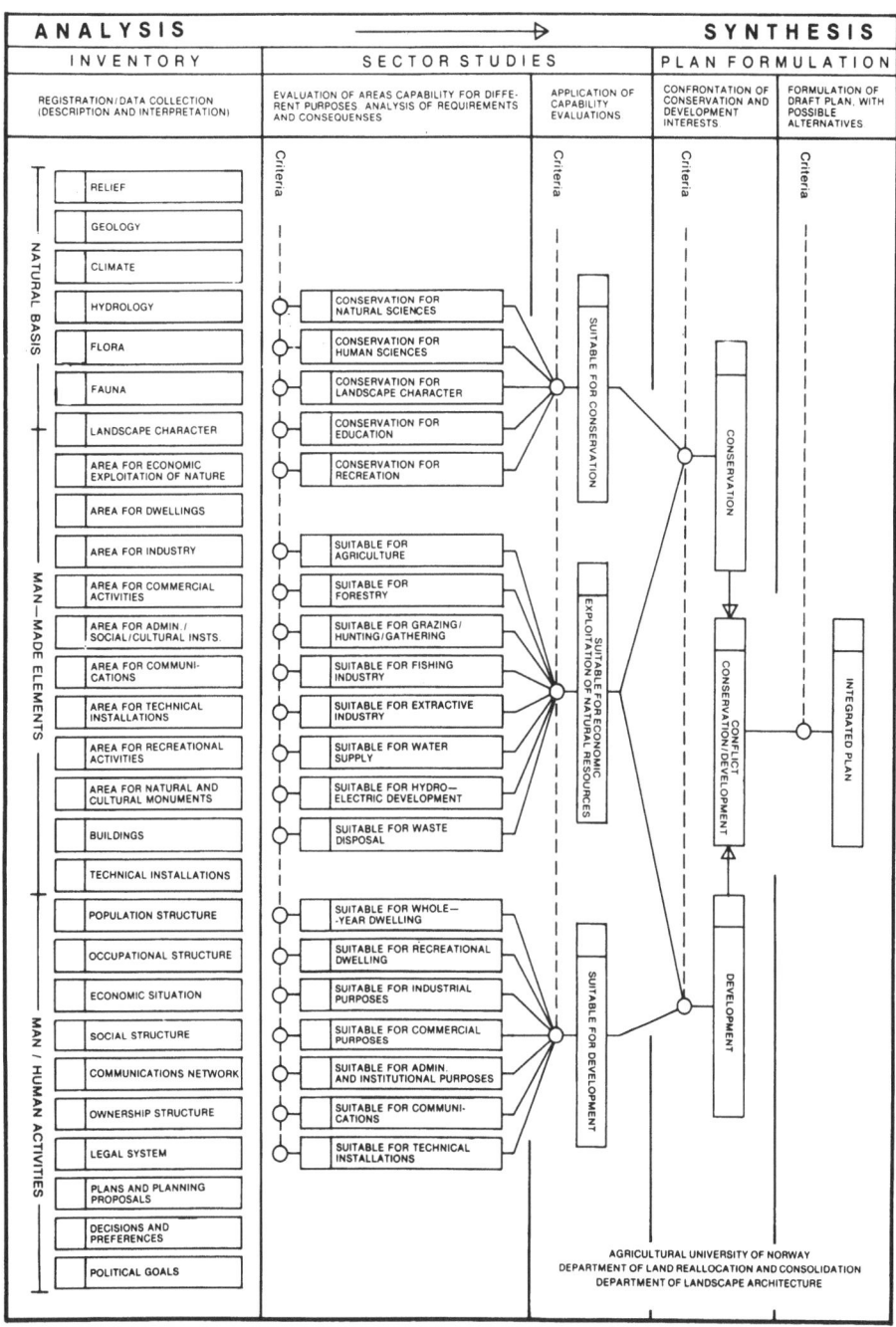

Figure 1: A Norwegian scheme for land-use planning, based on the confrontation between conservation and development, showing the inventory of descriptive data, the features of land capability and the way an integrated plan is developed. (Langdalen, 1975).

and themes in land use planning. The latter part of the procedure
- the conflict and priority discussion - is shown in Figure 2. The
suitability mapping procedure which is involved results in cartographic problems, which are illustrated in Figure 3.

LAND CLASSIFICATION SYSTEMS AS PLANNING TOOLS

Representative maps are needed for a broad understanding of land
classification systems in Norway. Until now, land use planning has
been based upon series of topographical maps, ranging from scales of
1:250 000 and 1:50 000 (for regional planning) down to 1:5 000 (for
municipal planning). A complete national map plan is being developed
by the Ministry of the Environment, approval in principle having been
given by Parliament. The project includes a series of thematic maps
that use transparency techniques and numerical data systems.

Economic maps and derived maps

Black and white economic maps of scale 1:5 000 (or 1:10 000) have
two main uses in land use planning: they are sufficiently detailed
to show property and ownership, as well as land condition and actual
use and the potential for agriculture and forestry. When completed,
the Norwegian economic map series will cover all productive areas
below the forest line.

Several smaller scale maps with aggregated class values, ranging
from 1:20 000 to 1:250 000 can be derived from economic maps. A new
land capability map of scale 1:10 000 has recently been prepared; it
is based on the economic map and a special vegetation inventory
above the forest line. This map surveys land use, plant production
potentials and ecological conditions. Mapped units are digitized
and may form part of a comprehensive registry system.

Thematic maps

Thematic mapping refers to a number of disciplines, which already
provide maps and other data in various scales and classification
systems, sometimes on an ad hoc basis. A number of public and private
institutions may be involved, both central and regional. Generally a
high level of coordination is needed for thematic mapping and some
essential work is not done for want of a responsible agency. A case
in point is biological mapping, for which there is an obvious need,
but for which no central organization is responsible.

The new demands for data, coupled with new opportunities opened
by computer technology have been responsible for the establishment
of a coordinated programme aimed at standardized systematic map series
and registries. In a report soon to be published, a group of
specialists has given production priority to the following themes:

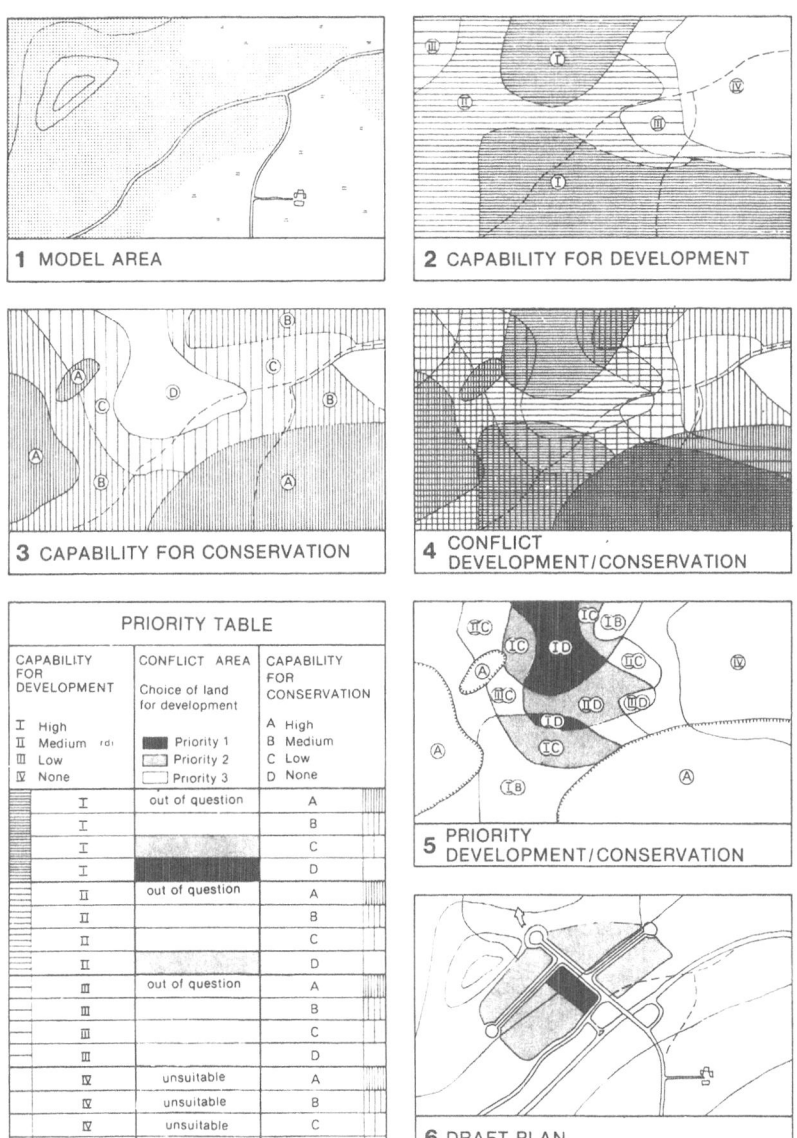

Figure 2: A theoretical example of plan formulation for an area of land under consideration for development. Following inventory work in the area (1) graded classifications are prepared of land suitable for development (2) and conservation (3) which are confronted in a conflict map (4). Systematic discussion of conflict categories (priority table) leads to the choice of land for development, expressed on the priority map (5) which is a basis for the plan proposal (6). (Langdalen, 1975).

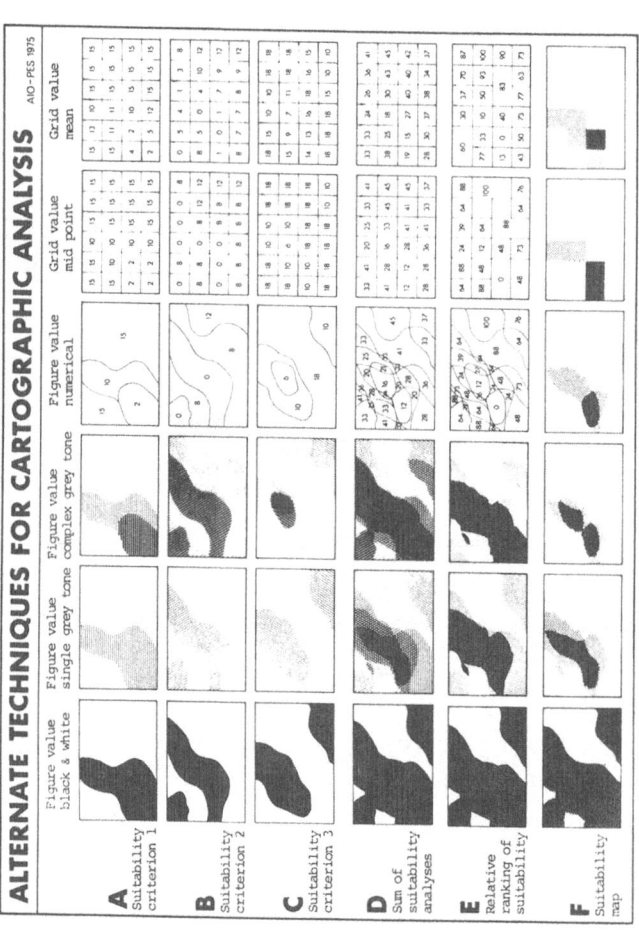

Figure 3: A comparison of techniques for cartographic analysis based on the methodology shown in Figures 1 and 2. Each column represents a particular technique which results in the different suitability maps at its base. (Oterholm and Skrovseth, 1977)

geology, natural hazards, vegetation, and fresh water resources.
Land resources in general, as well as climate are scheduled for
further research to improve mapping and presentation methods.

Local planning has shown geological and vegetation maps to be
key aspects of suitability analyses. Vegetation mapping started in
the 1970s, but is still in an experimental stage in three different
mapping institutions. Conformity in mapping and printing techniques
is overdue.

Land accounting

Early in the 1970s, experimental work started on a project aiming
at a complete resource accounting for Norway. An important part of
this was to establish land accounts, which will supply data in terms
of land balances and statistics of annual change in land use.
Priority has been given to the production of comprehensive land use
statistics for urban areas. Based on maps and aerial photographs,
land use information for 1955, 1965 and 1975 is now available. The
data collection is based on point sampling, using 100,200 or 300
metre grids, totalling 135 000 points for 247 urban areas (i.e. all
urban areas with more than 1 000 inhabitants). A similar technique
has been used to obtain national land use statistics based on 6 920
sample points. The land accounts are intended to present information
about actual, as well as planned land use and land capability.
Besides reducing costs, the point sampling method makes it possible
to link together various types of information on land use, such as
soil, vegetation, altitude and physical structure.

The national land use statistics are based on a 12 x 12 km grid
in mountain areas, and 6 x 6 km below the forest line. Many types
of data are collected, describing land use, geology, vegetation,
terrain, landscape, geographical location, etc.

Natural geographic regions in Norden

A Nordic land classification project was initiated in 1974 by
"Nordisk Ministerrad" (an Internordic political body) to provide a
mutual ecological base for physical planning, and especially to
identify a priority list of Nordic landscapes to be protected. A
working group of experts divided "Norden" (Denmark, Finland, Sweden
and Norway) into 60 natural regions, using vegetation types and
geomorphology as the basic criteria. This classification has been
further developed through a project entitled "Representative nature
types and endangered biotopes in Norden", whose efforts will soon
result in the publication of reports on vegetation and geomorphology,
both supplementing the regional classification, but not leading to
any change in the original regional definitions.

Another working party under the same Nordic political body is in the process of publishing an internordic guide on ecology and municipal planning.

Description and classification of landscapes

In the last few decades, there has been a great deal of demand for methods to describe and classify the quality of landscapes. Some interesting methods have been developed, especially at British and German universities. While some of these have been of mainly academic interest, others have found practical use in land use planning. Unfortunately none of these have been useful in Norwegian planning. It may be that these methods are too time consuming in a Norwegian context, as well as the fact that they do not reflect regional differences. In a country with huge unpopulated areas and drastic contrasts in landscape character, a universal, practical method has been hard to find. At present the Ministry of Environment is testing a simple methodology developed by US National Forest Service, based on two main principles:

(i)   a regional division of the country into various types of land-scape so as to classify landscape quality within comparable units, thus avoiding comparisons between amenity values in very different environments;
(ii) a classification of amenity values within each region, based on specified criteria.

The regional division of Norden into 60 natural geographic regions may prove to be a useful framework for this project, which is part of a comprehensive planning scheme for Norwegian water systems.

CONCLUSIONS

Land classification systems are tools to be used to improve both development planning and protective planning. Ecological classification is essential for land use analysis and sound decision making. The high complexity of modern planning requires advanced methods of analysing information and describing alternate solutions. It is a challenge to the natural and social sciences to develop methods of unveiling and confronting the direct and indirect interests involved in a given land area under planning.

To serve the planning process, the classification systems of the different disciplines must be coordinated to satisfy the requirements of mapping techniques and data registries, which are necessary to compare interests, compose solutions, and assess their impacts.

Classification systems must meet the requirements for continuous planning on the local, regional and national levels, as well as for short term ad hoc planning purposes.

To be usable in the planning process, each specialist discipline has to develop a classification language which is understood by the planners, decision makers and the public.

GENERAL REFERENCES

Langdalen, E., 1975, Conservation of the Natural and Cultural Land-
    scape as an Integral Part of the Municipal Survey Plan,
    Proceedings of the Symposium on Ecology and Planning, National
    Swedish Environment Protection Board, Stockholm
McHarg, I.L., 1969, "Design with Nature", The Falcon, Philadelphia,
    USA.
Ministry of Environment, 1975, "Survey of Norwegian Planning
    Legislation and Organisation", Ministry of Environment, Oslo.
Nordic Council of Ministries, 1983, "Representative Types of Nature
    in Nordic Countries", Nordic Council of Ministries, Oslo.
Oterholm, A.I., and P. E. Skrovseth, 1977, "Testing of methods for
    land use analysis", Agricultural Univ. of Norway, Oslo.
U.S.Department of Agriculture, 1974, "National Forest Landscape
    Management, Vol. 2, The Visual Management System", Agricultural
    Handbook No. 462, U.S.Department of Agriculture, Washington,
    D.C.

24

THE CANADA LAND INVENTORY

L. C. Munn

Policy Research and Co-ordination Branch
Lands Directorate
Place Vincent Massey
Hull, Quebec, Canada

INTRODUCTION

On October 3, 1963 the Government of Canada approved the undertaking of a comprehensive land resource inventory. Since that time
much has been written about the Canada Land Inventory (CLI), particularly the development, organization and implementation stage of
the inventory. It took approximately 15 years to complete the
inventory phase of the CLI. It is only in the last 5 years that we
have begun to use and draw upon the information to address resource
issues. It is only in the past 2 to 3 years that we have started
to realize the importance of the CLI in resolving social, economic
and environmental issues in Canada.

The CLI information and data permeates all facets of land use in
Canada. CLI terminology has become widely accepted and is used by
scientists and planners alike. Politicians and citizens refer to
class 1, class 2 land etc., when discussing land values and land
issues - they have become household 'words'.

PRELUDE TO INVENTORY

The decades of the 1940's and 1950's were momentous ones for
Canadian agriculture (Maxwell, 1972). Science-based technology was
applied to the agricultural industry on a grand scale, labour
scarcity and costs made mechanization mandatory for viable farm
operation, and changes in market patterns and conditions demanded
increased crop specialization. Although these forces presented
opportunities to increase the efficiency, productivity and income

of farms, certain necessary conditions had to be met by a farm enterprise before these opportunities could be captured. The farm had to
be located on reasonably productive land that was suited to mechan-
ized operations; it had to have access to sufficient capital for
underwriting investments in machinery and other necessary farm inputs;
it had to contain sufficient acreage to spread the costs of labour
and machinery and thereby keep unit costs of these factors at comp-
etitive levels. Finally, the managers of farms had to have sufficient skill to apply technical advances including the wise use of
artificial fertilizers, pesticides and herbicides.

Unfortunately, many Canadian farms could not meet these conditions. As a result, large numbers of farms located on poor soils,
and even small farms on good soils, were unable to adjust to the
new conditions. Farm abandonment began and during the 1960's the
number of farms in Canada declined at the rate of 10,000 a year.
Many of the people who left farming were able to find alternative
employment in industries located in, or near towns and cities but
inevitably some remained without jobs with poverty emerging in
many of the country's rural areas.

The Canadian Government responded to this situation by passing
the Agricultural Rehabilitation and Development Act in 1961 (later
renamed the Agriculture and Rural Development Act). This Act,
known by the acronym "ARDA", enabled the federal government to undertake farm enlargement and consolidation programmes, land improvement schemes and land use adjustment projects in cooperation with
the provincial governments.

When the ARDA programmes were being formulated, it became
apparent that a rational approach to land use adjustment and planning could not be undertaken without knowledge of the land's capability to support agricultural and feasible alternative uses. It
was to provide this knowledge that the Canada Land Inventory
project was launched in October, 1963.

THE CANADA LAND INVENTORY (CLI) - OBJECTIVES AND ORGANISATION

The CLI was established as a cooperative federal-provincial
project for the purpose of classifying lands according to their
present use and by their physical capability (Environment Canada,
1970a) for sustained use in :-

(i)   agriculture
(ii)   forestry
(iii) recreation, and
(iv)   wildlife (waterfowl and ungulates).

The underlying objective was to provide a basis for land use planning.

The project has been completed and land use and capability data are available for the settled parts of Canada and the adjacent forest fringe (an area of approximately 2.5 million km^2, Figure 1). These data are provided in map form at two scales for each resource sector. Field sheets are prepared at 1:50,000 and published maps are produced at 1:250,000. In addition, because of the volume of data produced (twenty thousand 1:50,000 maps and twelve hundred at 1:250,000), a computerized data bank called the "Canada Geographic Information System" has been developed to facilitate analyses and retrieval of information.

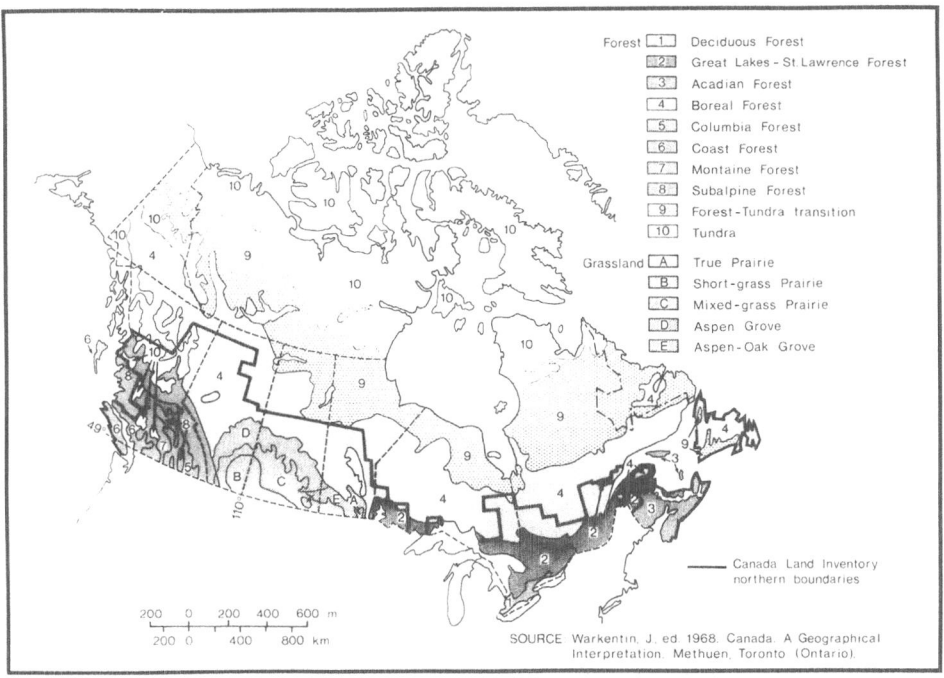

Figure 1 :   Map of Canada showing the limits of the area studied
             for the Canada Land Inventory and the principle veg-
             etation types.

The role of the Federal Government was primarily one of enabling and coordination. Thus, the Government of Canada financed all incremental costs incurred by the provinces during the conduct of the Inventory. It also directed the development of other national capability classification systems and provided coordination services to ensure their consistent application. The publication of the 1:250,000 CLI sector capability maps, and the provision of data processing facilities were also federal responsibilities.

Provincial governments were responsible for the planning and
conduct of the Inventory and the preparation of the capability maps
for all resource sectors except waterfowl. Inventory committees,
chaired by a provincial CLI Coordinator were established in each
province to direct this work which was carried out by numerous
federal and provincial resource agencies, university groups and
private consultants.

THE CLI LAND CAPABILITY CLASSIFICATION SYSTEMS (TABLE 1)

The biggest challenge to be met in launching the CLI project
was the development of nationally relevant systems of land capability classification. Initially, this appeared to be a formidable
task because of the diversity of Canada's geography and the number
and complexity of the factors determining the viability of a particular land use at a given time and location. Needless to say, the
CLI project would never have been conceived and undertaken in the
absence of the considerable knowledge and experience gained
through many years of natural resource surveys in Canada. Thus,
the planning which produced the CLI was, in turn, highly dependant
upon the scientific knowledge and methodology resulting from earlier
surveys of natural resources.

It quickly became apparent that the more dynamic socio-economic
determinants of land use could not be considered if a data base
having a value over a reasonably long period of time was to be produced. For this reason, factors such as present land use and cover,
accessibility, tenure and market conditions were excluded from the
assessment of capability. Although these factors are critical in
determining the highest and best use of land at any given time and
location, it was decided that consideration of them should take
place only when land use plans were being prepared and implemented.
The elimination of these factors permitted lands to be evaluated
strictly in terms of the ability of their inherent physical characteristics to support specified uses under current technological
conditions and good management practices.

The five classification systems, four concerned with different
aspects of land capability and the fifth describing present use,
developed for the CLI project, while dealing with different land
uses, are similar in many respects (Environment Canada, 1970a).
All are interpretive classification systems in which soils or land
types are grouped into one of seven classes on the basis of their
capability for production. The class designation indicates the
degree of limitation or potential; class 1 lands have the highest
potential, class 7 the least. The classes are further divided into
sub-classes, the number varying in the different classifications -
the classification for agriculture, for example, has thirteen sub-
classes. The sub-classes usually indicate types of limitation

TABLE 1a: The systems of classification used in the Canada Land
           Inventory (CLI): Agricultural Land Capability, Forestry
           Land Capability.

           SOURCE: Environment Canada. 1978. CLI: Objectives,
                    Scope and Organization.
                    Canada Land Inventory Report No. 1.
                    Lands Directorate. Ottawa, Ontario.

CLI AGRICULTURAL LAND CAPABILITY CLASSIFICATION

Classes	Description
	Soils in this class have or are...
1	No significant limitations in use for crops.
2	Moderate limitations that restrict the range of crops or require moderate conservation practices.
3	Moderately severe limitations that restrict the range of crops or require special conservation practices.
4	Severe limitations that restrict the range of crops or require special conservation practices or both.
5	Very severe limitations that restrict their capability to producing perennial forage crops, and improvement practices are feasible.
6	Capable only of producing perennial forage crops and improvement practices are not feasible.
7	No capability for arable culture or permanent pasture.
0	Organic soils (Not placed in capability classes).

Subclasses	Description	Subclasses	Description
C	Adverse climate	P	Stoniness
D	Poor structure	R	Bedrock
E	Erosion	S	Mix of D, F, M, and/or N
F	Low fertility	T	Topography
I	Inundation	W	Wetness
M	Low moisture	X	Cumulative limitations
N	Salinity		

CLI FORESTRY LAND CAPABILITY CLASSIFICATION

Classes	Description
	Lands having...
1	No important limitations to the growth of commercial forests. Productivity usually greater than 7.8 m^3/ha/yr.
2	Slight limitations to the growth of commercial forests. Productivity usually 6.3–7.7 m^3/ha/yr.
3	Moderate limitations to the growth of commercial forests. Productivity usually 5.0–6.3 m^3/ha/yr.
4	Moderately severe limitations to the growth of commercial forests. Productivity usually 3.6–4.9 m^3/ha/yr.
5	Severe limitations to the growth of commercial forests. Productivity usually 2.2–3.5 m^3/ha/yr.
6	Severe limitations to the growth of commercial forests. Productivity usually 0.8–2.1 m^3/ha/yr.
7	Severe limitations which preclude the growth of commercial forests. Productivity usually less than 0.8 m^3/ha/yr.

Subclasses	Description	Subclasses	Description
A	Aridity	Y	Pattern of D and R
C	Climate	E	Erosion
H	Low temperature	F	Low fertility
U	Exposure	I	Inundation
M	Soil dryness	K	Permafrost
W	Soil wetness	L	High carbonate
X	Mix of M and W	N	Toxicity
Z	Mix of rock/organic	P	Stoniness
D	Rooting depth	S	Cumulative limitations
R	Bedrock		

TABLE 1b: The systems of classification used in the Canada Land
            Inventory (CLI): Recreational Land Capability, Ungulates
            Land Capability.

            SOURCE: Environment Canada. 1978. CLI: Objectives,
                      Scope and Organization.
                      Canada Land Inventory Report No. 1.
                      Lands Directorate. Ottawa, Ontario.

CLI RECREATIONAL LAND CAPABILITY CLASSIFICATION

Classes	Description
	Lands in this class have...
1	Very high capability for outdoor recreation.
2	A high capability for outdoor recreation.
3	A moderately high capability for outdoor recreation.
4	Moderate capability for outdoor recreation.
5	Moderately low capability for outdoor recreation.
6	Low capability for outdoor recreation.
7	Very low capability for outdoor recreation.

Subclasses	Description	Subclasses	Description
A	Angling	O	Upland wildlife
B	Beach	P	Cultural landscape
C	Canoe tripping	Q	Landform
D	Deep water	R	Rock formation
E	Vegetation	S	Ski hill
F	Rapids, falls	T	Thermal spring
G	Glacier view	U	Yachting
H	Historic site	V	Viewing
J	Gathering/collecting	W	Wetland viewing
K	Camping	X	Miscellaneous
L	Landform	Y	Family boating
M	Upland waterbodies	Z	Man-made feature
N	Lodging		

CLI UNGULATES LAND CAPABILITY CLASSIFICATION

Classes	Description
	Lands in this class have...
1	No significant limitations to the production of ungulates.
2	Very slight limitations to the production of ungulates.
3	Slight limitations to the production of ungulates.
1W, 2W, 3W	Classes 1, 2, and 3 significant for winter range.
4	Moderate limitations to the production of ungulates.
5	Moderately severe limitations to the production of ungulates.
6	Severe limitations to the production of ungulates.
7	Limitations so severe that there is no ungulate production.

Subclasses	Description	Subclasses	Description
A	Aridity/drought	I	Inundation
C	Climate	M	Soil moisture
Q	Snow depth	N	Adverse soils
U	Exposure/aspect	R	Shallow soils
F	Fertility	T	Topography
G	Landform mix		

TABLE 1c: The systems of classification used in the Canada Land
            Inventory (CLI): Waterfowl Land Capability, Present Land
            Use (Circa 1967).

            SOURCE: Environment Canada. 1978. CLI: Objectives,
                      Scope and Organization.
                      Canada Land Inventory Report No. 1.
                      Lands Directorate. Ottawa, Ontario.

CLI WATERFOWL LAND CAPABILITY CLASSIFICATION

Classes	Description
	Lands in this class have...
1	No significant limitations to the production of waterfowl.
2	Very slight limitations to the production of waterfowl.
3	Slight limitations to the production of waterfowl.
1S, 2S, 3S and 3M	Classes 1, 2, 3 important for migration or wintering.
4	Moderate limitations to the production of waterfowl.
5	Moderately severe limitations to the production of waterfowl.
6	Severe limitations to the production of waterfowl.
7	Such severe limitations that almost no waterfowl are produced.

Subclasses	Description	Subclasses	Description
A	Aridity	J	Low marsh edge
B	Water flow	M	Soil moisture
C	Climate	N	Poor soil/water
F	Fertility	R	Shallow soils
G	Landform mix	T	Topography
I	Inundation	Z	Water depth

PRESENT LAND USE CLASSIFICATION (CIRCA 1967)

Classes	Description	Symbols
Urban		
1	Built-up area	B
2	Mines, quarries, sand and gravel pits	E
3	Outdoor recreation	O
Agricultural Lands		
1	Horticulture, poultry and fur operations	H
2	Orchards and vineyards	G
3	Cropland	A
4	Improved pasture and forage crops	P
5	Rough grazing and rangeland (Areas of natural grasslands and woodland grazing)	K
Woodland		
1	Productive woodland	T
2	Non-productive woodland	U
Wetland		
1	Swamp, marsh or bog	M
Unproductive Land		
1	Sand	S
2	Rock and other unvegetated surfaces	L
Water		Z

(wetness, steepness of slope, stoniness, etc.) except in the recreation classification where sub-classes indicate particular advantages
(Table 1a, b and c).

   All systems are national in character. That is, all areas
placed within the same class for each of the respective land uses,
have the same degree of limitation or potential, regardless of where
they are located in the country (Figure 2).

Soil Capability for Agriculture - the classifications were made by
federal and provincial soil survey organizations (Environment Canada,
1969a, 1976). Their analyses were based on soil survey maps and
reports prepared, for many of the settled areas of Canada, by the
National Soil Survey which was established in 1921. In this system,
soils designated as classes 1, 2 and 3 are capable of sustaining the
production of common cultivated crops adapted to a region. Class 4
soils are marginal for sustained arable culture and class 5 soils
are capable of supporting permanent pasture and hay production only.
Class 6 soils are suited only for 'wild' pastures which are not
intensively managed while class 7 soils are considered incapable of
supporting any form of agricultural activity.

Land Capability Classification for Forestry - consists of seven
classes based on the capability of the land to produce a given volume
of a stated species on a mean annual basis over the rotation of the
forest stand (Environment Canada, 1970b). Thus, class 1 land must
be capable of producing more than $8 \text{ m}^{-3} \text{ ha}^{-1} \text{ yr}^{-1}$ on a sustained
basis under good management. With increasing limitations and
corresponding reductions in gross yields of $1.5 \text{ m}^{-3} \text{ ha}^{-1} \text{ yr}^{-1}$, the
capability classes become progressively lower until class 7 is
reached with an annual yield of less than $0.75 \text{ m}^{-3} \text{ ha}^{-1} \text{ yr}^{-1}$. Lands
designated as class 7 are incapable of supporting 'commercial'
forests. These classifications were made by the provincial forest
management services.

   The land type, which is defined as land on a particular landform
segment and having a fairly homogenous combination of soil (parent
material), topography, moisture and climatic characteristics, is
the basic unit for which use-capability ratings are made for forestry.

Land Capability Classification for Outdoor Recreation - while consisting of seven classes like the other systems, differs from them
in that lands are classified by their positive features enabling
sustained recreational use, rather than by their limitations (Environment Canada, 1969b). The basis of the classification is the
quantity of recreation (measured in visitor days or hours) that may
be sustained per unit area per year under perfect market conditions.
Thus, lands with a capability to sustain intensive use, such as
sandy beaches and slopes suitable for skiing, are rated higher than
lands which will support less intensive uses.

Figure 2
Canada Land Inventory Mapping
Map Sheet Kitchener 40 O,P

The recreation classification programme was the responsibility
of the provincial parks and recreation departments with, in some
provinces, the help of private consultants.

Land Capability Classification for Wildlife - the units of land
allocated to the different classes of an ungulate classification
system are often identical to those identified in the land capability
classification for forestry (Environment Canada, 1969c). But, of
course, the types of limitation differ. Winter range is an important
facet of the classification for ungulates.

The waterfowl classification system is based on assessments of
the nature and amounts of wetlands, permanency of water and nature
and frequency of bordering vegetation. In addition to considering
dietary and breeding requirements, attention was given to areas used
by birds during migration.

Provincial wildlife agencies classified land for ungulates; the
federal Wildlife Service was responsible for waterfowl land classifications.

Present Land Use - in addition to mapping land resource capabilities
it was considered essential to determine the extent and nature of
existing land uses, particularly agricultural. Using a classification devised by McClellan et al (1968) mapping was done on 1:50,000
National Topographical System map sheets or if they were not available on the most suitable alternative, scales being adjusted when
being reproduced photographically. The 1:50,000 map sheets were
later generalised to 1:250,000 for input into the Canada Geographic
Information System and overlay analysis with other CLI data and
related information.

Land uses were mapped from aerial photographs with frequent
ground checks, the smallest areas designated being 2.6 ha.

THE USE OF CLI

Since the early '70s, when most of the data for the CLI had been
obtained, the information has become widely used throughout Canada,
both as part of the increased awareness of land as a resource, and
in the reactions of governments, at all levels, to problems of land
availability.

Figure 2 shows an area of the Great Lakes - St. Lawrence Forest
region (see Figure 1) in the southwestern part of the province of
Ontario, as it was mapped for the 5 CLI capabilities and land use.
Although mapped independently, some broad geographic correlations
are evident in the maps. The unit of organic soils near Luther Lake
shown on the land capability for agriculture map is also recognizable
on other maps - as low capability for forestry, as lower capability

than surrounding lands for ungulates, but prime capability for waterfowl. Because of the wetness, it was not cleared for agriculture,
but has been left in forest, as shown on the land use maps.

The land use map in this case indicates that, by current practices, a relatively good allocation of land use has occurred. Historically, agriculture has provided the highest social economic value
to society, and therefore the high capability land is mostly in agriculture. It is mostly class 1 and where the capability is reduced
it is mainly due to adverse topography or excessive soil moisture.
Islands in Luther Lake, although of high agricultural capability,
are in rough pasture probably due to difficulties of access.

Forestry capability is generally high but most of the area was
cleared for agriculture. Most forested areas are now too small to
have significant commercial importance.

Many of the water courses and wetlands remained in natural vegetation including the perimeter of the lake. However, the large
amount of clearing has resulted in a much smaller wildlife (deer)
population than the area is capable of supporting.

Luther Lake with its surrounding wetlands and vegetation represents an important staging area for waterfowl during migration and
therefore has a high capability rating for waterfowl.

The lake has a moderate recreation experience for angling, camping
and viewing of wetland wildlife, waterfowl and a wetland ecological
system. Land use planning should encourage development of a long-
term passive recreation use capitalizing on the above capabilities
and integrating it with use of the agricultural resources.

A Myth of Unlimited Resources: Educating the Public to Land Concerns

"Canada has long been viewed, both at home and abroad, as a
breadbasket of immense potential" (McCuaig and Manning, 1982).

There is a common misconception that Canada's supply of good
land is virtually inexhaustable. Conditions of climate, topography,
soil, and moisture availability limit the amount of land useful for
most purposes to a much smaller area. In a country which encompasses
922 million hectares of land, it is difficult to impress on most
people that good land is indeed a limited and extremely valuable
resource. Surely in a nation of this size there is no shortage of
good agricultural land ? However, analyses of the Canada Land Inventory data reveals that 86 per cent of Canada's land has no capability
for agriculture. About 2 of the 14 per cent with agriculture capability, is marginal for intensive agriculture and only suitable for
rough grazing. The balance of agricultural land amounts to something just in excess of 105 million hectares. Of this land, only